Rebecca Salter

Rebecca Salter *into the light of things*

Edited by Gillian Forrester

With essays by
Achim Borchardt-Hume
Richard Cork
Gillian Forrester
Sadako Ohki

Yale Center for British Art
in association with
Yale University Press
New Haven and London

This publication accompanies the exhibition
into the light of things: Rebecca Salter, works 1981–2010
organized by the Yale Center for British Art, New Haven,
and on view 3 February–2 May 2011.

Exhibition curated by Gillian Forrester.

Designed by Geoffrey Winston, Graphics with Art.

Set in InterFace Typographic Edition

Printed in Italy

Library of Congress
Cataloging-in-Publication Data

Salter, Rebecca, 1955–
Rebecca Salter: into the light of things / edited by Gillian Forrester.
p. cm.
Published to accompany an exhibition held at the Yale Center for British Art, New Haven, Conn., Feb. 3–May 1, 2011.
Includes index.
ISBN 978-0-300-17042-9 (alk. paper)
1. Salter, Rebecca, 1955—Exhibitions.
I. Forrester, Gillian. II. Yale Center for British Art. III. Title.
N6797.S25A4 2011
709.2—dc22
2010049514

Photography and Image Credits

Every effort has been made to credit the photographers and sources of all illustrations in this volume; if there are any errors or omissions, please contact Yale University Press so that corrections can be made in any subsequent edition.

Unless otherwise noted all works are from the collection of the artist. All her works are © Rebecca Salter. Unless otherwise noted, all photography of Salter's works, including the *Calligraphy of Light* and *Delta* installations at St. George's Hospital, London, is by FXP Photography, London.

© 2010 Agnes Martin/Artists Rights Society (ARS), New York: fig. 32; Art Archive/Sylvan Barnet and William Burto Collection: fig. 30; Christopher Gardner Photography: fig. 43; © Belvedere, Vienna: fig. 17; © 2010 Brice Marden/Artists Rights Society (ARS), New York: fig. 53; © 2010 Estate of Mark Tobey/Artists Rights Society (ARS), New York: fig. 42; © 2010 Franz Kline Estate/Artists Rights Society (ARS), New York: fig. 52; Geoffrey Winston: figs. 1, 10, 64, and 65; © Gerhard Richter: fig. 36; Gillian Forrester: fig. 8; © Jasper Johns. Licensed by VAGA, New York, NY: fig. 35; © MUMOK, Museum Moderner Kunst Stiftung Ludwig Wien, on loan from Sammlung Ludwig, Aachen: fig. 35; © Museum of Modern Art/licensed by SCALA/Art Resource, NY: fig. 32; Rebecca Salter: figs. 7, 19, 20, 67–69, and pp. 28–9, 46–7; Richard Caspole, Yale Center for British Art: fig. 23, plates 45, 99; Schloss Sanssouci, Postdam/Alinari/Bridgeman Art Library: fig. 31; © St. Bartholomew's Hospital Archives: fig. 54; © Tate, London 2009: fig. 16; Tate, London/Art Resource, NY: fig. 36; Yagi Akira: fig. 38; © Yale University Art Gallery: figs. 37, 41, 42, 44, 49, 52, and 53.

Jacket, endpapers, and section dividers are details from *into the light of things*, plate 154.

Contents

Director's Foreword and Acknowledgments

This publication and the exhibition to which it is related result from a long and highly productive collaboration that began in 2003. In September of that year, Rebecca Salter, the British, London-based, abstract artist, traveled to the rural town of Bethany, Connecticut, to take up a three-month residency at the Josef and Anni Albers Foundation. As Gillian Forrester, the Center's Curator of Prints and Drawings and the curator of the exhibition, recounts in her essay in this publication, the experience of working in nature, away from the busy metropolis and in an environment so sympathetic to artistic creativity, was transformative for Salter's practice. The artist's sojourn in Bethany also had important outcomes for the museum community at Yale. Rebecca regularly visited the Yale Center for British Art, and she generously invited staff members from the Center to make field-trips to her studio to view work in process. Friendships, which have proved to be both enduring and productive, were forged. In fact, the following year funds from the Center's Friends of British Art enabled the institution to purchase *Bethany Squares*, an outstanding suite of drawings, which Rebecca had made during her residence, a work that is both an homage to Josef Albers' chosen format of the square and a summation of her own experience in Bethany. Rebecca also presented the Center with a sketchbook made in the English Lake District, in memory of her father, and a very natural collaboration between the artist and the institution began to take shape.

Since Rebecca's formative stay at the Josef and Anni Albers Foundation, its Director, Nicholas Fox Weber, and his colleagues Oliver and Anne Barker, and Brenda Danilowitz, have continued to provide support and encouragement for this collaboration, and we are grateful to the Foundation for offering further hospitality in Bethany to Rebecca during the run of the exhibition. The friendships made during Rebecca's time in Bethany with colleagues at Yale and the Foundation have been nurtured through visits to Salter's London studio and her regular return visits to New Haven. Deeply impressed by Rebecca's work, and with a powerful sense of a meeting of minds and shared sensibilities, Gillian Forrester began to consider the idea of organizing a survey exhibition, a notion that was warmly received by her colleagues at the Center. Rebecca has described her practice as "involved with the attempt to capture stillness in movement, a stillness with potential, not a passive quiet." Gillian was struck by the understated complexity of her works, which eschew brash strategies but are nonetheless radical in intent and form, demanding the kind of hard looking and aesthetic appreciation that is, Achim Borchardt-Hume argues in his illuminating essay for this book, "One of painting's all the more urgent ethical mandates today." Borchardt-Hume also explores Rebecca's contention that she is concerned with "making an object," rather than a two-dimensional surface. This project is intended to be as much a reconsideration of the nature of painting and its relationship with drawing as a survey of a single artist's work.

Salter spent six formative years in Japan from 1979 to 1985. On one of her visits to New Haven, while undertaking research for a book on Japanese popular prints, she was introduced to Sadako Ohki, the Japan Foundation Associate Curator of Japanese Art at the Yale University Art Gallery. Sadako generously shared with Rebecca some extraordinary objects from the Gallery's rich Japanese collections, and curator and artist began a fruitful conversation that has continued to this day. When Gillian and Rebecca began to formulate the Center's exhibition, it seemed natural to invite Sadako to participate.

Sadako proposed curating a separate exhibition in the Ruth and Bruce Dayton Gallery of Asian Art at the Yale University Art Gallery that would take Rebecca's work as a starting point for exploring the complex relationship between Japanese and Western practice. Jock Reynolds, Henry J. Heinz II Director of the Art Gallery; Susan Baker Matheson, the Molly and Walter Bareiss Curator of Ancient Art and Chief Curator of the Gallery; and the institution's curatorial committee agreed to the proposal with alacrity.

The exhibition, entitled *Rebecca Salter and Japan*, runs concurrently with the Center's own show. Two of Salter's recent works, a painting and a drawing, are in dialogue with fifteen paintings, drawings, and ceramics by Japanese and American artists drawn from the Art Gallery's holdings and from the collection of Peggy and Richard M. Danziger, LL.B. 1963. Notable collectors of Japanese art, they have donated important objects to the Gallery in recent years and very generously have supported the outstanding exhibition curated by Sadako in 2009 on the tea culture of Japan. We greatly appreciate their contributions to this exhibition, and we are grateful to them for so generously allowing the Center to reproduce images of some exquisite objects from their collection in this publication and in the brochure accompanying the two associated exhibitions. Salter has observed that her project in Japan was to "finesse the bridging of the two cultures," absorbing and understanding difference over a protracted period of time, rather than merely to superimpose discrete cultural practices onto one another. Sadako's exhibition and essay raise important questions about the complex relationship between Eastern and Western artistic practice, extending the discussion initiated by the Solomon R. Guggenheim Museum's groundbreaking and provocative exhibition in 2009, *The Third Mind: American Artists Contemplate Asia, 1860–1989*.

We are deeply grateful to Jock Reynolds and his colleagues at the Gallery for their support of this project from its inception, and particularly to Sadako for crafting such an elegant and thought-provoking exhibition, and for contributing her illuminating essay to this publication. David Ake Sensabaugh, the Ruth and Bruce Dayton Curator of Asian Art at the Gallery, has generously provided indispensable support and advice, as has Ami Potter, the department's Museum Assistant. We also would like to thank other members of the Art Gallery's staff, including Jill Westgard, Kate Ezra, Pamela Franks, Suzanne Boorsch, Lisa Hodermarsky, Jennifer Farrell, Chris Sleboda, Diana Brownell, Jeffrey Yoshimine, Clarkson Crolius, and John ffrench and his colleagues in the Visual Resources Department, for supporting the exhibition and its related programs in many ways.

The exhibition of Rebecca's work at the Yale Center for British Art, along with the accompanying book and programs, could not have been brought to completion without the willing and generous help of others, and we are much indebted to a large number of friends and colleagues both at Yale and elsewhere for their advice, encouragement, and practical assistance. We are particularly indebted to Laura and James Duncan (Yale College, Class of 1975), who have long been highly valued supporters of the Center and the Art Gallery, and of Rebecca, and whose help and encouragement have been invaluable in realizing the project.

We also are deeply grateful to individuals in the United States, Japan, and Britain, who have generously lent important works to the Yale Center's exhibition: Richard and Janet Caldwell, Diane Davies,

Gill and Steve Evans, Howard Foote, Bruce and Aphrodite Garrison, Fanchon and Howard Hallam, Vicky and Terry Klein, David Klingensmith, Andrew Lambirth, Holly Lennihan and Robert Cox, Lea Babcock Scherer and Jeffrey A. Scherer, and Delia Smith. Additionally, we are indebted to those lenders who have chosen to remain anonymous. We also would like to thank Howard Scott, who has organized a complementary exhibition of Rebecca's new work at the Howard Scott Gallery in New York, which will run concurrently with our exhibition, and Amanda and Brian Beardsmore, who have hosted a series of solo exhibitions of Rebecca's work at the Beardsmore Gallery in London, for their assistance with securing loans and help in innumerable other ways.

Achim Borchardt-Hume and Richard Cork have contributed insightful essays to the publication, which, in concert with those by Gillian Forrester and Sadako Ohki, provide a penetrating account of Rebecca's eclectic practice and wide range of intellectual concerns. We are thankful to Belinda Harward, the arts officer at St. George's Hospital, London, for agreeing to be interviewed by Richard Cork for his essay on *Calligraphy of Light*, Salter's reconception of the hospital's reception area.

We are delighted to include in the exhibition and publication a series of woodblock prints derived from four drawings made by Rebecca, entitled *Quadra*, and printed by the Kyoto workshop of Satō Keizō just months before the exhibition opened. Carved by master craftsmen Fujisawa Hiroshi and Kitamura Shōichi, and printed by Nakayama Makoto, the prints are masterpieces of the medium, translating Salter's originals with extraordinary skill and sensibility. Rebecca, who learned to make woodblock prints from Professor Kurosawa Akira in Kyoto in the 1980s, received funding between 2006 and 2008 from The Great Britain Sasakawa Foundation to interview printmakers, papermakers and other craftsmen in Japan. Her ultimate goal is to create an historic archive of their working lives, which will function as a poignant memorial to artistic practices in danger of extinction. Salter is indebted to many Japanese craftsmen who have generously shared their knowledge with her over the past three decades. We give special thanks to Sashiyama Takeshi, who coordinated the project; to Satō Keizō and his team; and to Claire Cuccio, who has painstakingly documented the process of carving and printing the woodblocks, and whose illuminating photographs are displayed in the exhibition.

In concert with Gillian, the Center's Associate Curator and Head of Exhibitions and Publications, Eleanor Hughes, together with her colleagues Imogen Hart, Craig Canfield, Anna Magliaro, Diane Bowman, Andrea Wolk Rager, and Jo Briggs, has overseen the logistical details of the project with characteristic energy, efficiency, and aplomb. Sally Salvesen, the Center's publisher with Yale University Press, London, has worked tirelessly with this team to ensure that this volume has come so successfully to fruition. We are deeply grateful to Geoffrey Winston for his elegant and thoughtful design of the book, the product of much painstaking labor. Peter White of FXP Photography, London, has undertaken much of the book's superb photography. His work has been augmented by that of Melissa Fournier, the Center's Associate Museum Registrar, and her colleagues Richard Caspole and Kurt Heumiller.

Many others enabled and ensured the realization of this project in countless ways. The Center's Registrar, Timothy Goodhue, and his superb colleagues, notably Corey Myers, handled so many details of the loans and the traveling of the exhibition. We are especially grateful to the staff of the Department of Prints and Drawings for their tireless help and forbearance, and especially John Monahan, David Thompson, Adrianna Bates, and our student assistant, Lucy Fitz Gibbon. We owe thanks to Theresa Fairbanks-Harris, Dong-Eun Kim, Mary Regan-Yttre, and Mark Aronson, who have undertaken the complex conservation and framing requirements for the exhibition. Under the direction of Richard Johnson, the Center's installation team, working with Lyn Bell Rose and Elena Grossman, have brilliantly executed the exhibition, responding sensitively to Rebecca's refined aesthetic. Lyn and Elena have also designed the elegant brochure which serves both exhibitions. Additionally, we have benefited from the knowledge and advice of our curatorial colleagues: Scott Wilcox, who has done so much to foster the Center's relationship with Rebecca and proposed the purchase of *Bethany Squares*; Cassandra Albinson, who read and commented on Gillian Forrester's essay; Elisabeth Fairman; and Angus Trumble. Rebecca Szantyr and Romina Tortoriello also undertook invaluable research for the project. We also are grateful to Beth Miller, who has been an indefatigable advocate of Rebecca's work and supporter of the project from the outset, as well as her colleagues Amy McDonald, Kaci Bayless, Julienne Richardson, and Amelia Toensmeier in the Department of Advancement and External Affairs. Martina Droth, Lisa Ford, Imogen Hart, Jane Nowosadko, Linda Friedlaender, Cyra Levenson, and Jennifer Kowitt have created a program of accompanying events that will greatly amplify and illuminate the exhibition and publication, and to them we express our gratitude.

Special thanks are also due to the following people, who have assisted Gillian Forrester in innumerable ways: John Curley, Merlin Forrester, Ruth Forrester, Sophie Forrester, Susanna Lawrence, Morna O'Neill, Holly Shaffer, Guilland Sutherland, and Ian Warrell. She also is very grateful to Rebecca Salter and Geoffrey Winston for their gracious hospitality in London, and to Rebecca for acting as a superb *cicerone* in Tokyo and Kyoto, a life-changing experience, as indeed has been the whole project.

Rebecca Salter would like to add her heartfelt thanks to all of the above, and particularly to Gillian, without whose thoughtful insight and curatorial flair this project would have been impossible. She also would like to thank her husband and the designer of this book, Geoffrey Winston, as well as Muffin de Wardener, and her friends and family for their sustained and loving support. Her experience of living and working in Japan, as this book demonstrates, has been a profound influence on her life and practice, and she thanks her many Japanese friends who have enabled her understanding of their country especially Nishioka Tsutomu, Yamazaki Nobuko and Nishioka Fumi, Yagi Akira and Yagi Sakiyo, Ōtsuka Ennosuke and Aoyama Shinju, the late Tada Michitarō, and the Isomura family.

Finally, we hope that through Rebecca's splendid exhibition at the Center—the first monographic show of her work—as well as through the accompanying book and the associated exhibition at the Yale University Art Gallery, viewers will discover in her works the same exquisite resonances that we have come to experience over our years of friendship. It is to Rebecca that we owe our deepest thanks, and to her we offer our warmest appreciation and affection.

Amy Meyers
Director, Yale Center for British Art

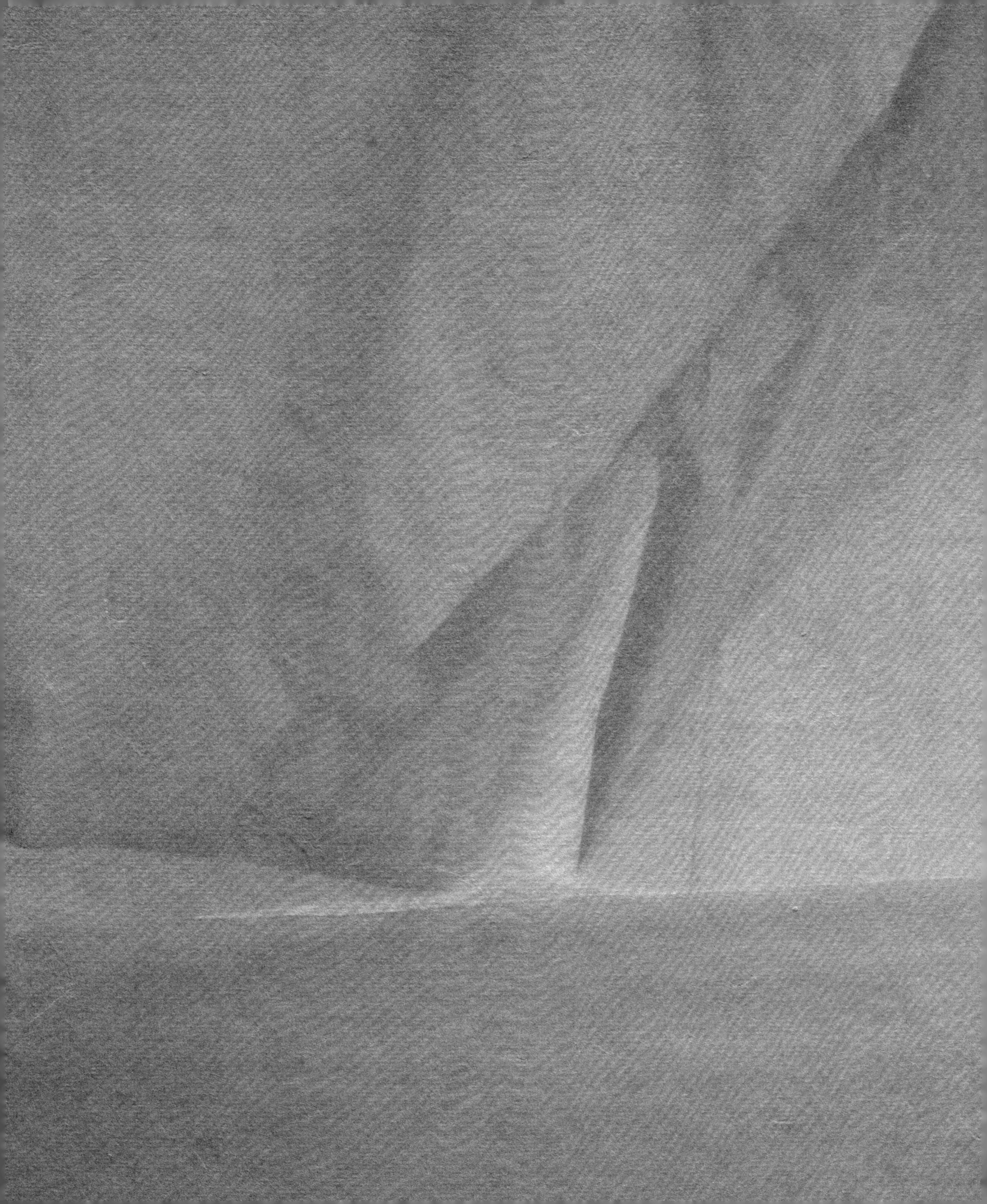

From Without to Within

Gillian Forrester

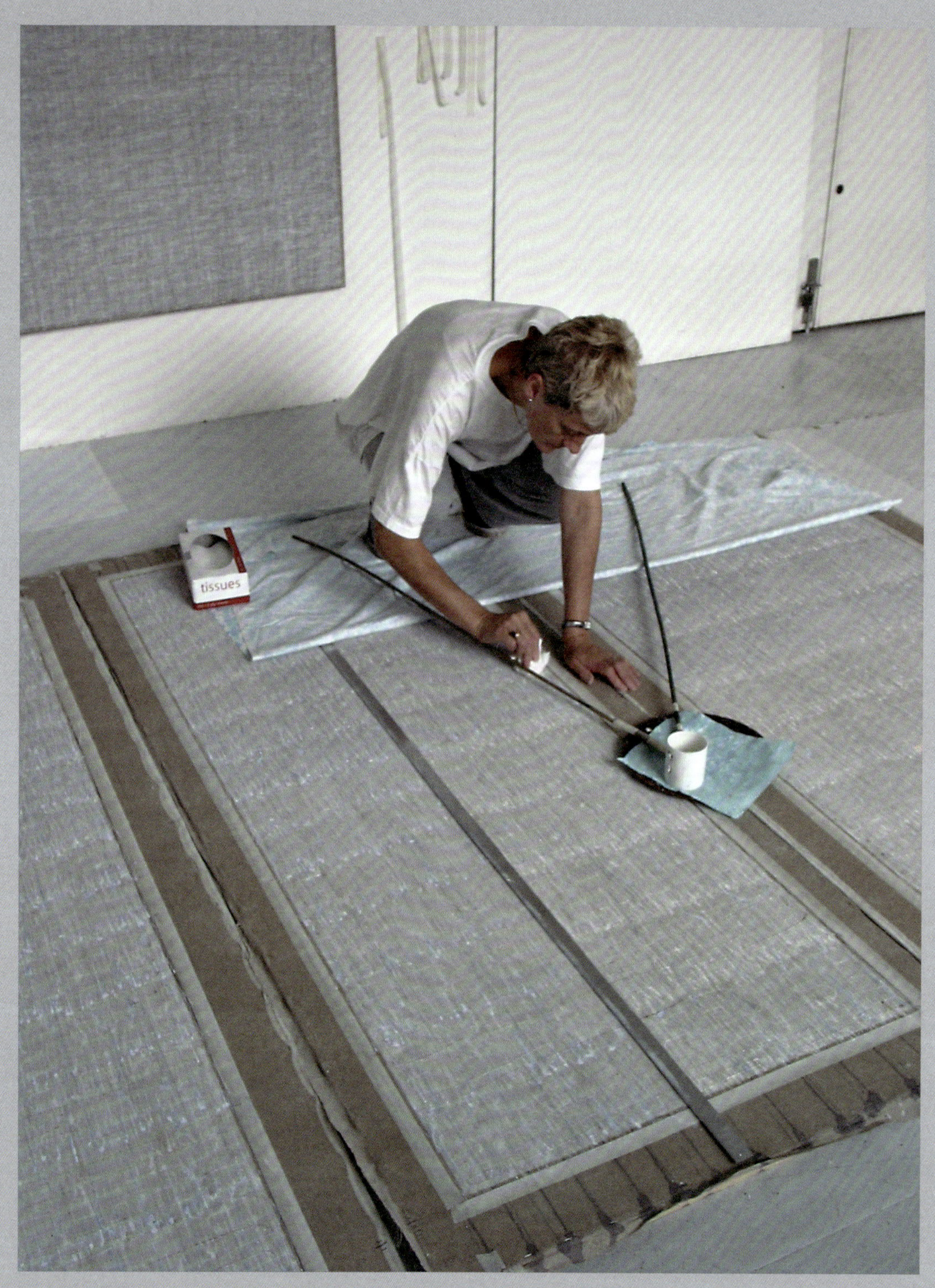
tissues

From Without to Within

Gillian Forrester

Close your bodily eye, that you may see your picture first with the eye of the spirit. Then bring to light what you have seen in the darkness, that its effect may work back, from without to within.

Caspar David Friedrich

BORDER WALKING

In the early twenty-first century it is considered more or less critical to the success of anyone connected with the art world—whether artist, curator, collector, or dealer—to travel extensively, either for exhibitions of one's own work or that of others, or to make site-specific works, receive prizes, take up residencies, and visit art fairs. This incessant motion seems to function as a contemporary analogue both to the movements of those in the service of imperial agendas in the eighteenth, nineteenth, and early twentieth centuries and to counter-culture journeyings in search of exotic locales in the 1960s, though present-day art-world travelers, who greet each other like erstwhile colonial administrators or hippies as they crisscross the globe, are more likely bound for Basel, Miami, and Venice than Calcutta or Marrakech. Rebecca Salter's travel practices are distinctive, however; she travels sparingly in order to conserve her time in the studio and tends to eschew fashionable art-world destinations. Achim Borchardt-Hume, in his essay in the present volume, describes Salter as a "border walker" who transits between cultures, genres, and modes of engagement. I would like to explore Borchardt-Hume's astute characterization of Salter by providing an account of her own extensive travels and their significance for her practice, as well as the possibilities for journeying, even pilgrimage, which her works offer both to the receptive viewer and to the artist herself as she creates them.

Salter describes herself as an abstract artist, a term in many respects accurate, but also a broad one, which she may have chosen as a strategy for tacitly discouraging other categorizations that might circumscribe both her practice and our understanding of her work. Two other commentators on Salter's work—Charlotte Klonk in an article from 1991, and more recently Anna Moszynska—have both proposed persuasive readings of Salter's works as landscapes, and in this essay I will develop this proposition, arguing for the centrality of landscape in Salter's practice from 1981, the year she embarked on her exploration of making works in two dimensions.[1] Landscape is indeed far from constituting a limited and limiting category for Salter, offering nothing less than *Erlebniskunst*, or a framework in which the entire world can be delineated.[2]

This fundamental aspect of Salter's practice persuades me to place her within a European Romantic tradition, although, as I will show, her aesthetic and physical trajectory, which has oscillated between West and East, indicates a far more complex history than my categorization might at first suggest. Salter's fascination with Japan has never been driven

fig. 1: Rebecca Salter working in her London studio, 2005

by the desire for "otherness" that has historically attracted Western artists to Asia, and her distinctive hybridity seems to stand outside of postcolonial discourse. Mapping as it does the internal topographies of human mind and spirit, Salter's highly reflexive work seems preoccupied with the exploration of a poetics rather than a politics, yet her project also has its own highly freighted politics, exploring as it does the central question of what it means, both for artist and viewer, to make art in a postmodern world.

MAKING THINGS

Born in 1955, Salter had a conventional middle-class upbringing and early education; she attended a girls' grammar school in Cheltenham, followed by a degree course in ceramics at Bristol Polytechnic, from which she graduated in 1977.[3] Salter's formal education proved to be foundational for her mature practice, though her unpredictable career trajectory might be read as a subtle subversion rather than a logical outcome of these formative if often frustrating years. Art school was not an unusual path for young women in Britain in the 1970s, but in a cultural climate where few female artists succeeded in establishing successful careers, it was typically regarded as a preamble for the domestic sphere, rather than a strategic career choice.[4] British art colleges in the mid-1970s were overwhelmingly masculine environments, where the heroic aesthetics of abstract expressionism and modernist monumental sculpture were the predominant models. Salter, like many of her female contemporaries (and indeed her contemporaries who were neither white nor heterosexual), faced formidable obstacles in her quest to construct her identity as an artist and function in a male-dominated milieu.

At grammar school Salter was discouraged from applying to art school on the grounds (fallacious, as her later work has eloquently demonstrated) that her drawing skills were deficient and that her chances of becoming a painter would be slight. She persevered, though was sufficiently influenced by this critique when she arrived at Bristol that she chose a specialization in ceramics, a somewhat marginalized field that straddles, often uneasily, the boundaries of art and craft. Ceramics also occupies an ambiguous space in terms of gender; although some of its most influential practitioners in the mid–twentieth century were (and indeed still are) male and often—as in the case of Bernard Leach—espoused a self-consciously heroic masculine style, the medium's capacity for extreme refinement and delicacy has also inevitably tethered it to notions of female practice, as exemplified by the work of the influential Austrian émigré potter Lucie Rie.[5] Salter's decision can be seen as a strategy for negotiating the complexities of gender politics and creating a space in which she could forge her own practice unimpeded by the expectations of her teachers and peers. Since early childhood, Salter had enjoyed "making things," and the hands-on element and the tactility of the clay also attracted her to ceramics.

Despite strenuous efforts by students and staff in the late 1960s to radicalize the teaching of art in British colleges, manifested most notably in the landmark sit-in at Hornsey College of Art in 1968,[6] the curriculum remained rigidly compartmentalized in accordance with the modernist paradigm, and students had few opportunities for broadening their practices by working across media and disciplines and engaging with theoretical issues.[7]

At Bristol, however, Salter was fortunate to have an enlightened teacher, George Rainer, who encouraged his students not to be circumscribed by their choice of medium, and the freedom to experiment she enjoyed under his tutelage proved foundational for her practice. Rainer's understanding of Asian art and aesthetics, which then were receiving little attention from British scholars and artists, helped to foster her own burgeoning interest in the field. As Sadako Ohki recounts in her essay in the present publication, during her time at Bristol Salter first encountered, in reproduction, *Genji Monogatari* (Genji narrative picture scrolls), which became the topic of her undergraduate thesis. The possibilities offered by the scroll form for constructing narratives and mapping space and time, which were so markedly distinct from Western paradigms, were to be highly influential much later when she turned to working in two dimensions.

Salter graduated from Bristol in 1977, a landmark event she commemorated by depositing her degree-show works in the River Avon. This gesture, readable either as destructive or regenerative in impulse, seems to have constituted a repudiation of the pedagogic strictures Salter encountered at Bristol but was also, more significantly, an early manifestation of her preoccupation with transience. Loss is critical to her practice, if simultaneously constituting a threat to its commercial success. Salter's abandoned Bristol works were large clay sculptures, which she deliberately fired at low temperatures to leave them in a fragile state, thus subverting the conventional transformation from common mud to permanent object that underpins the making of ceramics. Salter routinely destroys works she finds unsatisfactory, while retaining some which may be instructive from a technical point of view. Initially I was shocked by this ruthlessness but now understand it as a fundamentally Romantic impulse to experience loss, as underscored by her recent avowal of the "pleasure of bereavement" which follows such a cull.[8] Longing is also central to Japanese aesthetics, *wabi*, meaning the cultivation of imperfection and aging, derives from the word *wabishi*, meaning "want."[9] Salter also believes passionately in the necessity for the artist to acknowledge and learn from her failures: "I've always thought that it is really, really important to talk about things that don't work. When I was at art school there were no women artists; I never went to a lecture by a woman artist. There was a series of talks by men who would give you a sanitized career trajectory...."[10] Salter was to find her philosophical outlook validated less than two years later when, funded by a Leverhulme Scholarship and unencumbered by the freight of her student work, she traveled to Japan to study ceramics at the Kyoto City University of Arts. The expression "failure is the origin of success" is a central principle of creative endeavor in Japan, and the freedom to make mistakes proved to be highly productive for Salter.

EAST

Salter's decision to go to Japan was in many respects logical given her focus on ceramics but was nonetheless an unconventional one for a young British artist. In the late nineteenth century, the Aesthetic movement in Britain had appropriated formal qualities of Japanese art, primarily via *ukiyo-e* woodblock prints, but interest had waned by the 1920s, and the main nexus of contact between Britain and Japan in the visual arts was the field of decorative arts.[11] During Salter's student years British ceramics was still dominated by Bernard Leach, who trained in Kyoto in the early twentieth century and was a lifelong advocate of Japanese cultural practices, describing himself as "a courier between East and West."[12] Although Leach often has been designated "the father of British studio pottery," a moniker which still has currency, his work has not been universally admired.[13] Salter shared her teacher George Rainer's distaste for the patriarch's self-consciously artisanal pots and incessant proselytizing; she opted to study with the celebrated avant-garde potter Yagi Kazuo, whose sustained campaign to resituate ceramics outside of the narrow categorization of craft ran counter to Leach's passionate espousal of the Folk Craft movement.[14] Aside from the field of ceramics, there was a popular vogue in Britain for essentialized Eastern philosophy, epitomized by the American writer Robert M. Pirsig's bestselling life manual *Zen and the Art of Motorcycle Maintenance* (1974) and an interest in Japanese aesthetics among architects and designers, but British artists for the most part did not share their American contemporaries' fascination with Asia.[15] In the 1970s, before the advent of low-cost transcontinental air travel, Japan seemed remote and of marginal importance in comparison with New York, which was considered the hub of the international art scene.

Salter is essentially a solo traveler who, by her own admission, has never wished to belong to a group, and the prospect of developing her practice in isolation from the Western art world was enticing. Salter's self-sufficiency proved to be advantageous, since just two months before her departure she learned that Yagi Kazuo had died at the age of sixty-one, leaving her without a teacher in Kyoto. While this unfortunate event likely would have seemed catastrophic for a Japanese student, who would be habituated to a traditional pedagogic system based on hierarchy and mentorship, Salter found her situation exhilarating, and the absence of structured tuition provided a space in which she could experiment. Kyoto, as Sadako Ohki notes, was (and is) deeply conservative, and the radical intervention of Yagi and his fellow members of the avant-garde *Sōdeisha* (Crawling in Mud Association) notwithstanding, lineage and tradition still underpin the local production of ceramics to a great extent.[16] Salter's outsider status in Japan enabled her to participate in Kyoto's artistic, intellectual, and social worlds without being circumscribed by their complex codes. Japan's patriarchal society was unsympathetic to promoting indigenous women artists, who often felt compelled to move to New York in order to gain recognition for their work (as in the cases of Yoko Ono and Yayoi Kusama, for example). Salter's liminal status as female foreigner, however, set her outside the complex gender politics of Japanese society and art world, where she assumed the identity, as she has put it, of a "third sex," far from invisible, yet able to work with relatively few constraints.[17]

Salter also arrived at a moment when the traditional system of apprenticeship or mentoring had been partially superceded by the introduction of Western-style art schools, which in turn precipitated a crisis of confidence in traditional Japanese painting and a resurgence of interest in crafts. For Salter, this was a refreshing change from Britain, where the polarities between art and craft were still firmly entrenched, and the handmade was often denigrated vis-à-vis painting as an inferior form of production.[18] The inherently Western concept of the artist as creative genius is alien to traditional Japanese artistic practice, where stress is laid on the importance of mastery of skill through repetition, attention to detail, and sheer diligence.[19] Salter, who characterizes herself as a maker of "things,"[20] and, by her own admission, has been hard-wired with a Protestant work ethic since birth, found this artisanal model highly congenial. It is also critical for her that her works are well-made, though she acknowledges that this pursuit of the highest production values can be an endgame, resulting in the creation of an object that might be exquisitely crafted but devoid of vitality.[21] Salter admits to a tendency to "fiddle" with her works, and continually grapples with the challenge of determining the point when a work is finished.[22]

In Kyoto, Salter was also able to establish her reputation by showing her ceramics regularly. While there were few art dealers with galleries, commercial spaces that could be rented by the week abounded, even though, as Salter has wryly noted, the eclectic mix of offerings in such galleries meant that one's show might well be sandwiched between displays of *ikebana* and macramé.[23] Unlike recession-torn Britain, Japan's economy was booming, and Salter readily found a market for her work. Having originally intended to come to Japan for two years, she remained there for six. Although her experience of the visual was to be critical for the development of her practice, Salter has noted that the most significant experience of her sojourn in Japan was learning the language; she become fluent in reading and writing within a year and voraciously read Japanese texts on art, architecture, and aesthetics that were unavailable in English, as well as newly translated works on postmodernism and semiotics that were widely read in Japan before they became popular in Britain. Salter was also commissioned by the scholar Tada Michitarō to translate his publications and lectures, which explored Japanese notions of the relationship between space and time, and sacred and secular sites; much of her later work can be understood as a negotiation between these binary states.[24]

It is inevitable that one should want to read Salter's complex relationship with Japan, which is still evolving over thirty years after her initial arrival, in terms of postcolonial discourse. Edward Said's groundbreaking polemic *Orientalism* was published in 1978, the year before Salter's departure for Japan, and though his focus, controversially, was on the Western encounter with the Middle East, his theoretical standpoint has inevitably been grafted onto debates regarding the relationship between the West and Asia. Salter's engagement with Japan can be most productively viewed, not as a Saidian encounter with the exotic, nor even in terms of J. J. Clarke's ameliorative position of "affirmative enlightenment," but rather using the model adopted by Alexandra Munroe for her groundbreaking exhibition and publication, *The Third Mind*. Munroe examined the relationship between artistic practices in Asia and America in terms of Takeuchi Yoshimi's formulation of "Asia as method," a scenario in which Western cultural and political practices can be inflected by Eastern ones while both retain their own value-systems.[25] Salter has observed that her project in Japan was to "finesse the bridging of the two cultures," absorbing and understanding difference over a protracted period of time, rather than merely to superimpose discrete cultural practices onto one another.[26]

Although Salter was enjoying a successful career as a ceramicist in Japan, she made a radical decision to abandon the medium. The artist had become frustrated by the inherent

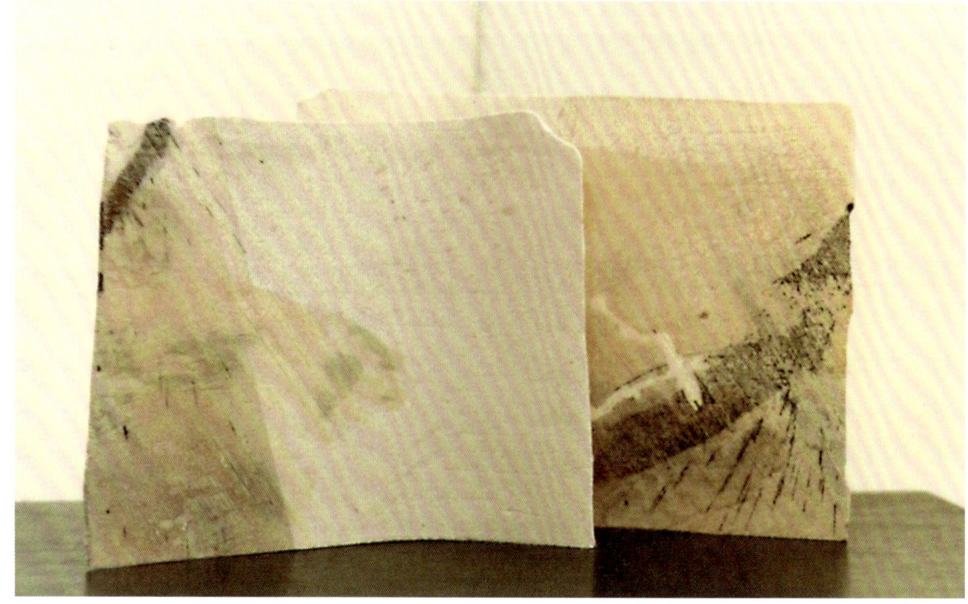

fig. 2: *Ceramic*, 1981
low-fired ceramic backed with Japanese paper
dimensions and location unknown

unpredictability of making kiln-fired objects and felt the need to exert greater control over the outcome of her works.[27] Salter's final ceramic works, known now only through an installation photograph, were slim pieces, more or less drawings on clay, which also betray a desire to work in two dimensions (fig. 2). In the same year, after studying calligraphy and woodblock printing, Salter began to make drawings and woodblock prints. Much of the quiddity of these distinctive works, which synthesize European and Japanese art-making practices and sensibilities, derives from the use of Japanese paper, prized highly by Salter for its soft surfaces, absorbent qualities, and subtle coloration. As Junchirō Tanizaki notes in a text on Japanese aesthetics, *In Praise of Shadows*, "Western paper is to us no more than something to be used, while the texture of Chinese paper and Japanese paper gives us a certain feeling of warmth, calm and repose. ... Western paper turns away the light, while our paper seems to take it in, to envelop it gently, like the soft surface of a first snowfall. It gives off no sound when it is crumpled or folded, it is quiet and pliant to the touch as the leaf of a tree."[28]

Salter's earliest two-dimensional work, made in 1981, was a suite of woodblock prints on paper, which she had stained with persimmon juice, imbuing it with an appearance of

fig. 3: *Untitled B118*, 1981
mixed media on Japanese paper
24⅜ × 37⅜ in. (62 × 95 cm)

fig. 4: *Untitled A57*, 1982
mixed media on Japanese paper
24¼ × 37¾ in. (64 × 96 cm)

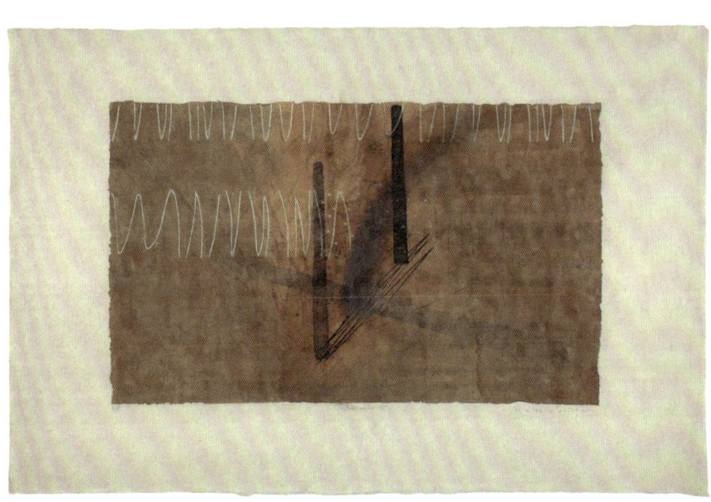

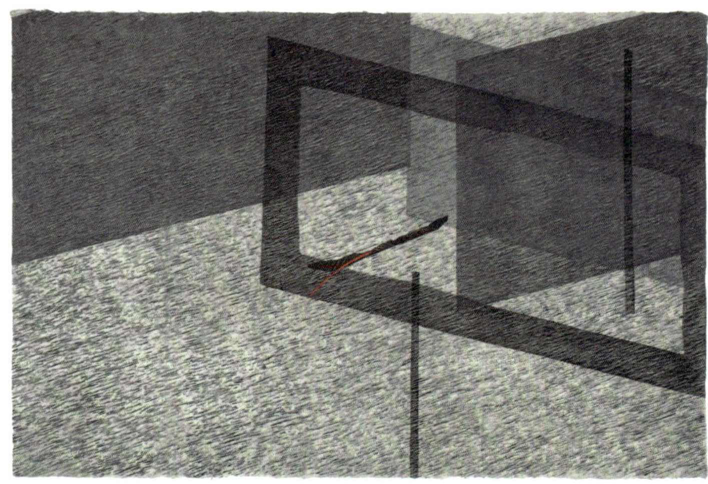

fig. 5: *Untitled B161*, 1984
mixed media on Japanese paper
25¼ × 63 in. (64 × 160 cm)

agedness characteristic of the aesthetic of *wabi* (see fig. 3). As Ohki recounts below, Salter was preoccupied with *ma*, or Japanese conceptions of space and time, and the prints represent an attempt to create what she has termed an "odd space."[29] In a series of woodblock prints made later in the same year (fig. 4), the intervals, created through the placement of vertical and horizontal lines and rectangular structures, embody a more sophisticated rendering of spatial and temporal elements. Salter then moved on to making hybrid works combining woodblock and drawing by printing on the back of absorbent Japanese paper and distressing the color that seeped through with a brush using *sumi* (fig. 5; see also figs. 39, 40). As Ohki has noted, these early works on paper seem to allude to scrolls and open books.[30] Salter was commissioned to design a book while she was in Japan (fig. 6),[31] and the diptych format that she favors (a subject to which I will return) has strong affinities with the double-page spread of a book. The horizontal format of the hybrid works, in particular the panoramic *Untitled B161* (fig. 5), irresistibly suggests to me the emergence of Salter's engagement with landscape and a working-out of the relationship between the interior spaces and the outer world, which seems a central concern of her practice.

fig. 6: Narumi Hiroyuki, *Boku sorekkiri*, designed by Rebecca Salter, 1983

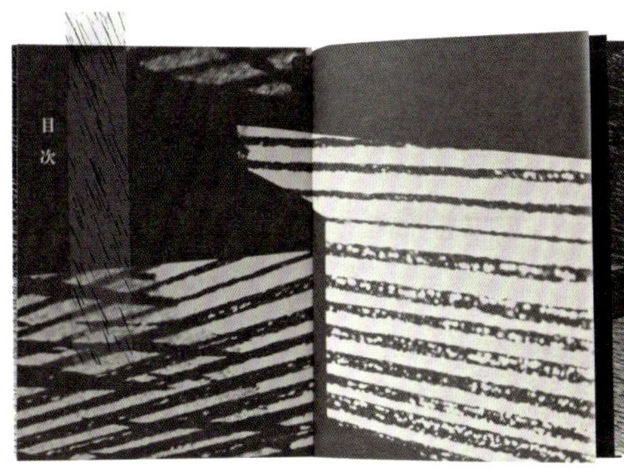

Salter, the border walker, is fascinated by the porousness of boundaries between spaces in Japan, where domestic environments are continually reconfigured for different purposes using screens, and gardens are traditionally designed to be contemplated from viewing stations or apertures in buildings (figs. 7, 8). This preoccupation with liminality plays out both through Salter's choice of subject matter and her hybrid working methods. Most crucially, Salter regards her works as "spaces and places," which she inhabits while she is making them, and has described how she experiences the powerful sensation of "disappearing," akin to a Zen meditative state, during the creative process.[32]

A NEW SPIRIT IN PAINTING?

In 1985, and not without misgivings, Salter left Kyoto for good and returned to London. In 1979, the year of her departure for Japan, Margaret Thatcher had taken office as prime minister and was now enjoying her second term; the intervening six years had witnessed

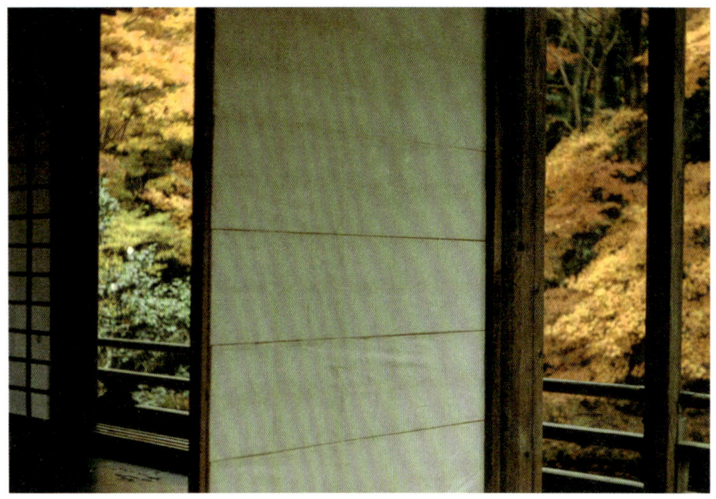

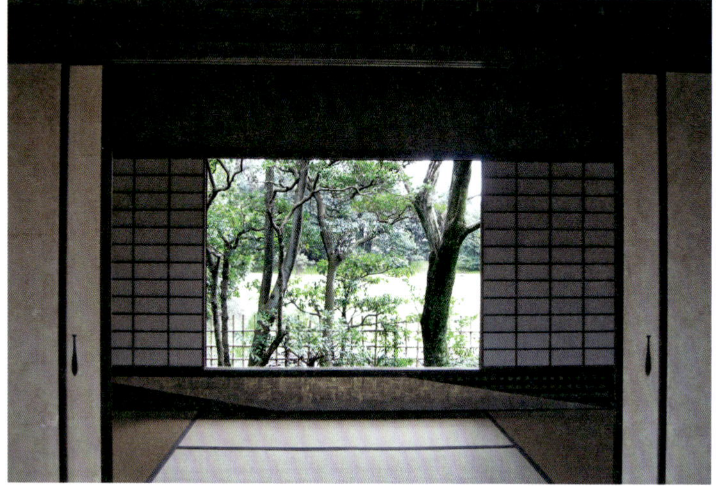

fig. 7: Kyoto interior: Sanzen-in, 1983

fig. 8: View from tea house at Katsura Palace, Kyoto, 2010

a strategic and a steady decline of governmental support for intellectual and creative endeavor, and the weak economy was not conducive to private and corporate collecting of art and support of public cultural institutions.[33] The prevailing taste in art, in keeping with the political tone, was conservative, and conceptual art, which had flourished in the 1970s, was being displaced by a focus on the more traditional media of sculpture and painting. In 1976, when Salter was still studying at Bristol Polytechnic, R. B. Kitaj had organized for the Arts Council of Great Britain *The Human Clay*, a polemical exhibition of the work of British figurative painters, in response to a perceived lack of appreciation for figuration. In his introduction to the exhibition's catalogue, he coined the now ubiquitous term "School of London."[34] Kitaj owned, "[I] felt very out of sorts with my time,"[35] but his polemical exhibition was to prove influential, laying the ground for the Royal Academy's more ambitious and high-profile exhibition *A New Spirit in Painting*, organized by Christos M. Joachimides, Norman Rosenthal, and Nicholas Serota in 1981. The triumvirate, who prefaced the catalogue by noting, "We are in a period when it seems to many people that painting has lost its relevance as one of the highest and most eloquent forms of artistic expression," selected thirty-eight painters from Britain, the United States, and Europe, including a significant number from West Germany.[36]

Although the exhibition was indubitably invigorating, exposing the British public to some remarkable works by unfamiliar painters, its message for emerging artists was not encouraging. Some commentators, mainly in retrospect, hailed the show as pathbreaking, but others drew attention to the fact that the majority of the painters featured were over fifty, and two (Picasso and Philip Guston) were dead, undermining the exhibition's claims to represent the cutting edge.[37] Abstraction (presumably strategically) was sidelined, with Britain represented by the single artist Alan Charlton, inducing Richard Shone in his *Burlington Magazine* review to note that "non-figurative art has not been understood or practised with much distinction in England."[38] All the featured painters were men (and, indeed, white). Despite some claims made at the time for the expanded opportunities available for women artists, as Gill Perry has noted, they remained marginalized until the emergence of the Young British Artists in the late 1980s.[39] The practitioners of the "new British sculpture" in the 1970s and the early 1980s were also predominantly male, including Tony Cragg, Richard Deacon, Bill Woodrow, and Richard Long.[40] Media divisions still prevailed in art school curricula until Jon Thompson, principal of Goldsmiths College, introduced a radical cross-disciplinary "open studio" curriculum in the late 1980s,[41] fostering the new "Sensation" generation of artists whose aesthetic of subversion and brash self-promotional strategies were to transform the British and international art scene.

Salter's move was in some respects predictable, since the Japanese art world had considerable limitations, and London would provide a larger arena in which to make and show her work; however, a more fundamental motivation for uprooting herself from an environment which had proved congenial and productive was a need for discomfort, which is critical to her practice. Salter's typical pattern of working is to make a series of objects over a period of time using a specific format and/or technique but, when habit and comfort set in, to abandon that trajectory and explore new concerns. She confesses to having a low boredom threshold (not altogether convincingly, since her methods are always self-consciously laborious), and avers that the only way of maintaining her interest is "to keep the risks high."[42]

fig. 9: *Untitled C129*, 1986
mixed media on Japanese paper
34½ × 48 in. (88 × 122 cm)

In keeping with this strategy, Salter returned to London in the knowledge that she would find an inhospitable cultural climate. The move from Kyoto to London ruptured the productive trajectory of her work, and now in a radically different environment she was forced to reimagine her artistic identity. Salter continued to make drawings on Japanese paper (fig. 9), working in the bedroom of a rented basement apartment and then in a diminutive studio in Old Street. She maintains that it should be possible for an artist to work in any environment, but these dense and convoluted images (which she has preserved, fortunately, considering them instructive) lack the vitality and spatial complexity of her Kyoto prints/drawings. Indeed, they suggest a sense of claustrophobia, and the dark reddish coloration, reminiscent of Mark Rothko's late works, is a radical departure from the muted, *shibui* palette she had explored in Japan.[43] Salter dislikes such "off colors," as she terms them, and has since only used them for the grounds of her canvases, systematically erasing or obscuring them in the course of making a painting; her use of such pigments in these transitional works implies an almost masochistic desire to subvert her own aesthetic. Salter's move to a more congenial studio in the Space complex in Dalston in 1989 proved beneficial, but she had already extricated herself from the impasse in which she found herself, once again through the stimulus of traveling.

A WHOLE WORLD OF IMAGES IN THE WATER

In the summer of 1988 Salter made her first trip to the Lake District in the northwest of England, staying near Skiddaw. The north of England has exerted a powerful hold on the imagination of British writers and artists from the eighteenth century onward (late in his career, J. M. W. Turner said of the border outpost of Norham Castle that it "made him a painter"), and the rugged landscape, extraordinary quality of light, and constantly mutating weather made a deep impression on Salter. She revisits the area every year in the fall with

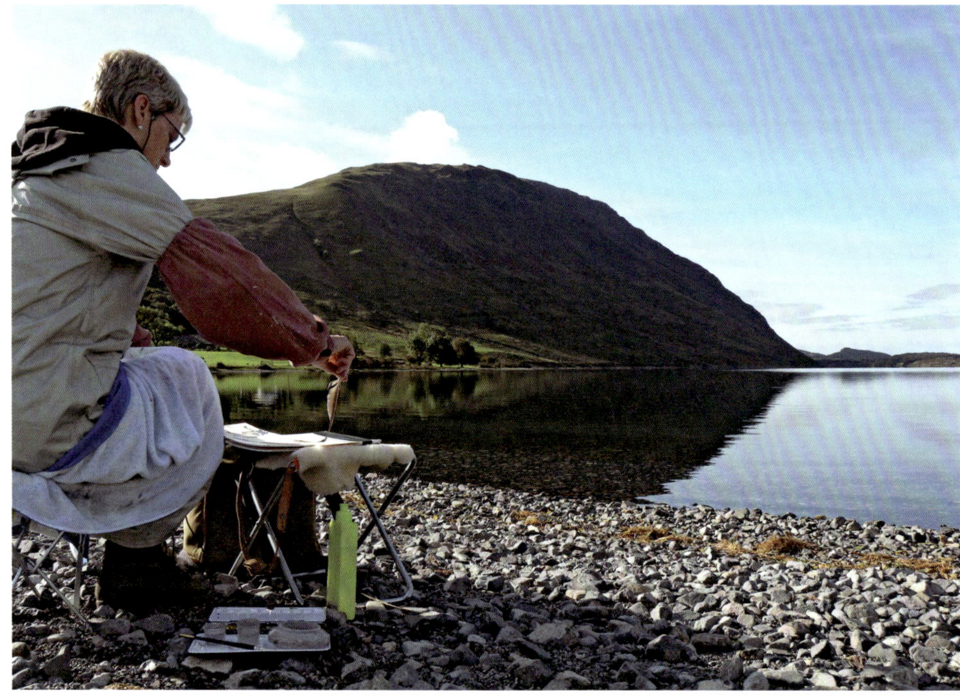

fig. 10: Rebecca Salter sketching at Wastwater, 2008

her husband, Geoffrey Winston, and since 2007 has stayed in a remote farmhouse opposite the Screes at Wastwater, an area described by William Wordsworth as "long, narrow, stern and desolate."[45] Wordsworth's contemporary Samuel Taylor Coleridge found Wastwater more exhilarating, writing in his notebook on 4 August 1802, "When I first came, the lake was a perfect mirror—and what must have been the glory of the reflections on it! The huge facing of rock, said to be half a mile in perpendicular height, with deep ravines and torrent-worn, except where the pink-striped Screes came in as smooth as silk—all this reflected, turned into pillars, dells, and a whole world of images in the water."[46] Wastwater is virtually unchanged today, remaining too remote for cellphone and even radio reception; Salter finds its austerity highly sympathetic and a rich source of inspiration (figs. 10, 11).

On every trip Salter brings a single thirty-three-page sketchbook made from Banks Cream, a paper with a softly toned coloration which subtly modulates as the light changes (Salter dislikes working on harsh white surfaces). She sets herself the objective, which is always met, of filling it in the course of six days. Salter also prepares herself for such expeditions by bringing watercolor and drawing implements (which are often eccentric; recently she has favored using the tips of sharpened wooden stirrers filched from coffee shops) but frequently finds them inappropriate for the circumstances in which she finds herself and is forced to improvise with whatever comes to hand.[47] The unpredictability of the weather in the Lake District tests the plein-air artist's ingenuity, and Salter recalls how on a very wet day on a sketching expedition, "[I] actually worked with the rain; I let the sketchbook get wet and just drew into the rain."[48] Such uncertainties are exhilarating for Salter, who, as Ohki suggests, is constantly striving to maintain equilibrium between her competing desires to allow chance to operate in her work and to control outcomes.[49] Salter's new tools, such as

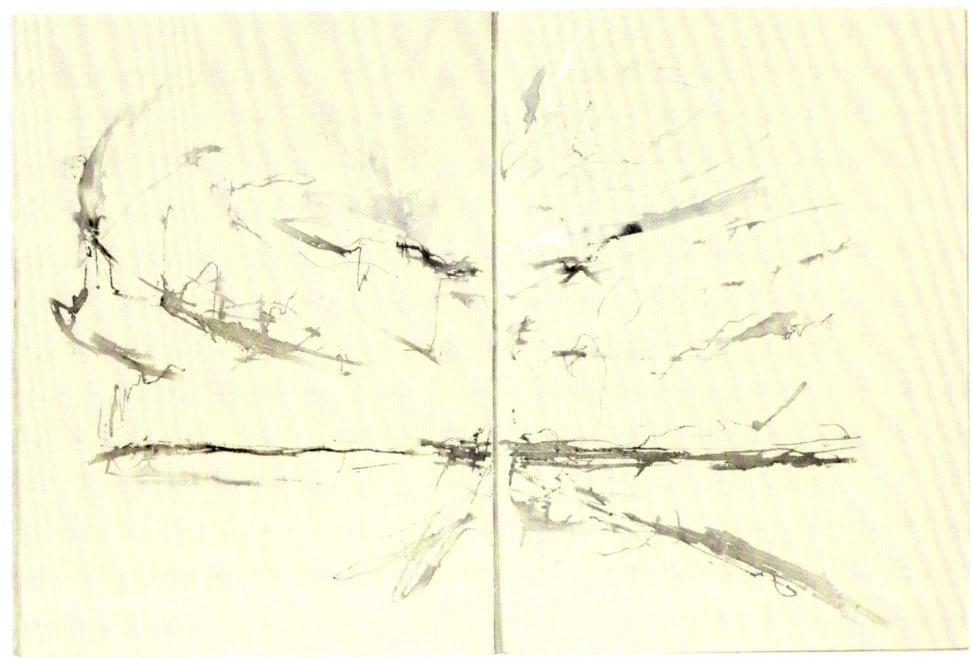

fig. 11: *Wastwater Looking East*
spread from Lake District sketchbook
2008
watercolor on paper
12½ × 9½ in. (31.1 × 24.1 cm)

the feathers she used for her Wastwater sketches in 2008 and 2009, often become incorporated into her practice after she returns to the studio. She characterizes this constant process of experimentation and adaptation as "that eternal quest to keep drawing alive."[50] Salter will also use scraps of paper while her drawings in the sketchbook are drying, but notes, "It's never as exciting as doing it in a book. I think part of it is that you can't take the pages out. ... I really like the high stakes of the book. I like the fact that nobody can have them, and I like the fact that they are in the order in which I make them. ... I don't ever do sketchbooks in the studio apart from working out sums and measurements."[51]

Salter's modus operandi as a practitioner of landscape is distinct from that of some of her contemporaries such as Andy Goldsworthy, Hamish Fulton, and Richard Long, whose methods typically involve mapping and reconfiguration of the physical terrain. (Long has recently noted, "I enjoy the simple pleasure of ... passing through the land and sometimes leaving (memorable) traces along the way.")[52] Salter leaves no mark on the landscape, and although she would not necessarily characterize herself as a "green" artist, her practice implies a profound respect for the environment. In some respects Salter's practice of immersing herself in the landscape for short periods at regular intervals, followed by processing the experience in the seclusion of the metropolitan or urban studio, is much more congruent with the conventions of a Romantic landscape tradition exemplified by artists such as Turner, John Constable, and Caspar David Friedrich. Once she has returned to London, however, Salter's method is radically different from that of her predecessors, who used on-site sketches as a basis for finished works; though her brief but intense sojourns in the landscape clearly function as potent sources of inspiration for her studio works, she never refers to her sketchbooks or to her photographs again, as if the very act of drawing or photographing outdoors seems to have served to lodge the experience in her mind. This is a process that she admits not to understand herself, but simply concludes, "I never cross-reference anything. It's as if I feed it in and shut the door."[53] Her studio is strikingly bereft of visual materials other than her own works-in-progress, and the notion of having the archetypical artist's pinboard, laden with art-historical referents, is horrifying to her.[54] Given the place of the Lake District in the Romantic imaginary, it is inevitable that Salter's response to the landscape would be mediated to a certain extent by her knowledge of art history, but her eschewal of the work of others once she enters the hermetic space of the studio enables her to purge her mind of associations and inhabit her own "whole world of images."

FRAGMENTS FROM THE FUTURE

Salter's first visit to the Lake District inspired a new series of drawings on Japanese paper, which are powerfully suggestive of the piercing light, ever-changing atmospheric conditions, and expansive terrain of the northwest of England. Though owing much to her earlier exploration of calligraphy, these confident works are characterized by a more gestural mark making and richer surface textures than are evidenced in the works she made in Japan and signal a regeneration of her practice. After making the drawings, Salter cut them up and reorganized the pieces, mounting them on a backing sheet, a working method which she was to sustain for a decade (fig. 12). In 1990 Salter began to make paintings, which she also cut up and reassembled. There were practical considerations for the move to canvas, as the large sheets of Japanese paper she was using for drawings were unwieldy and costly to frame; however, her decision to eschew the "seductive softness," as she has put it, of handmade paper also speaks to her need for discomfort and new technical challenges. Salter finds the surface of oil paintings harsh and unsympathetic, but by scraping the canvas and applying layers of acrylic mixed with natural pigments, thereby avoiding shiny binders, she mimics the silky texture of paper. The challenge with which she grappled in her first canvas, *Untitled F112* (fig. 13), was to saturate the canvas sufficiently so that it would not fray when cut, while simultaneously creating a soft surface. In the same year that she began to paint on canvas, Salter made the transition from Japanese to Western paper.

Salter's method of cutting and reassembling has some affinities with the aleatory "cut-ups" made in the 1960s by the Beat Zen writers and artists, whose interest was in the operation of chance;[55] but as ever with Salter, the relationship between hazard and control is a complex one. As Charlotte Klonk notes in "Towards the Problem of an Aesthetic of Nature in Contemporary Art," the artist organized her fragments "deliberately, but not intentionally."[56] In this illuminating article, written after a studio visit in 1991 and shortly after Salter's transition to painting, Klonk perceptively linked Salter's practice, along with that of the artist Nicky Hirst, to a concept of nature explored by Goethe in his late novel *Wilhelm Meisters Wanderjahre*, seeing a concern on the part of both artists to maintain a delicate equilibrium between objectivity and subjectivity.[57] She observes, "These women explore nature's processes, not in an objectifying way (which would lead to naïve realism) but empathetically by tracing natural operations within the course of their work processes, and within their own creative activity."[58] Klonk traces the lineage of their practices back to the moment before art and science became polarized in the mid-nineteenth century, when the

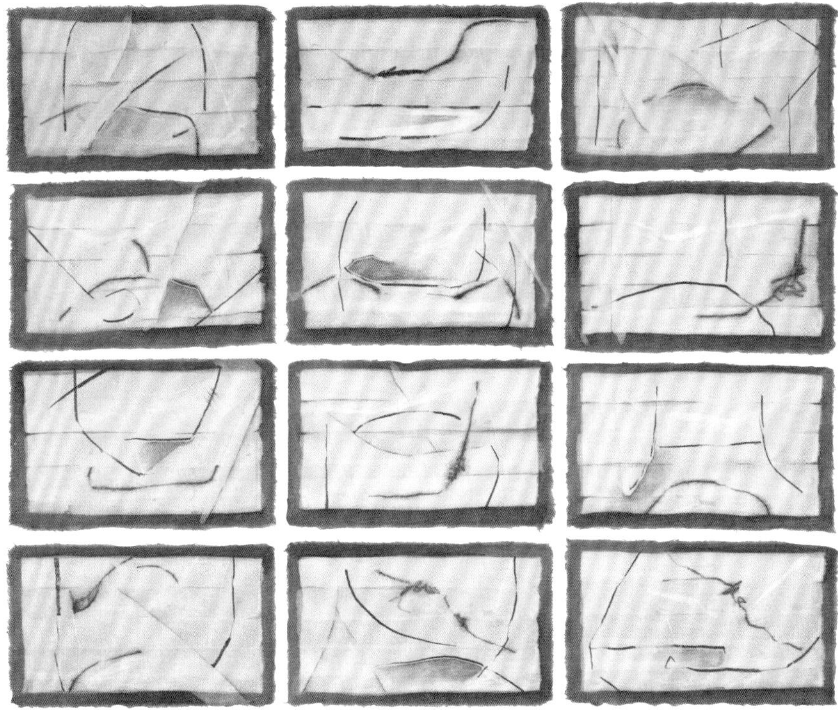

fig. 12: *Untitled D58*, 1988
mixed media on Japanese paper
18½ × 23¼ in. (47 × 59 cm)

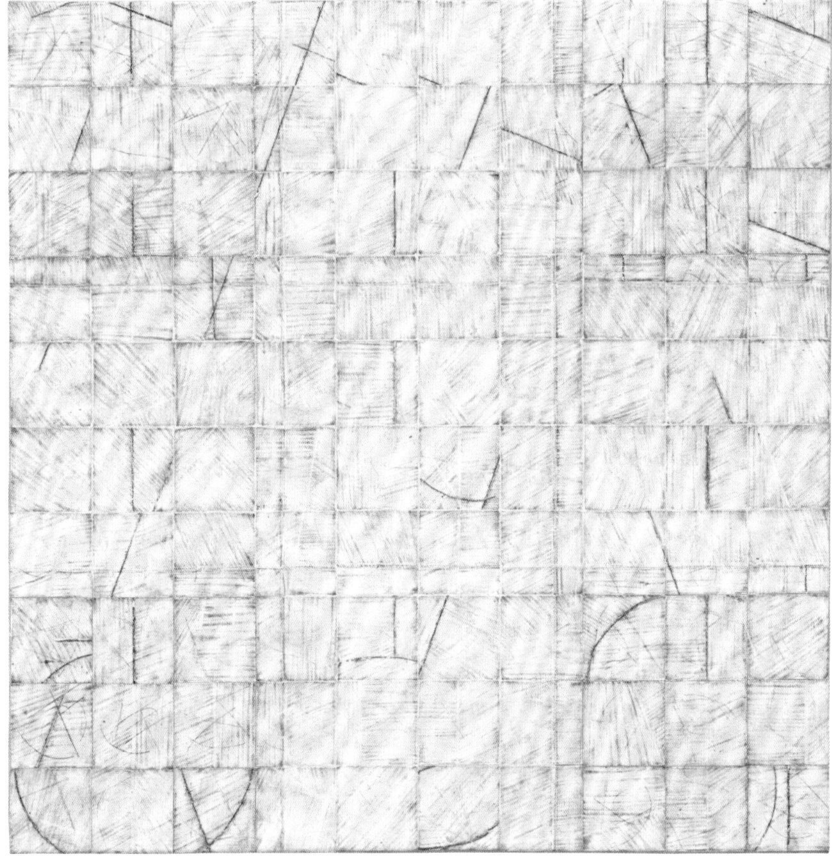

fig. 13: *Untitled F112*, 1990
mixed media on canvas
30 × 30 in. (76.2 × 76.2 cm)

fig. 14: *Untitled J48*, 1995
mixed media on paper
29⅞ × 44⅛ in. (76 × 112 cm)

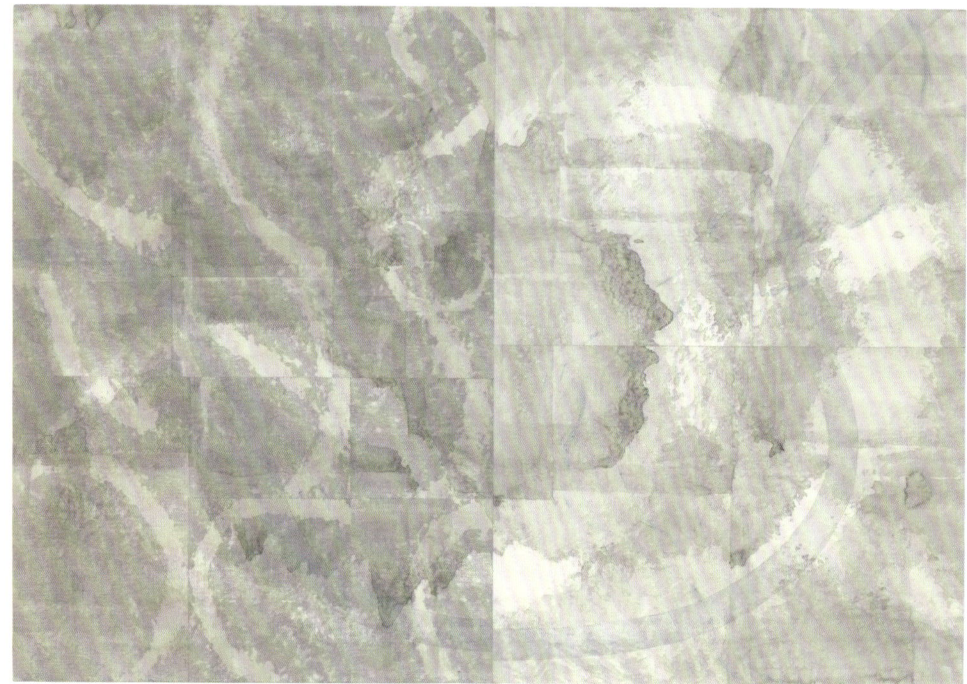

former was "relegated to the merely subjective and science took over the factual ground."[59] For Klonk, Salter's project is Romantic in impulse, yet, in an era where technology and nature are in contention, it is also highly relevant to modern life, mapping a path which mediates the subjective/objective divide (fig. 14).

Salter's "cut-ups" can also be seen as analogous to the German Romantic philosopher Friedrich Schlegel's theorizing of the function of the fragment, both "from the past" and "from the future" in artistic creation.[60] For Schlegel, making a whole, finished work from fragments was central to the Romantic endeavor, yet the inherent difficulty of such a project paradoxically meant that such a work would inevitably be doomed to remain incomplete. The creation of an autonomous and finished whole is critical to Salter's project, however, and her capacity for resolving the fragmentary nature of experience may be the source of the calming effect that viewers attest to experiencing in the presence of her works. Nevertheless, completion is always hard-won; Salter estimates that her large cut-up painting *Untitled J1* (fig. 15) took between two and three months to execute. The stakes were high as she would only be able to determine whether the work had succeeded after she had applied the last fragment, and, since she used a strong adhesive, mistakes could not be rectified.

LOOKING FOR A PERFECT SPACE

Although Salter preserves the pigments she scrapes from her canvases during the laborious layering sequence (see fig. 33), she never documents the progress of her works, and the stages of creation remain hidden though still present under the surface. Salter prefers to think of her art making in terms of a "passage" rather than a process, an evocative word which aptly conveys the journey that she makes to create each work.[61] As I have noted above, she also thinks of her works as "places to be" while she is making them, recalling the artist Agnes Martin's remark that she was "looking for a perfect space" when making her own work.[62] Salter's pieces are also places to wander freely; her method is essentially improvisational, and, although her works are distillations of her experience, she never begins

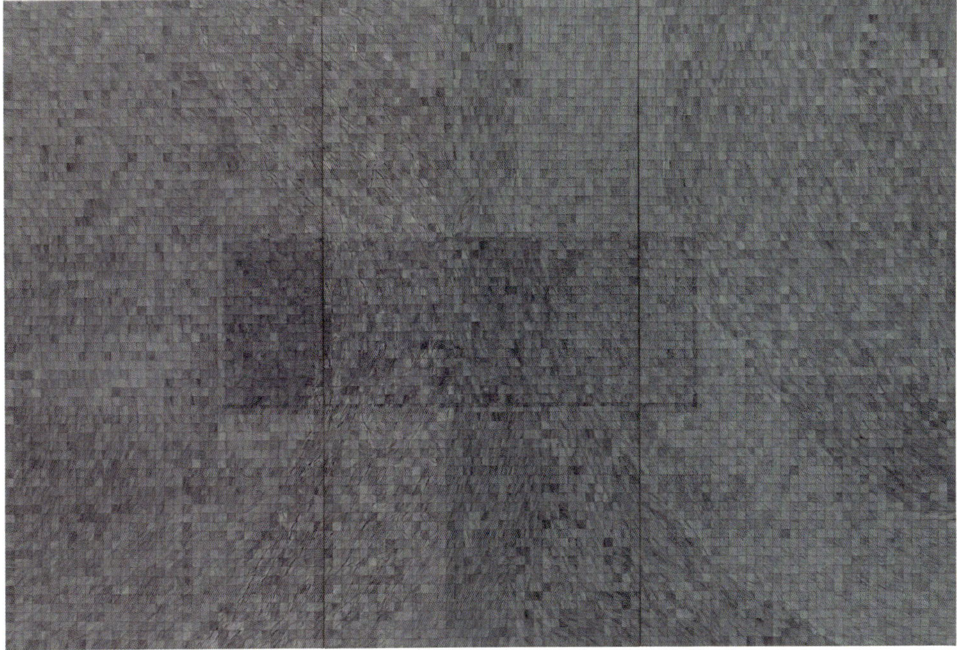

fig. 15: *Untitled J1*, 1994
mixed media on canvas
72 × 108 in. (182.9 × 274.3 cm)

with an image in mind.[63] Her works offer an analogous experience for the viewer; as Klonk noted, the experience of contemplating Salter's works is akin to finding "oneself on an extensive voyage within the picture with no beginning and end."[64] Though she might reject the analogy (Salter distances herself from the performative model of the heroic virtuoso artist) this intensely experiential method has affinities with that of Turner, whose self-absorption and capacity for engaging with his works when he was creating them was legendary (fig. 16).[65] Like Turner, Salter scrutinizes her works at close quarters while she is making them but does not sit and contemplate them as, famously, was Mark Rothko's practice.[66]

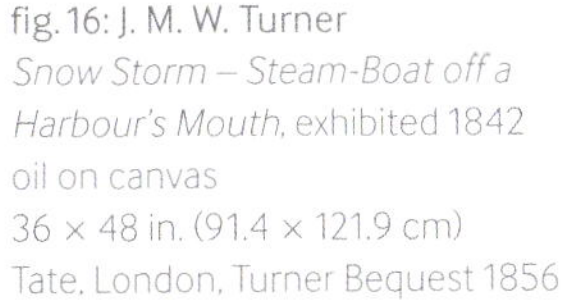

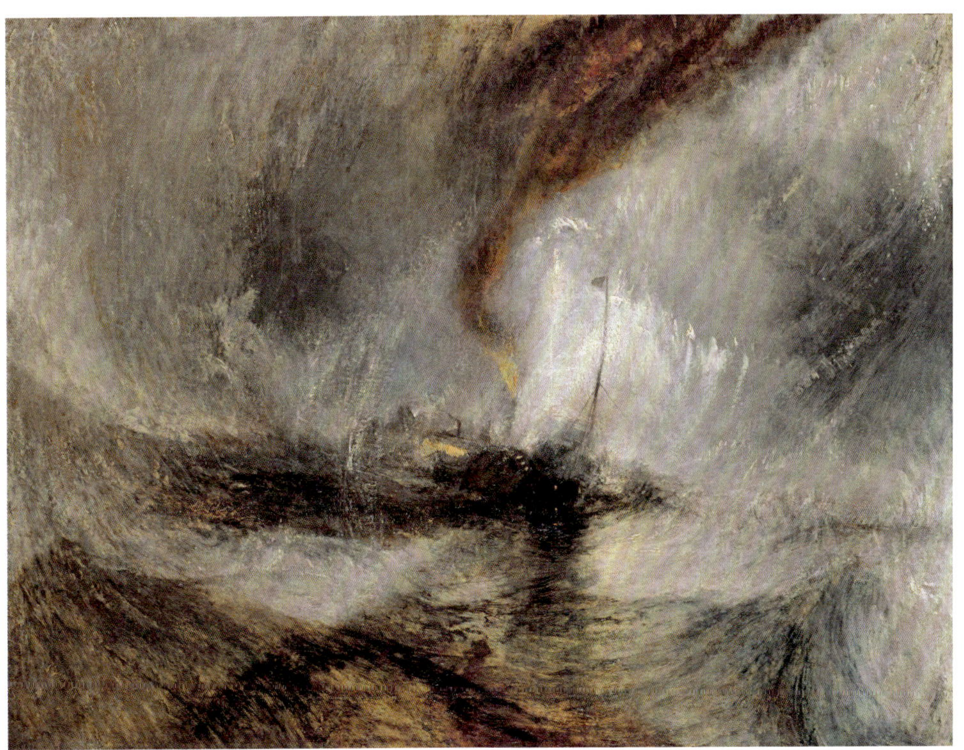

fig. 16: J. M. W. Turner
Snow Storm – Steam-Boat off a Harbour's Mouth, exhibited 1842
oil on canvas
36 × 48 in. (91.4 × 121.9 cm)
Tate, London, Turner Bequest 1856

fig. 17: Caspar David Friedrich
View Through an Open Window in the Artist's Studio, 1805–06
sepia on paper
12⅜ × 9¼ in. (31.4 × 23.5 cm)
Belvedere, Vienna

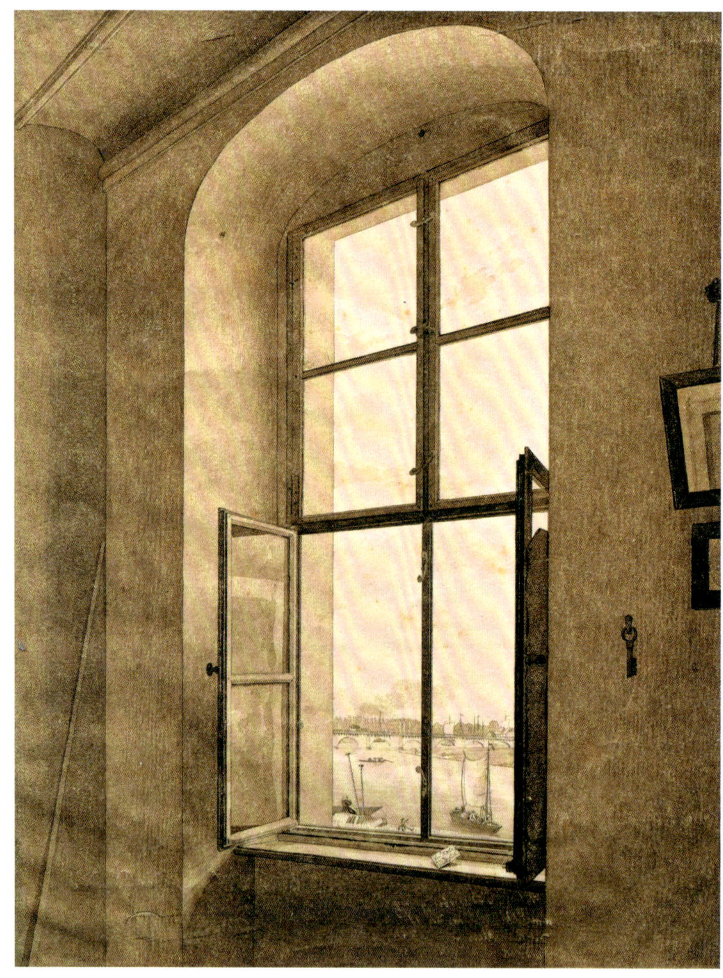

In 1992 Salter purchased a space in a studio complex occupying a former industrial site in Finsbury Park, London. Although she feels that it should be possible for her to work anywhere and is resistant to fetishization of the studio,[67] the move to the more spacious and lighter space indisputably had a profound impact on her work, which became more expansive and luminous. Despite her profound engagement with landscape, Salter has never considered moving to rural isolation and feels stimulated and grounded by living and working in such a busy, diverse, metropolitan neighborhood. The studio satisfies Salter's need for seclusion, yet its north-facing window, which overlooks a busy railway line, connects her with the external world, providing a state of equilibrium between inner and outer worlds, which was also a keynote of Romantic practice (fig. 17). The trees visible from the window, which change their leaves with the seasons, provide a temporal framework as well as an index of the processes of nature.

MANHATTAN OF THE DESERT

In the late 1990s Salter abandoned making her cut-up canvases and drawings. Her original impulse had been to create interstices between the fragments, "spaces to be" for the viewer, but, as Charlotte Mullins has noted, Salter had come to feel that "the internal edges caused by the cuts were restricting the eye from entering the work."[68] The meticulous crafting of the works had suddenly become a constraint, and the works now seemed overdetermined to Salter, disrupting the delicate equilibrium of objectivity and subjectivity, to use Klonk's

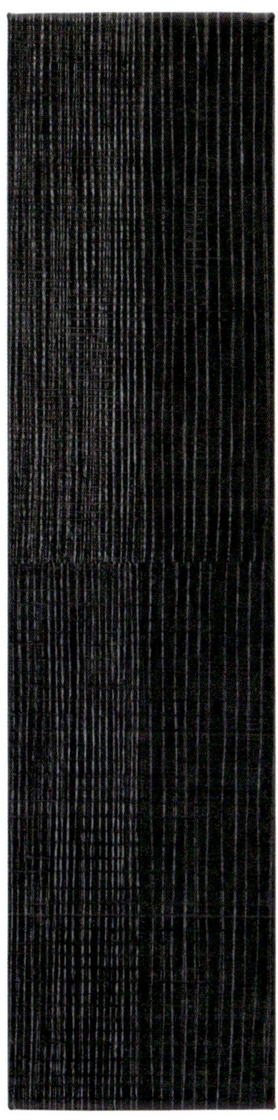

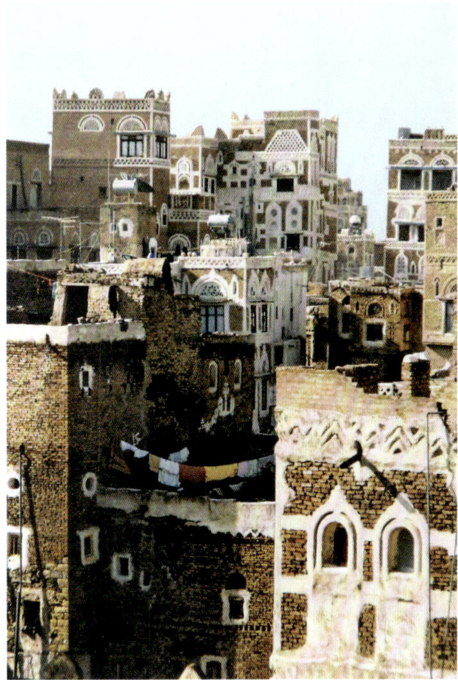

fig. 18: *Untitled BB32*, 2001
mixed media on canvas
28 × 7 in. (71.1 × 17.8 cm)
Private collection, Cambridge, UK

fig. 19: Sana'a, 2001

formulation. When a method of working or format has outlived its purpose for her, Salter ruthlessly abandons it and moves on to something new, typically stimulated by her travels.

In 1998 Salter visited Yemen for the first time and was fascinated by the organic relationship of landscape and architecture in the cities of Sana'a and Shibam (known as the "Manhattan of the desert"), where she stayed. Salter returned in 1999, 2000, and the summer of 2001 to stay in Sana'a with a friend's mother and was intrigued by the hermetic internal spaces of domestic architecture, which are clearly delineated by function, unlike those she had encountered in Japan. The sexes are strictly segregated, but Salter, again as an outsider and occupying the status of the "third sex," could move relatively freely between the gendered spaces. In the aftermath of the events of September 11, 2001, a return to Yemen for Salter would be extremely difficult. It appears in her inventory of longing as the "ultimate lost place," but its traces survive in her works, most overtly in the series of vertical paintings she made in 2000 and 2001, a new format, which references the tall buildings and doors of Sana'a and Shibam (fig. 18). It is also possible to read in the texture of lines incised into the grounds of these works an allusion to the white tracery that is a distinctive feature of the Yemeni architecture (fig. 19).

BETHANY

Of all Salter's travels, with the exception of her formative years in Japan, her three-month residency at the Josef and Anni Albers Foundation in Bethany, Connecticut, in fall 2003 has been arguably the most transformative for her practice. Although just a thirty-minute drive from New Haven, rural Bethany feels far from the bustle of city life, and the Foundation's two studio houses provide artists with a tranquil and secluded environment in which to work. Salter has admitted that she viewed her visit with some apprehension; although making periodic forays to the countryside is a critical component of her method, she feels most at home in the metropolis and was uncertain how she might function effectively away from her studio. The residency is a commonplace of the artist's, writer's, and scholar's life today, but

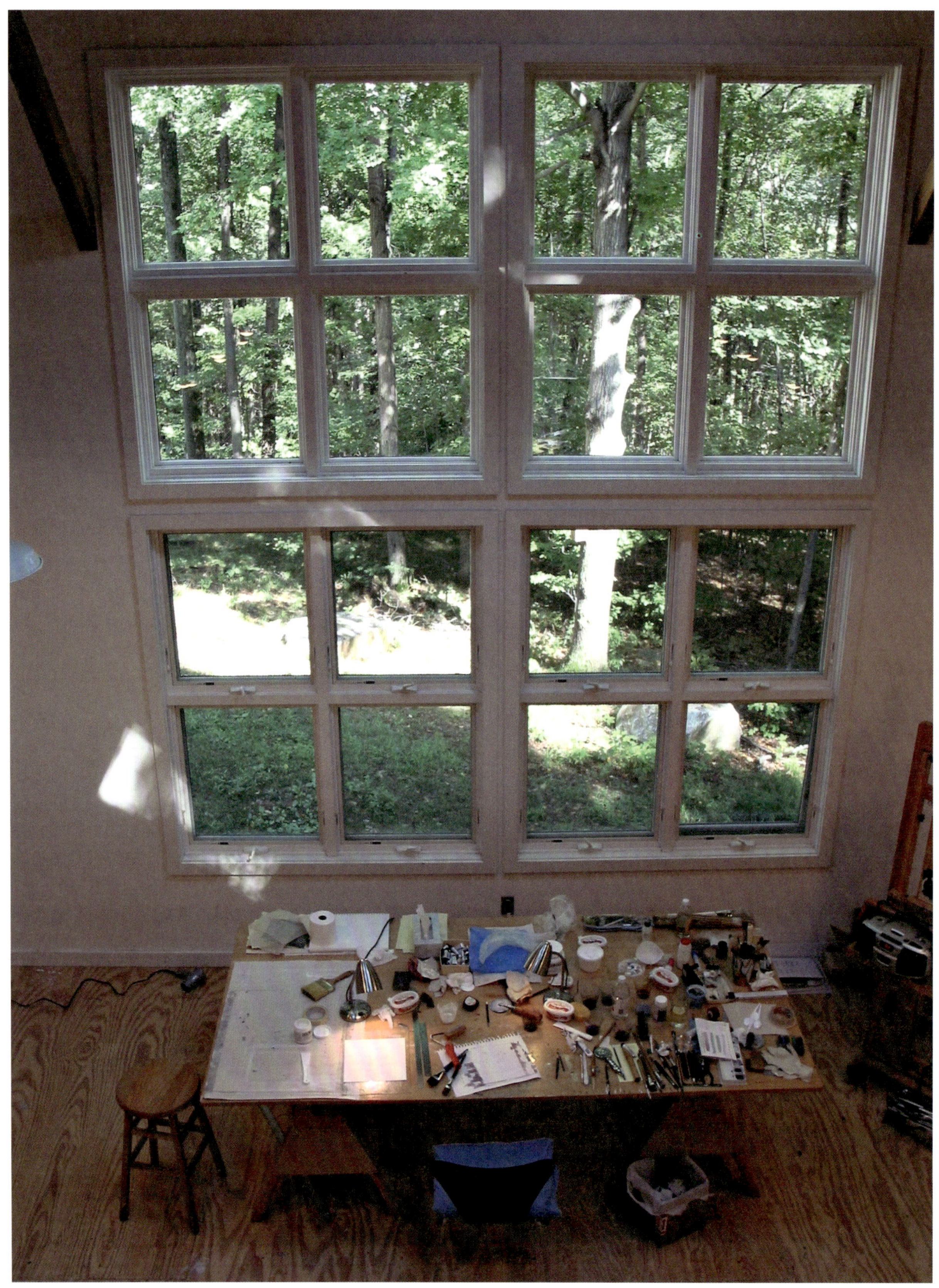

fig. 20: Interior of Rebecca Salter's studio, Josef and Anni Albers Foundation, Bethany, Connecticut, 2003

working in an unfamiliar environment, often without the habitual materials and tools of one's trade, can be practically challenging and destabilizing to one's sense of identity.

Salter had envisaged that her work at Bethany might take the form of an homage to Josef Albers's practice. In readiness she brought with her a set of square pieces of aluminum to use as supports, but she quickly recognized that this period of self-imposed exile from London would provide her with a space to evaluate and recalibrate her own practice. Salter focused on drawing, punctuating hours in the studio with walks in the Foundation's wooded grounds (fig. 20). She finds the over-lush spring and summer landscapes overpowering, preferring the more muted colors (and, I would suggest, the elegiac mood) of fall and winter, so the timing of her stay at Bethany, between September and December, was propitious. The opportunity to contemplate the natural world in a much more sustained manner than she had previously enjoyed resulted in a series of drawings that vividly evoke natural forms while simultaneously representing a return to a Japanese-inflected calligraphic line (figs. 21, 22). Salter's emotional register is always finely calibrated, but the drawings do possess a quiet exuberance, which indicates how liberating she had found her Bethany sojourn. Salter later noted, "I think that I really discovered, what I knew all along, was that I'm really about drawing. I think painting is the wrong word. ... In Japanese, draw and paint is the same word, which probably has its origins in the word to 'scratch.' Painting comes out of calligraphy because they both use the brush. It's in the language because it comes from writing. That's what I think I do. I draw pictures."[69]

During her time at Bethany Salter also made regular field trips to New Haven, visiting the Yale Center for British Art and the Yale University Art Gallery. The friendships forged during that period with colleagues at Yale and the Foundation have been nurtured through meetings in London and Salter's regular visits to New Haven. In 2004 the Center acquired *Bethany Squares*, a suite of drawings she had made during her residency. The work, both homage to Albers's square and a summation of her experience at Bethany, functions as a

fig. 21: *Bethany 8*, 2003
mixed media on paper
19¼ × 25¼ in. (49 × 64 cm)

fig. 22: *Bethany 11*, 2003
mixed media on paper
19¼ × 25¼ in. (49 × 64 cm)

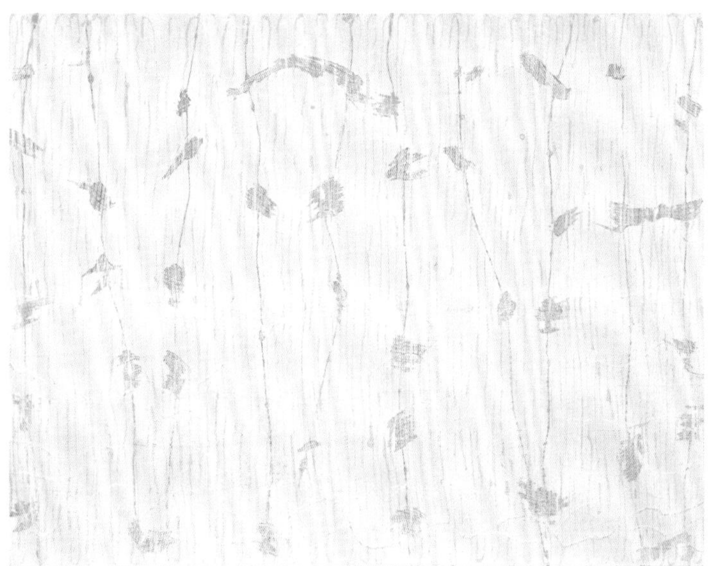

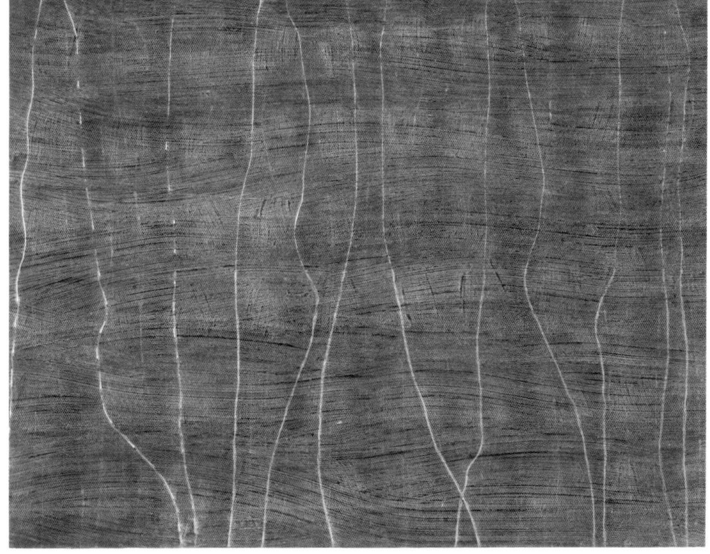

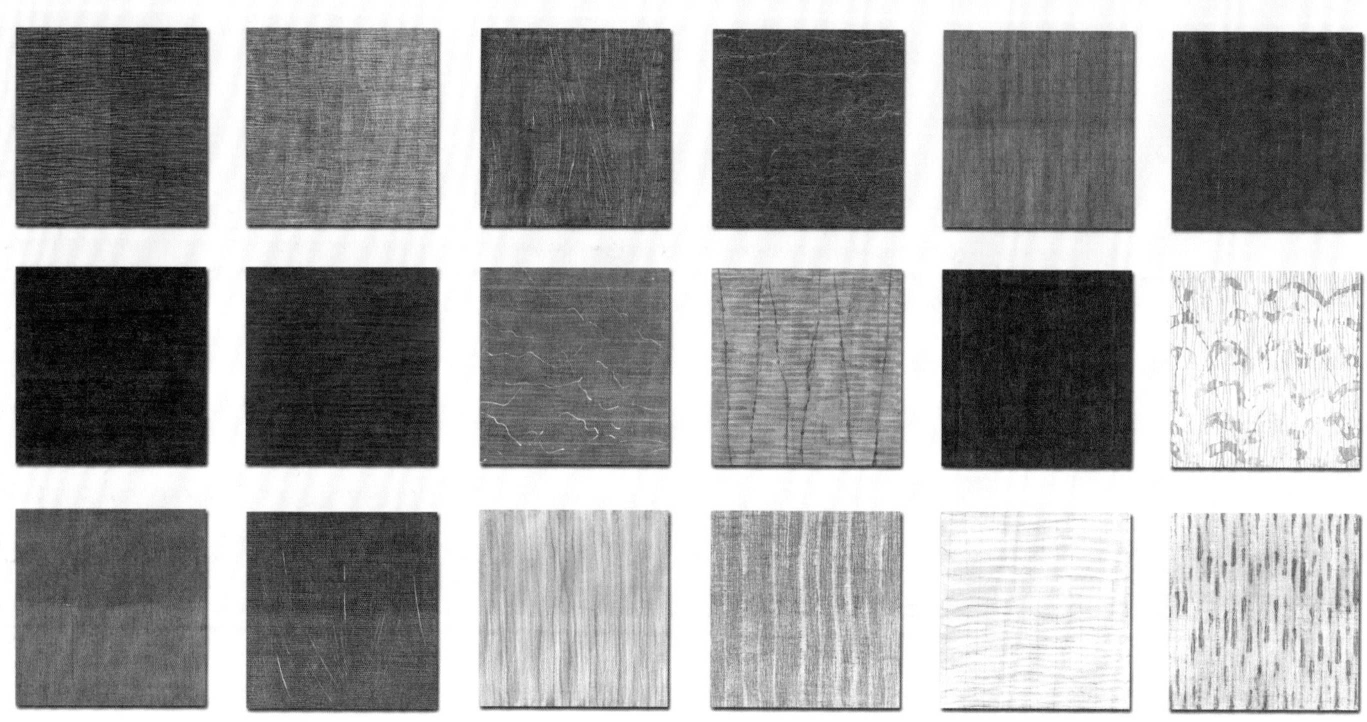

fig. 23: *Bethany Squares*, 2003
mixed media on paper mounted on aluminum
each square: 9 × 9 in. (23 × 23 cm)
Yale Center for British Art, Friends of British Art Fund

series of variations that invites musical analogies (fig. 23). In 2005 Salter donated one of her cherished Lake District sketchbooks to the Center in memory of her father, who had died that year.

Salter has continued to investigate the concerns that preoccupied her at Bethany, most overtly in a series of large square and rectangular drawings begun in 2006 (figs. 24, 25). Her work of the past four years also demonstrates a renewed engagement with Japan, stimulated in part when she received funding between 2006 and 2008 from Great Britain Sasakawa Foundation to interview printmakers, papermakers, and other craftsmen in Japan. Her ultimate goal is to create an historic archive of their working lives, which will function as a poignant memorial to artistic practices in danger of extinction.

THE HUMAN MEASURE

Since the 1990s Salter has enjoyed a series of successful shows at commercial galleries, and her work has been acquired for museum collections. Over time she has attracted a dedicated group of patrons, but Salter, an anomaly at a moment when a media presence is considered almost a prerequisite for an artist's success, does not crave fame. She regards the income from sales as a means of enabling her to continue to do what she loves most, to make work. Salter is, however, far from impervious to the affect of her works. Owners and viewers frequently comment on their therapeutic value, and Salter observes that "creating a sense of calm is absolutely fundamental to me while I am doing it, and for people who live with the work."[70] As Mullins notes, Salter's "sense of scale has always been based on the human measure—how far she can stretch, the size of the arc she can manage with one wrist movement while painting a line."[71] Her self-conscious eschewal of the monumental format is a telling indication of her empathetic philosophy.

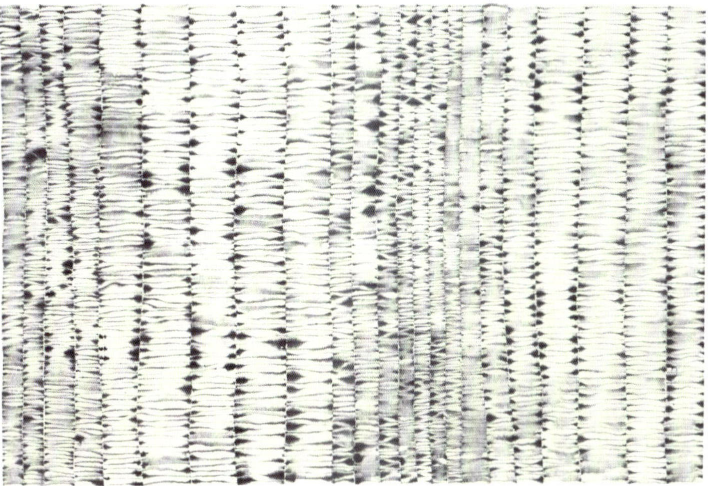

fig. 24: *Untitled MM2*, 2008
mixed media on paper
39¾ × 59⅞ in. (101 × 152 cm)

fig. 25: *Untitled RR35*, 2009
mixed media on paper
39⅜ × 59⅞ in. (100 × 152 cm)

fig. 26: *Calligraphy of Light*, 2009
recycled glass and bamboo panel
main entrance, St. George's Hospital, London
(in collaboration with Gibberd, architects)

The diptych is a constant in Salter's work, a format associated with religious painting.[72] Salter, whose father was an Anglican minister, rejects any specific Christian reading of her works, claiming more prosaically that dividing a canvas or large sheet of paper mitigates the daunting challenge of creating a single work. The binary format does seem meaningful to me on a more philosophical level, however, suggesting points of departure ranging from William Blake's dialectic of the "contrary states of the human soul" to Carl Jung's formulation of the "light" and "shadow" sides of the psyche. While conforming to the Romantic conception of *Erlebniskunst* as "infinite whole," Salter's binaries offer more complex possibilities for her reinscription of the world.[73] Her palpable concern for enriching human experience made Salter a natural recipient of the commission in 2006 to reconceive the reception area of St. George's Hospital in south London (fig. 26).[74] Uniting her central concerns of color, line, light, and space, this liminal place, completed in 2009, has achieved the perfect mediation between inside and outside worlds.

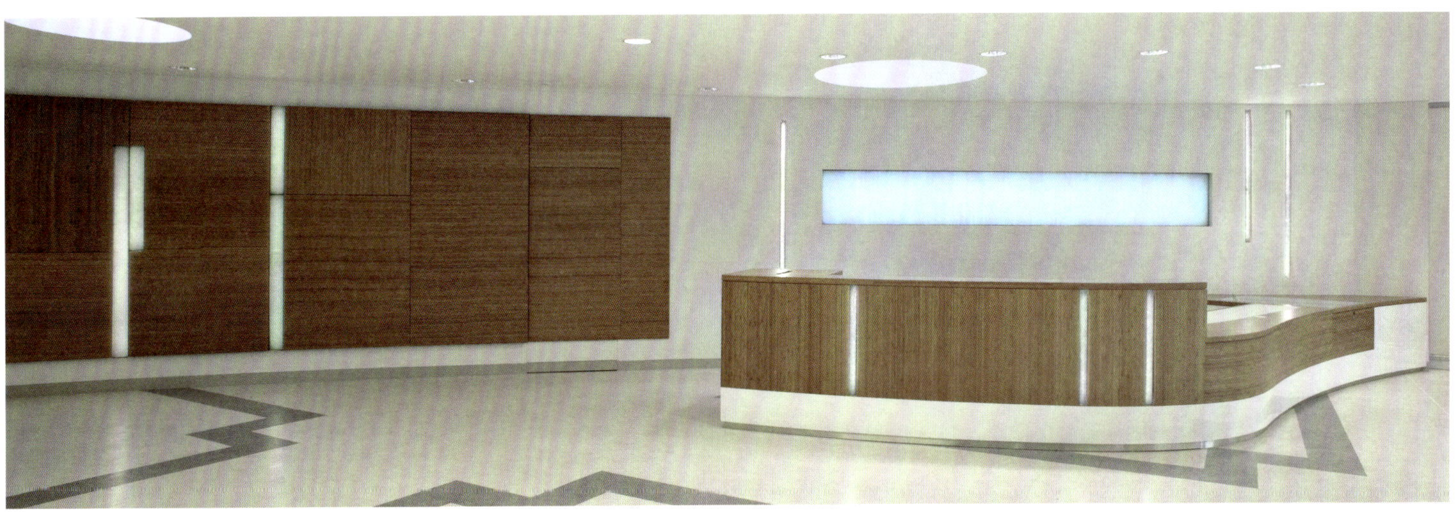

notes: From Without to Within

I am very grateful to Rebecca Salter for generously, and fearlessly, sharing so much about her philosophy and working practices over the course of the past seven years; this voyage of discovery has been a privilege and pleasure for me. I would also like to thank Cassandra Albinson for her insightful comments on this essay, and Jay Curley, Morna O'Neill, Sadako Ohki, Romina Tortoriello, and Geoffrey Winston for their contributions.

The epigraph for this essay is from Caspar David Friedrich, "Remarks on a collection of paintings, mostly by living or recently deceased artists," quoted in William Vaughan, Helmut Börsch-Supan, and Hans Joachim Neidhardt, *Caspar David Friedrich, 1774–1840*, exh. cat. (London: Tate Gallery, 1972), 103.

1 Charlotte Klonk, "Towards the Problem of an Aesthetic of Nature in Contemporary Art: Nicky Hirst and Rebecca Salter," *Scroope* (Cambridge Architectural Journal) 3 (June 1991): 11–14; and Anna Moszynska, introduction to *Rebecca Salter: Bliss of Solitude*, exh. cat. (London: Beardsmore Gallery, 2006), n.p.

2 For *Erlebniskunst*, see Hans-Georg Gadamer, *Truth and Method*, rev. trans. Joel Weinsheimer and Donald C. Marshall (New York and London: Continuum, 2004), 49–69.

3 Salter's was the last generation to attend such grammar schools, as the institution was dismantled, at least nominally, by the Labour government's Education Act of 1976, which set out to establish a more egalitarian comprehensive school system. The West of England Art School had been incorporated into Bristol Polytechnic, formerly Bristol Technical College, in 1970, as a result of a national initiative to incorporate art into the curricula of the newly created polytechnics.

4 For an account of what he terms the "faultline of endemic sexism" in Britain's art schools in the postwar period, see Paul Wood, "Between God and the Saucepan: Some Aspects of Art Education in England from the Mid-Nineteenth Century until Today," in Chris Stephens, ed., *The History of British Art: 1870 to Now* (New Haven and London: Yale Univ. Press, 2008), 178. For the reception of women artists in Britain in the 1970s, see Gill Perry, ed., *Difference and Excess in Contemporary Art: The Visibility of Women's Practice* (Oxford: Basil Blackwell, 2004).

5 For Leach, see Emmanuel Cooper, *Bernard Leach: Life and Work* (New Haven and London: Yale Univ. Press, 2003). For Rie, see Tony Birks, *Lucie Rie*, 2nd rev. ed. (Yeovil, UK: Alphabet and Image, 1998). The issue of ceramics and gender is much more complex than my binary account might suggest. For example, the aesthetic of refinement was also a hallmark of Rie's close associate and fellow émigré Hans Coper, who self-consciously positioned his practice against Leach's. See Margot Coatts, ed., *Lucie Rie and Hans Coper: Potters in Parallel* (London: Herbert, 1997); and Cyril Frankel, *Modern Pots: Hans Coper, Lucie Rie and Their Contemporaries; The Lisa Sainsbury Collection* (Norwich, UK: Univ. of East Anglia Press, 2002).

6 See Lisa Tickner, *Hornsey 1968: The Art School Revolution* (London: Frances Lincoln, 2008).

7 See Richard Cork, "The Art College Malaise," in *Everything Seemed Possible: Art in the 1970s* (London and New Haven: Yale Univ. Press, 2003), 370–75; and Wood, "Between God and the Saucepan," 186.

8 Artist's statement, 24 July 2010 (unpublished), sent to the author. Salter's avowal of the importance of loss also underpins the destructive nature of her working methods. As Achim Borchardt-Hume observes, "Salter recognizes erasure and subtraction, and the aggression involved therein, to be elements as valid to the creative process as the addition and construction of new layers"; see below, p. 44.

9 For the importance of *wabi* for Salter's work, see pp. 57–58 in the present publication.

10 Conversation with the author, 6 October 2009, artist's studio, Finsbury Park, London. It is conceivable that the pots that Salter abandoned in the River Avon may have been repurposed by the nomadic British artist Richard Long. See Anna Gruetzner Robins, "'Ain't Going Nowhere': Richard Long; Global Explorer," in Steven Adams and Anna Gruetzner Robins, ed., *Gendering Landscape Art* (Manchester: Manchester Univ. Press, 2000), 166. For a comparison of Long and Salter's radically differing practices as landscape artists, see below, p. 15.

11 See Tomoko Sato and Toshio Watanabe, eds., *Japan and Britain: An Aesthetic Dialogue, 1850–1930*, exh. cat. (London: Lund Humphries, in association with Barbican Art Gallery and Setagaya Art Museum, 1991). See also the webpage for Watanabe's *Forgotten Japonisme* project on the University of the Arts Research Centre for Transnational Art, Identity and Nation website, http://www.transnational.org.uk/projects/17-forgotten-japonisme, accessed 21 July 2010.

12 Quoted in Cooper, *Bernard Leach*, 250.

13 For a characterization of Leach as "the father of British studio pottery," see his artist page on the British Council website, http://collection.britishcouncil.org/collection/artist/5/18557, accessed 21 July 2010. Edmund de Waal has more recently, and controversially, noted the considerable limitations in Leach's knowledge of Japanese cultural practice. See Edmund de Waal, *Bernard Leach* (London: Tate, 1997); and the transcript of John Tusa's interview with Edmund de Waal, http://www.bbc.co.uk/radio3/johntusainterview/dewaal_transcript.shtml, accessed 20 July 2010. See also Morag Shiach, *Discourse on Popular Culture: Class, Gender and History in Cultural Analysis, 1730 to the Present* (Cambridge, UK: Polity, 1989), 131–38.

14 For Yagi, see Louise Allison Cort and Bert Winther-Tamaki, *Isamu Noguchi and Modern Japanese Ceramics: A Close Embrace of the Earth*, exh. cat. (Washington, D.C.: Arthur M. Sackler Gallery, Smithsonian Institution, in association with Berkeley: Univ. of California Press, 2003), 156–86; and below, p. 51.

15 See Alexandra Munroe, ed., *The Third Mind: American Artists Contemplate Asia, 1860–1989*, exh. cat. (New York: Guggenheim Museum, 2009) for a penetrating exploration of this complex relationship. See also below, pp. 65–68.

16 See below, pp. 51–52.

17 Artist's statement, 18 April 2010 (unpublished), sent to the author. For the issue of gender, in particular the "feminized Orient," see Bert Winther-Tamaki, "The Asian Dimensions of Postwar Abstract Art: Calligraphy and Metaphysics," in Munroe, ed., *Third Mind*, 152; and p. 65 in the present volume.

18 Artist's statement, 18 April 2010 (unpublished), sent to the author.

19 See Fiona Robinson, "Rebecca Salter," *Axis* 7 (November 2007 to February 2008), http://www.axisweb.org/dlFULL.aspx?ESSAYID=96, accessed 10 August 2010.

20 Conversation between Achim Borchardt-Hume and Salter, 3 April 2010, quoted below, p. 42.

21 Artist's statement, 18 April 2010 (unpublished), sent to the author.

22 Conversation with author, 6 October 2009, artist's studio, Finsbury Park, London.

23 Artist's statement, 18 April 2010 (unpublished), sent to the author.

24 See below, p. 54.

25 Edward Said, *Orientalism* (New York: Pantheon, 1978); J. J. Clarke, *Oriental Enlightenment: The Encounter between Asian and Western Thought* (New York: Routledge, 1997); and Takeuchi Yoshimi, "Asia as Method," in *What is Modernity? Writings of Takeuchi Yoshimi*, trans. and ed. Richard F. Calichman (New York: Columbia Univ. Press, 2005), 249–65. For a succinct and illuminating synthesis of these methodologies, see Munroe, ed., *Third Mind*, 27.

26 Artist's statement, 18 April 2010 (unpublished), sent to the author.

27 See below, p. 52.

28 Junichirō Tanizaki, *In Praise of Shadows*, trans. Thomas Harper and Edward G. Steidensticker (London: Vintage, 2001), 17–18.

29 Conversation with artist, London, 6 October 2010.

30 See below, pp. 54–57, for a close reading of *Untitled B34*, *Untitled B119*, and *Untitled B120*, which are closely related to *Untitled B161* (fig. 5).

31 Narumi Hiroyki, *Boku sorekkiri* (*Just Me*) (Kyoto: privately published by author, 1983).

32 Artist's statement, 24 July 2010 (unpublished), sent to the author.

33 For critiques of Thatcher's project, characterized by Stuart Hall as "regressive modernization," see Raphael Samuel, "Mrs. Thatcher and Victorian Values," in Alison Light, ed., *Island Stories*, vol. 2 of *Theatres of Memory* (London and New York: Verso, 1998), 330–48; and Geoffrey Wheatcroft, *The Strange Death of Tory England* (London: Allen Lane, 2005).

34 R. B. Kitaj, *The Human Clay*, exh. cat. (London: Arts Council of Great Britain, 1976).

35 R. B. Kitaj, introduction to *The Human Clay*, n.p.

36 Christos M. Joachimides, Norman Rosenthal, and Nicholas Serota, *A New Spirit in Painting*, exh. cat. (London: Royal Academy of Arts, 1981), 11.

37 See, for example, Richard Shone, "A New Spirit in Painting at the Royal Academy," *Burlington Magazine*, 123, no. 939 (March 1981), 182–83, 185; and John Russell, "A Big Berlin Show That Misses the Mark," *New York Times*, 5 December 1982.

38 Shone, "A New Spirit in Painting," 185.

39 Richard Cork, "The Flowering of Women's Art," in *Everything Seemed Possible*, 26–27; and Perry, ed., *Difference and Excess in Contemporary Art*, 9–10.

40 The very talented Alison Wilding (born 1948) was an exception, and in the 1980s a number of women sculptors, including Rachel Whiteread and Shirazeh Houshiary, came to prominence.

41 See Perry, ed., *Difference and Excess in Contemporary Art*, 11.

42 Artist's statement, 24 July 2010 (unpublished), sent to the author.

43 The importance of *shibui* in Salter's palette is discussed by Ohki, below, pp. 58–59.

44 Walter Thornbury, *The Life of J. M. W. Turner, R.A.: Founded on Letters and Papers Furnished by His Friends and Fellow Academicians*, rev. ed., 2 vols. (London: Chatto and Windus, 1877), vol. 1, 139.

45 William Wordsworth and Alan Sedgwick, *A Complete Guide to the Lakes...* (Kendal, UK: Hudson and Nicholson, 1853), 116.

46 Ernest Hartley Coleridge, "S. T. Coleridge as a Lake Poet," *Transactions of the Royal Society of Literature*, 2nd ser., vol. 24 (1903): 23.

47 In 1991 Salter visited Samoa, having won the Royal Overseas League prize, which enabled her to visit a Commonwealth country of her choice. She recalls that she brought a sketchbook and set of watercolors, but on observing the landscape realized that "those colours don't exist in an English watercolor palette." Conversation with the author, 6 October 2010, London.

48 Conversation with the author, 6 October 2010, London.

49 See below, p. 52.

50 Conversation with the author, 6 October 2010, London.

51 Conversation with the author, 6 October 2010, London.

52 Wall text in a survey exhibition of Long's career at Tate Britain in 2009. See also Gruetzner Robins, "'Ain't Going Nowhere.'"

53 Conversation with the author, 6 October 2010, London.

54 Conversation with the author, 6 October 2010, London.

55 See Alexandra Munroe, "Buddhism and the Neo-Avant-Garde: Cage Zen, Beat Zen, and Zen," in Munroe, ed., *Third Mind*, 198–215.

56 Klonk, "Towards the Problem of an Aesthetic of Nature," 13.

57 Klonk, "Towards the Problem of an Aesthetic of Nature," 13.

58 Klonk, "Towards the Problem of an Aesthetic of Nature," 11.

59 Klonk, "Towards the Problem of an Aesthetic of Nature," 14.

60 Friedrich Schlegel, quoted in Joseph Leo Koerner, *Caspar David Friedrich and the Subject of Landscape*, 2nd. ed. (London: Reaktion Books, 2009), 32.

61 Artist's statement, 24 July 2010 (unpublished), sent to the author.

62 Quoted in Kathryn A. Tuma, "Enhancing Stillness: The Art of Agnes Martin," in Catherine de Zegher and Hendel Teicher, eds., *3 x Abstraction: New Methods of Drawing by Hilma af Klint, Emma Kunz, and Agnes Martin*, exh. cat. (New York: The Drawing Center and New Haven and London: Yale Univ. Press, 2005), 50.

63 Artist's statement, 18 April 2010 (unpublished), sent to the author. See below, p. 38.

64 Klonk, "Towards the Problem of an Aesthetic of Nature," 13.

65 See, for example, E. V. Rippingille's description of Turner painting at the British Institution on "Varnishing Days," quoted in John Gage, *J. M. W. Turner: A Wonderful Range of Mind* (New Haven and London: Yale Univ. Press, 1987), 92.

66 See Achim Borchardt-Hume, "Shadows of Light: Mark Rothko's Latest Series," in *Rothko: The Late Series*, ed. Borchardt-Hume, exh. cat. (London: Tate, 2008), 25.

67 Cf. Chuck Close: "I always could work anywhere ... I know so many artists for whom having the perfect space is somehow essential. They spend years designing, building, outfitting the perfect space, and then when it is just about time to get to work they'll sell that place and buy another one." Quoted in Joe Fig, *Inside the Artist's Studio* (New York: Princeton Architectural Press, 2009), 40.

68 Charlotte Mullins, "Painting Shadows," *Rebecca Salter*, exh. cat. (London: Hirschl Contemporary Art, 2002), n.p.

69 Robinson, "Rebecca Salter."

70 Artist's statement, 25 July 2010 (unpublished), sent to the author.

71 Mullins, "Painting Shadows," n.p.

72 See Mullins, "Painting Shadows," n.p.

73 Gadamer, *Truth and Method*, 61.

74 See below, p. 74.

Overleaf (left): Rebecca Salter, photographs taken during her residency at the Josef and Anni Albers Foundation, Bethany, Connecticut, 2003

Overleaf (right): Rebecca Salter, the artist's studio at the Josef and Anni Albers Foundation, Bethany, Connecticut, 2003

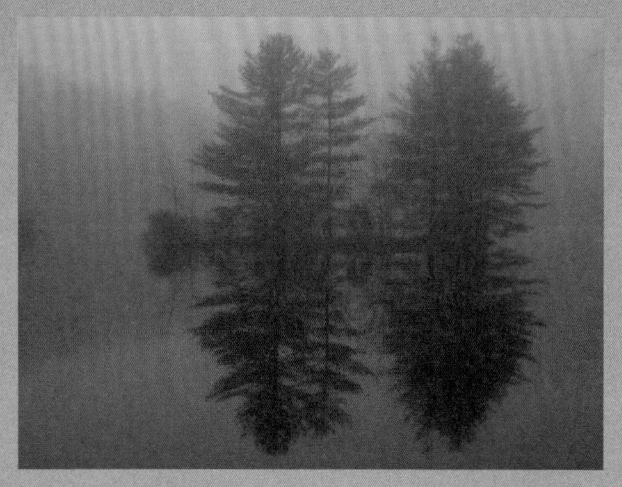

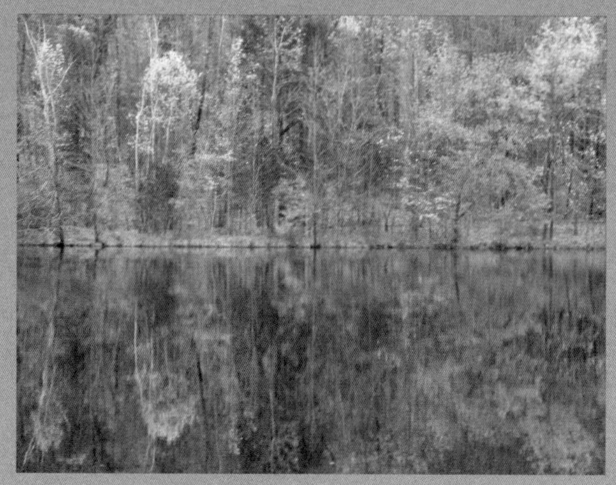

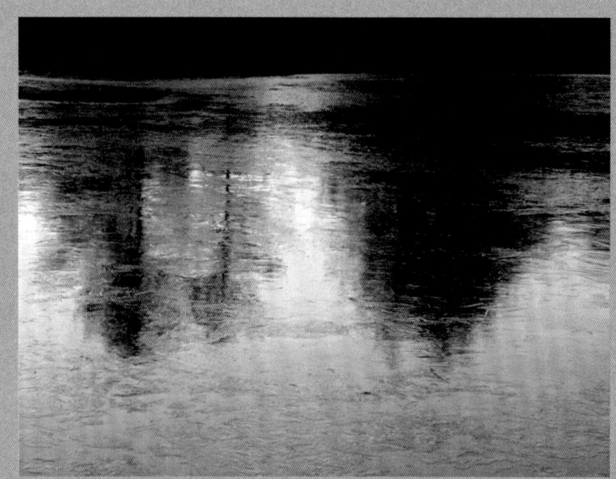

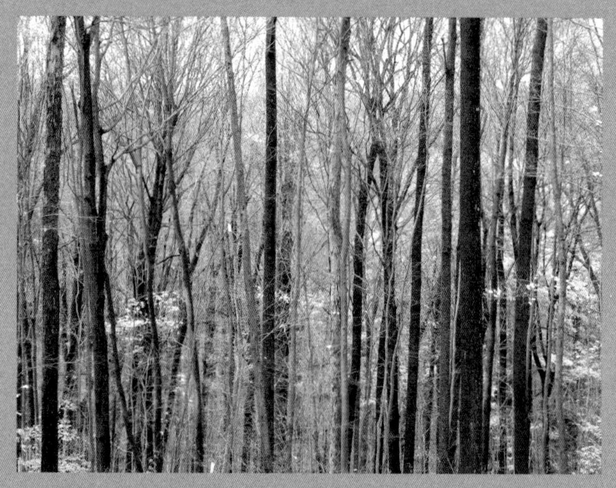

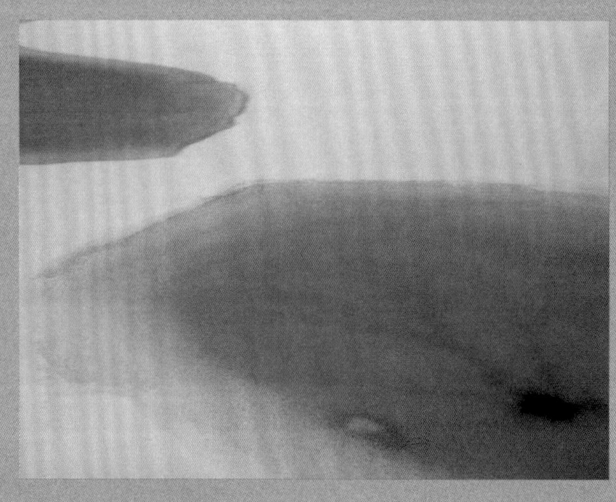

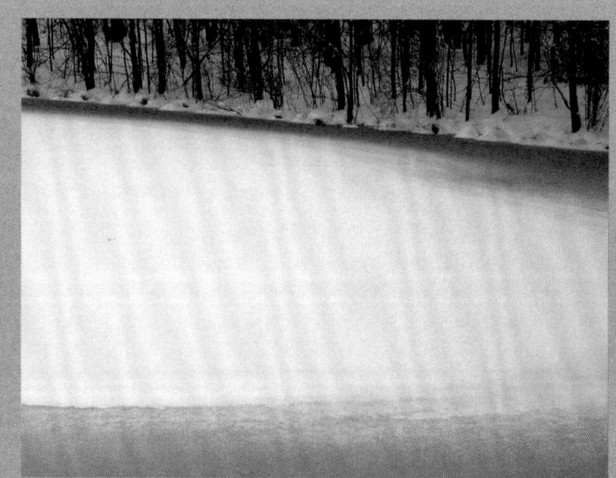

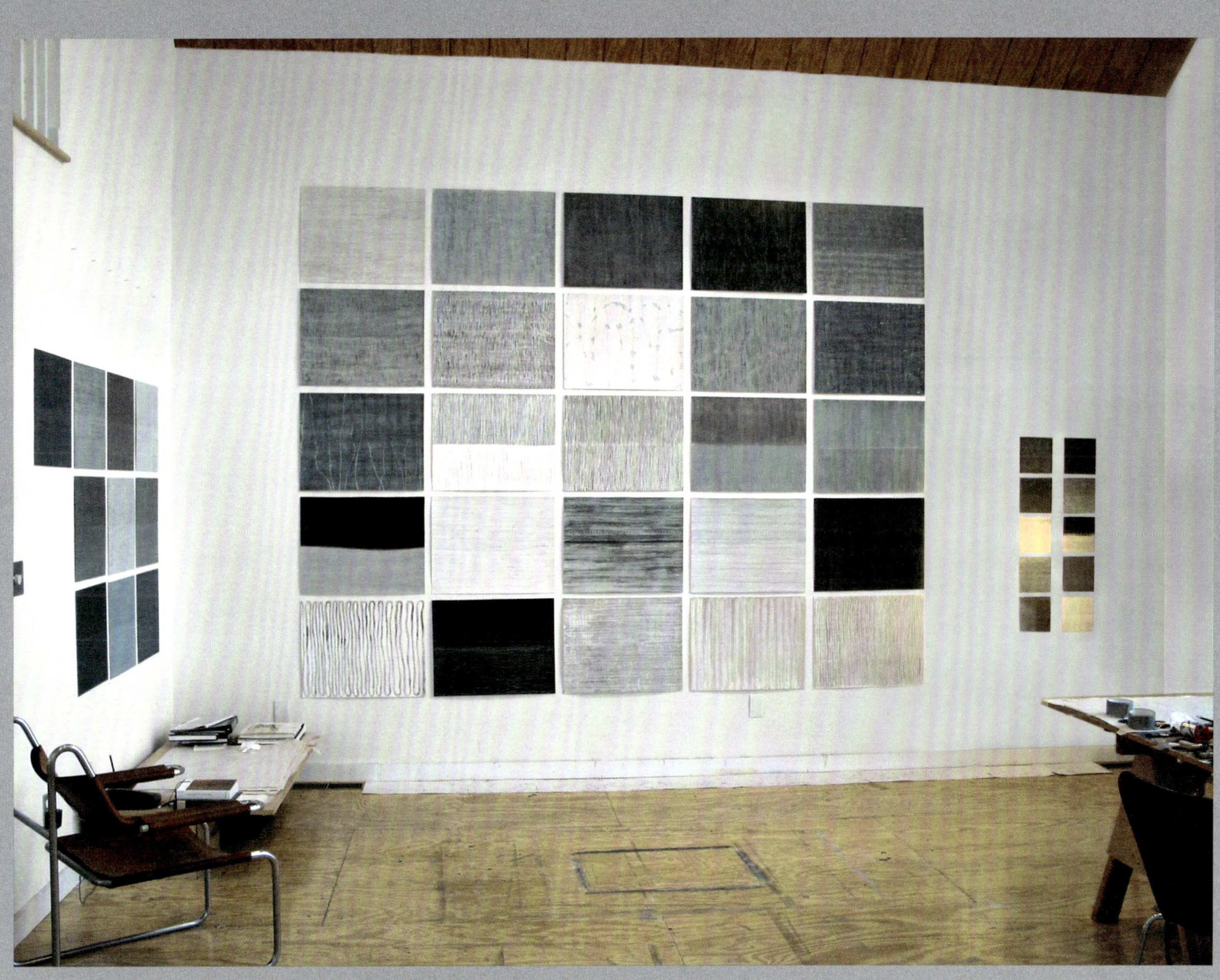

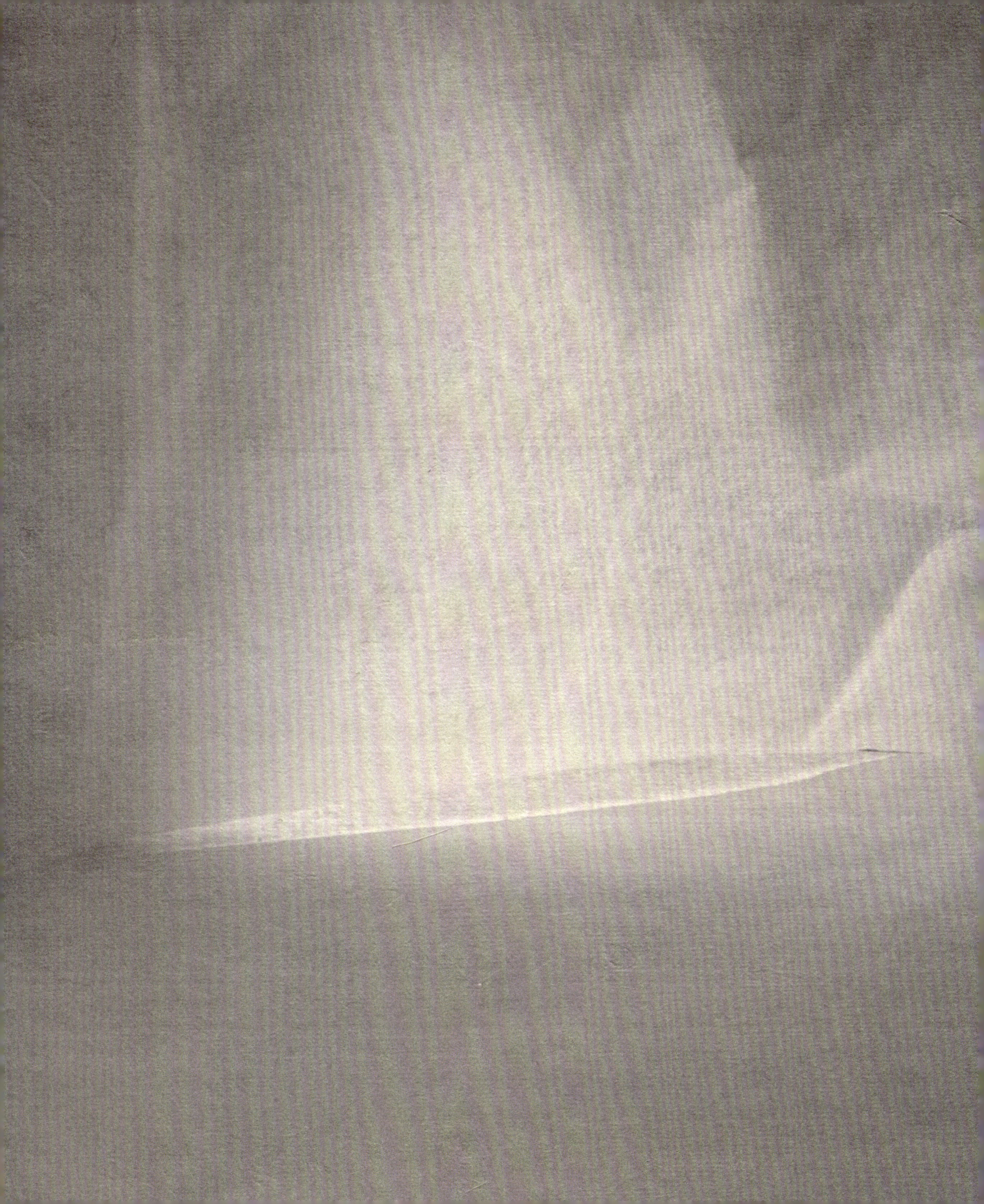

On the Surface of Things

Achim Borchardt-Hume

On the Surface of Things

Achim Borchardt-Hume

Before examining Rebecca Salter's distinctive approach to abstract painting, it may be helpful to consider what we mean when we speak of painting in general, and abstract painting in particular. By common understanding, a painting in Western art is made up of a support—most frequently canvas on a stretcher—covered in a medium such as oil or acrylic to present the viewer with an image. This image may represent an aspect of external reality, express a moment of inner feeling, or be abstract; that is, it may foreground its own internal logic of line and form. Thus defined, painting is one of Europe's most successful cultural exports as is evidenced not least by the recent boom of painting in territories where it has no historic roots, such as China or the Middle East. A Renaissance invention, painting is closely tied to the global conquest of an anthropocentric view of the world which promoted an emphasis on individuality and intellectual independence, the latter rising to become the engine powering the Enlightenment project of the past three centuries, culminating in twentieth-century modernism.

Easel painting, in combination with the development of central perspective, was uniquely qualified to give visual expression to this shifting engagement with the world. By constructing a space to be surveyed from a predetermined, fixed-point perspective, the Renaissance artist created the illusion that the world contained within the picture frame was fully at the command of the viewer. Mankind was confirmed as the master of the universe, contesting the all-seeing eye of divinity. Through its portability, easel painting emphatically asserted its autonomy from its immediate surroundings and held the inherent ability to detach itself from its point of origin. It could migrate, across time and space, in a way that was both a premonition of and a metaphor for the modern "un-place."

With its verticality and insistence on a frontal encounter, painting underscored uprightness as the physical condition elevating humankind over all other living species. Sigmund Freud, in *Civilization and its Discontents*, proposes that it was by walking upright that the genital (with its polar division of gender) superseded the anal (where no such distinction exists) and that, as a consequence of this new posture, the retinal came to replace the olfactory as the privileged sense with which to negotiate the world.[1] At the beginning of the twenty-first century, the visual reigns supreme, with a "virtual" reality composed of a flood of alluring images continuously expanding the human brain's evolutionary limitations for processing visual information. One of painting's all the more urgent ethical mandates today, therefore, is to provoke an in-depth process of aesthetic appreciation; that is, to insist that every particularity of the painted object is registered by the viewer. This extends from retracing the physical process of making to appreciating the lived time invested therein. It is through this experience that a rapport of profound commonality between the fellow human beings of painter and viewer is being established.

Abstract painting is uniquely suited to this task, as it avoids the trap of lulling the viewer into a false sense of security based on her ability to decipher an iconographic riddle

fig. 27: *Untitled RR31*, 2009
mixed media on linen
detail; see plate 14 /

Instead, it places the act of looking at the forefront of the encounter with the work of art. At the beginning of the twentieth century, Pablo Picasso's and Georges Braque's cubism radically altered the representational gambit by fracturing the picture surface to confirm pictorial reality as subject to its own logic. This led to two different models of abstraction: one took external reality as its starting point, stripping away all superfluous detail in the quest to reveal a Platonic essence (Piet Mondrian's geometric compositions, which evolved from his increasingly abstract depictions of trees and seascapes, are a prominent example); another devised compositional schemas as visual metaphors for notions of social progress (constructivism, for instance). Both were at heart utopian and consistently sought to transcend everyday reality, including the artistic materials employed (which by way of their physicality inevitably anchored even the most abstract composition to the very reality it sought to leave behind). The same was true of the third key historical model of abstract painting, which occurred in the climate of humanist crisis caused by World War II and the Holocaust. Abstract expressionism, in particular, infused the earlier European models of abstraction with an American sense of scale to transform the encounter of eye and canvas into one of body and canvas so as to create what Barnett Newman called "a place."[2] Cursory though this account may be, suffice it to say that, to this very day, abstract painting has continuously provided a rich terrain for philosophical speculation about the very concept of painting, both as a practice and as an object. In so doing it implicitly asks questions about the relationship of art and artist to viewer and the world at large.

All of the above, as we shall see, has a bearing on Salter's working practice, which does not so much endeavor to devise a conscious critique of abstract painting and its history as offer a space for reflection on the unspoken assumptions that circumscribe the ways in which it is being experienced. Salter begins her investigation from the somewhat precarious position of an artist who at heart is a border walker between different disciplines (art and craft), cultural reference points (Britain and Japan, the West and the East), and modes of engagement (making and thinking, skill and intuition). Looking at Salter's work and its manifold sources, especially those not forming part of the critical canon, thus affords an opportunity to look afresh at our experience of painting and the way it may help us to position ourselves in the world.

It may be instructive in this context to take a closer look at *Untitled RR31*, one of the most recent works in this exhibition; the way this painting may be "read"; and how this reading compares to the work's physical and conceptual genesis (figs. 27, 28). *Untitled RR31* in more than one way is typical of Salter's practice. It is colored in the artist's trademark muted grays, which, on closer inspection, reveal hints of other chroma, though only just. There is no discernible image as such, rather an overall pattern of striations that are spaced at regular—one is tempted to say melodic—intervals. Though it is possible to imagine these continuing beyond the confines of the canvas, they are comfortably contained by the field they inhabit. The lines' fluidity and soft outlines eschew any kinship with the hard edges and calculated geometry of the grid as it so frequently appears in much modernist painting. Rather than being determined by a preconceived "intellectual" design, the painting's final appearance is decided largely by the characteristic properties of the materials deployed: the

fig. 28: *Untitled RR31*, 2009
mixed media on linen
74¾ × 70⅞ in. (190 × 180 cm)
verso

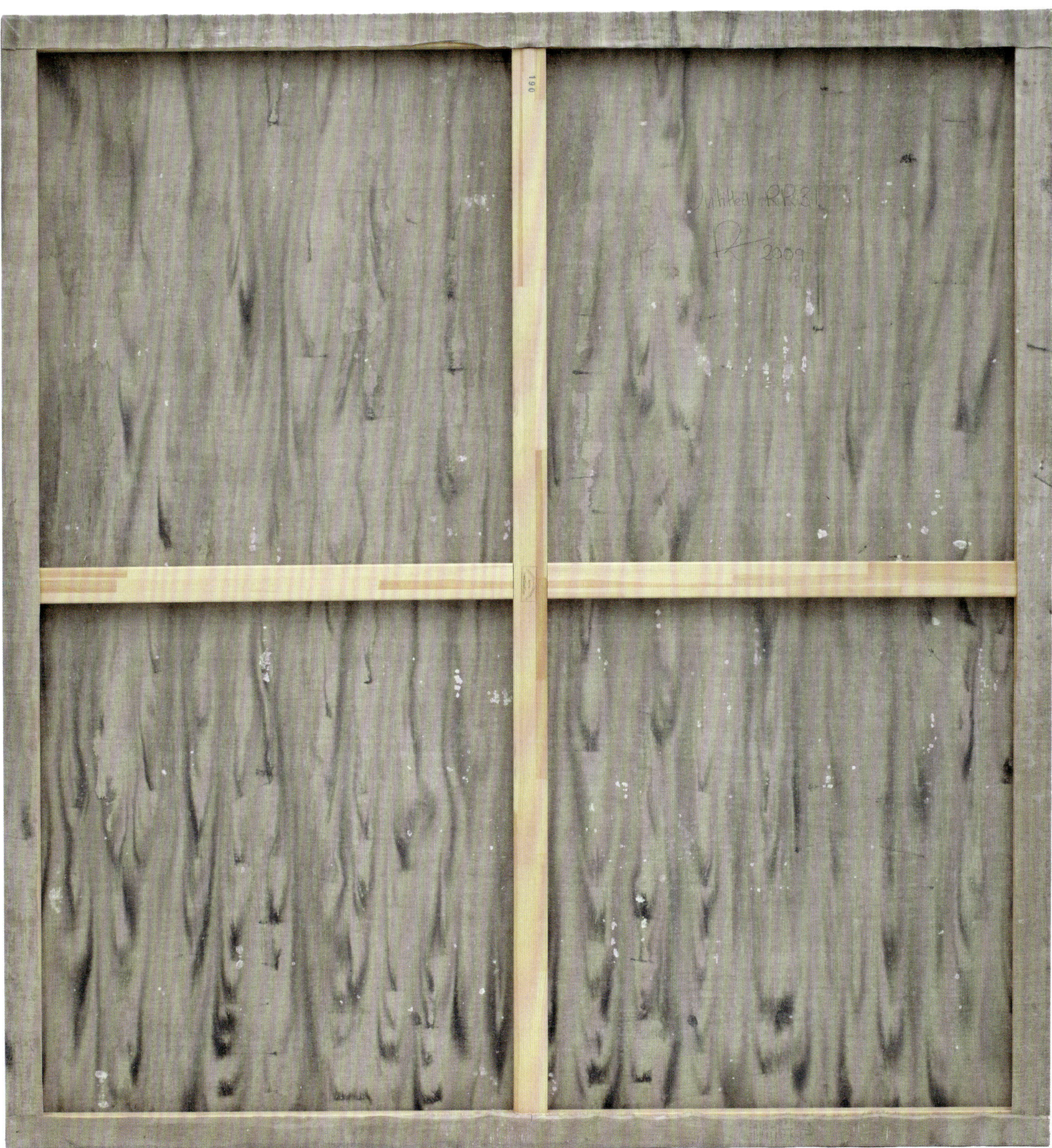

way colors bleed into one another; the degree to which they fuse with the ground; their density, translucency, and opaqueness. The surface seems mostly dry, papery, not unlike the skin of a mature adult, having left the gloss of youth behind. There is a very clear sense that in contemplating this painting we are invited to consider the result of a process informed by a set of skills and a familiarity with a wide range of materials, acquired over a prolonged period of time. This process entails clear decisions on behalf of the artist, one of which is to accept that part of it will inevitably be beyond her control.

It is interesting then to learn more about how *Untitled RR31* came into being. Like all of Salter's paintings, it was executed on a piece of canvas stapled to a wooden board so as to create the necessary resistance for the process of applying paint. Salter worked with the board flat on the ground rather than erect on an easel or fixed to the wall. The linen canvas was not primed but left untreated to retain its full absorbency. Salter applied an initial set of marks, after which she drenched the canvas in water. Only then did she place it upright to allow the fluid paint medium to run across the surface. Once the canvas was dry, Salter detached it from the board, turned it over, and continued to work on what had previously been the reverse. She then applied further layers of thin washes and highly diluted watercolor containing small amounts of pigment. Sometimes she would scrape these off again before applying the next wash. Throughout this process, Salter worked the surface with a variety of makeshift tools, from wooden ice-cream sticks to plastic ink dispensers, with the most conventional tool of the trade, the brush, being noticeably absent. Only at the point where she deemed the process completed—or perhaps it would be more accurate to say the point at which she felt there was nothing further to add—did she detach the canvas from the board and stretch it on a conventional wooden stretcher. The transformation of a stained and painted piece of cloth, worked from all sides, into a "painting" by way of fixing the canvas onto a stretcher marks the end of the process of "making." Once the work is stretched and hanging upright in the conventional manner of a painting on the wall, Salter does not work the surface any further.

Salter's strategy of making does away with a host of conventions that inform our cultural understanding of painting and how we relate to painting as a model. She works on the canvas not vertically (the orientation of humanist elevation) but on the ground (historically the position of the abject). She disbands the distinction between front and back, one of the defining parameters of painting and its mirror-like insistence on a frontal encounter. She surrenders control over the final appearance of the work to material processes she cannot necessarily determine; though rather than play with chance in the manner of Marcel Duchamp or Jean Arp, her allowance for chance effects is predicated on her intimacy with her materials, which allows her somewhat to preempt the result.

This capsizing of conventions traditionally governing the medium is rooted in Salter's unusual framework of reference. One of the legacies conceptual art has bestowed on painting over the past forty years or so is that painters, like artists working in any other medium, have to demarcate their chosen field of operation. Salter initially chose ceramics as her field in an attempt to avoid the confines of an overly determined and particularly a stereotypically male attitude to painting predicated on heroic gestures of self-expression and

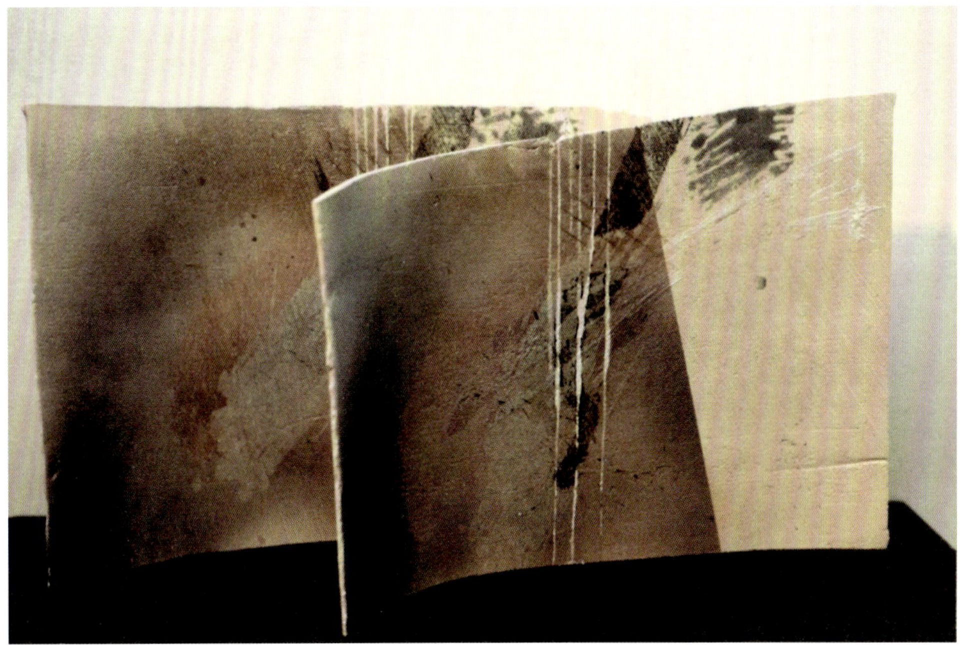

fig. 29: *Ceramic*, 1981
low-fired ceramic backed with
Japanese paper
dimensions and location unknown

the hunt for definitive formal solutions. However, she chose not to make functional objects but ceramics that took on the role of painted three-dimensional objects (fig. 29).

During the mid-1980s Salter transferred her field of inquiry to Japan, that is, to a cultural context whose very core is diametrically opposed to the anthropocentric paradigms on which Western painting is founded. Japan's polytheistic view of the world is reflected in its lack of preoccupation with the notion of a single, fixed center. As much is true of its art. Traditional Japanese depictions of landscape, for instance, are structured not by the optical laws of central perspective but by the harmonious coexistence of distinct elements in a fluid space. There is also an understanding of selfhood that is far more relational than that of the tyrannical "I" that is so central to Western conceptions of individualization and self-realization. A similar comprehension of everything as inevitably relating to something else also informs the way in which Salter organizes the constitutive elements of her paintings. Formal relationships—the way one line may correspond to another, or two color washes interact—function as analogies for the relationship between the individual and the world, symbolically represented here by the painted object.

Having abandoned ceramics, Salter turned to working with ink on Japanese paper. (Today works on paper still make for a substantial part of her output.) She began to take evening classes in calligraphy to learn more about the idiosyncratic qualities of the traditional materials deployed: the specially made papers, pigments, inks, and brushes (fig. 30). Her studies provided her with greater insight into the conventions of calligraphy, which, in Japanese aesthetics, is understood as evolving both in space (the movement of the arm and hand exerting pressure on the brush) and in time (the duration a single stroke occupies from the moment the brush touches the paper to the moment when they are separated). Educated viewers, such as Salter herself, are able to relive this process by following the line traced by the ink, its thickness and density, the firmness of its edges, and the intensity of its turning points. It is interesting to note in this context that the only black Salter uses in her work is Japanese calligraphic ink—watered down to different degrees, allowed to run, disperse, and fuse with traditional watercolors as the artist sees fit—and that this is the only

fig. 30: Jiun Onkō
Daruma, Edo Period (1615–1868)
hanging scroll, ink on paper
48⅝ × 21¾ in. (123.4 × 55.4 cm)
The Art Archive/Sylvan Barnet and William Burto Collection

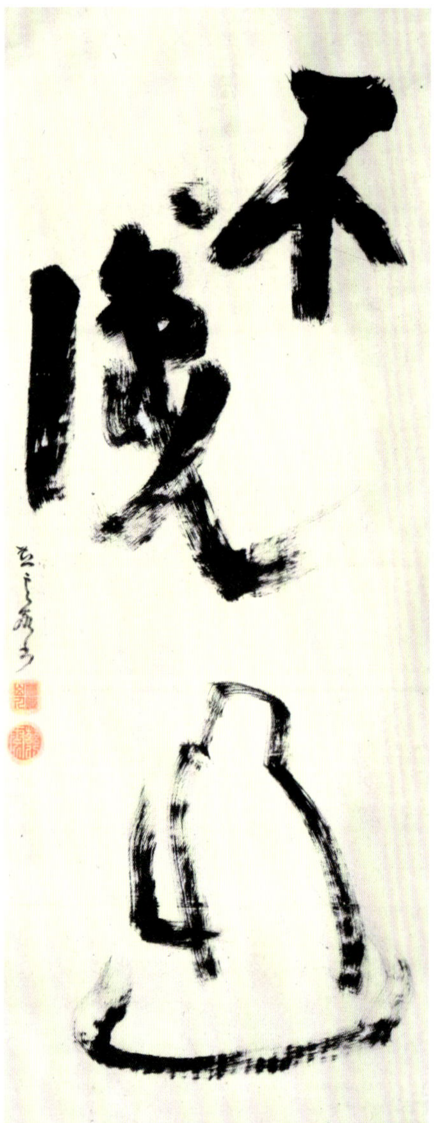

medium applied with a brush. In her paintings Salter effectively treats the canvas as though it were paper, and paper as equal to canvas, thereby putting into question traditional value judgments about the different media.

Salter chooses to exploit the challenges posed by her materials, such as the runniness of the ink and color washes, by varying the scale of her canvases and changing their orientation. *Untitled RR31* was turned 180 degrees so that the trails of the ink in a gravity-defying move now run upward. Unpredictable in their final appearance, Salter's paintings function in a way that is fundamentally different in motivation from, for instance, the color fields of Mark Rothko. Rothko presented himself as bearing witness to and making physically manifest an image which first existed in his mind.[3] The artist's struggle was thus one of approximation: how to come as close to the ideal effect desired and imagined as possible. In contrast, Salter's paintings unequivocally emerge on the canvas with the artist as much acting on her own accord as being forced to react to the (at times unpredictable) actions of her materials.

However, more than her handling of unusual materials, it is perhaps her understanding of surface and space that underscores Salter's investigation of conceptual models that

fig. 31: Caravaggio
The Incredulity of St. Thomas, 1602–03
oil on canvas
42⅛ × 57½ in. (107 × 146 cm)
Schloss Sanssouci, Postdam

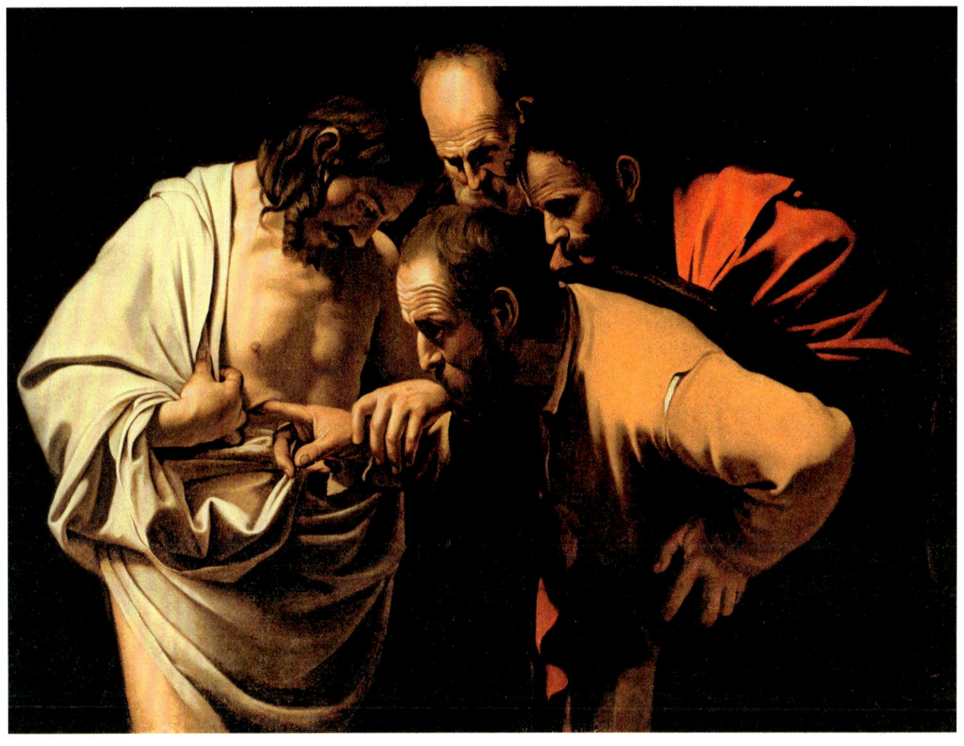

stem from beyond the conventional discourse around painting. Historically the canvas has been treated as a screen separating the reality of the painting (both as object and depiction) from what lies outside it. It is for this effect that the canvas traditionally is primed, thus reducing its absorbency and thereby its fusion with the paint medium. This treatment of the canvas as a passive screen readied to receive the markings of the painter surreptitiously translates into the relationship between painting and viewer, a relationship Salter describes as largely "confrontational, rather than conversational."[4] As viewers, our relationship to the surface—a relationship that is intrinsically at risk of being mistaken for "superficial"—and how this shapes our engagement with painting is emblematically captured in Caravaggio's dramatic depiction of the story of the doubting Thomas (fig. 31). Faced with the apparition of the resurrected Christ, Thomas does not trust his eyes and hence seeks reassurance via the sense of touch by placing his fingers into the savior's open wound. The retinal, reliant on surface only—that is, on what is visually presented to it—is shown here as untrustworthy, albeit in the most captivating and artful way. Triumphantly demonstrating the power of his imagination, Caravaggio the painter effectively places his fingers as much into the open wound of painting as the disciple Thomas did in the flesh of his master.

In the late-modernist debate around the ongoing legitimacy and continuing possibilities of painting, the surface of the canvas became an arena of particular contention. In postwar critical discourse, especially the writings of Clement Greenberg, painting was described as finding its teleological fulfillment in the assertion of the flatness of the canvas.[5] This rendition of flatness as painting's primary condition also played a central role in the rise of one of the favorite tropes of modernist painting, the grid. The parallel lines of the

fig. 32: Agnes Martin
Untitled #6, 1989
synthetic polymer paint and pencil on canvas
71⅞ × 72 in. (182.7 × 182.9 cm)
Museum of Modern Art, New York, Gift of the American Art Foundation

geometric grid in paintings otherwise as diverse as those by Piet Mondrian or Agnes Martin are located on a picture plane that is parallel to the image surface. Positioned thus, they deny and subvert the illusionist space that central perspective had constructed via a set of diagonals converging on a fixed point at the center of the image (fig. 32).

However, defining flatness and the break with illusionistic space as the logical conclusion to the evolution of modernist painting left later generations of artists with the dilemma of what to do other than engage in an ultimately unsatisfactory endgame of final gestures. Much late modernist painting by artists such as Frank Stella and Ad Reinhardt appears principally to be preoccupied with killing the ontological essence of painting, its ability to act as analogy for an engagement with life and the world. It may not be all that surprising, therefore, that younger generations of practitioners, frustrated by the circularity of this endgame, either abandoned painting altogether or took painting, to borrow Rosalind Krauss's term, into an "expanded field" through a heightened emphasis on its material constituents, such as support, frame, and painterly medium.[6] This latter option was sought out especially by voices hitherto marginal to the discourse around painting, including female, non-white, and openly gay artists.[7] It was all the more striking, therefore, that the critical discourse failed to respond, instead getting trapped in an endless recanting of the "death of painting."

Salter implicitly rejects this critical narrative by treating both surface and space in an altogether different way than the late modernists. To begin with surface, it may be helpful to think of it, not as that which hides something that lies behind, but as a continuum, only part of which may be visible at any one time though all of it is present. The body's own surface is such a continuum of the external and the internal, connected via various orifices. Every ceramicist, by squeezing clay between thumb and fingers, will be aware that the surface of an object such as a vessel is seamless and continuous, inside and out.

If painting traditionally has to negotiate this separation of physical surface (what we see with our eyes) and the meaning that may hide behind this surface (what the painting is about), Salter's treatment of the canvas dispenses with this dichotomy. Her paintings are objects in which front and back are not in a mutually exclusive relationship but are interchangeable during the process of making, when paint may seep through from one side to the other. At the same time, by abandoning brushes for scrapers and other tools, Salter elevates the tactile qualities of the surface to the level of its retinal effect. Or rather, she treats them as inseparable: what can be seen can be felt, and vice versa. Her paintings neither are in pursuit of metaphysical truths nor aspire to capture an otherworldly sublime. Instead, by way of the process of making and its decoding through the parallel process of perception, they ponder what it is to relate to the world. If anything, the surfaces of Salter's paintings act as visual metaphors for the balance between knowing and not knowing, between what can be controlled and what cannot. The surface, visually and physically, is the result of an event. Not an event in the sense of Jackson Pollock's dervish-like dancing across the canvas to express something of his tormented struggle for a universal truth, but rather an event in the sense of the quiet, day-to-day business of ordinary life.

fig. 33: Pigment scraped from three paintings, date unknown

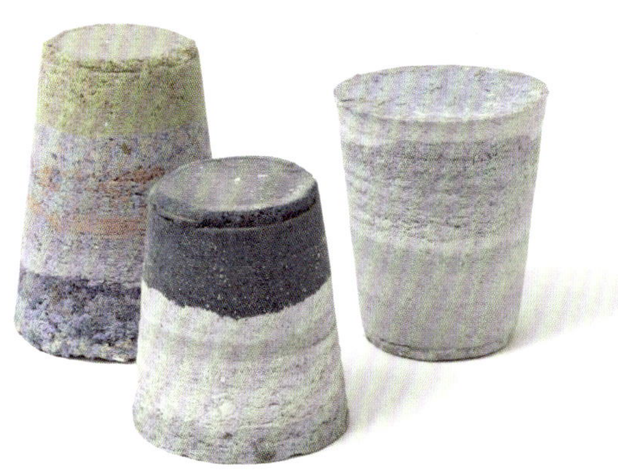

If surface in Salter's paintings does not act as a barrier or screen but as an active interlocutor, then space forms part of the same conversation. The space in a painting such as *Untitled RR31* is neither illusionary nor otherworldly. Rather it adheres to an optical and material logic that creates its own reality; that is, a painterly reality which adds to the familiar terrain of everyday reality without seeking either to represent or to transcend it. Space, a sense of what is in front and what behind, is conceived via the actual space of the surface, however thin this may be, through the build-up of layers and washes, the scraping back of pigment, and the reapplying of new strata. At times, Salter retains the sediments from this process (fig. 33). Akin to geological cross sections these time capsules of redundant matter provide an archaeological account of time evolving in space—the time it took to make the work, materialized in the evolution of the surface.

Surface and space are intertwined further through the way the painting continues around the edges of the stretcher. Whether painting the edges of a canvas emphasizes the autonomy of the painting as an object or ties it more directly to its architectural context has been subject to extensive debate, especially in the aftermath of abstract expressionism. Salter astutely expands on these polarized positions by observing that the integration of the edges of the canvas does not so much make an object of the painting but of the stretcher. This is an important distinction, particularly given that Salter's own paintings, as we have seen, are not painted on stretched canvas but instead subvert the conventional boundaries of front, side, and reverse. When asked how she would categorize her work, Salter suggests the most universal term to label an object: a "thing."[8] Overly generic though this may sound, "thing" may indeed be the most apt word to describe her paintings and the way they inhabit the world. Though, on the one hand, they function as aesthetic, autonomous objects, on the other, they are not detached from reality. Shaped neither by Salter's artistic subjectivity nor the search for metaphysical transcendence, they are generated by a materiality that firmly anchors them in the here and now, with the proviso that they make us see this here and now with a newly attuned sensitivity. This materiality becomes especially evident when Salter translates her understanding of surface and space from two dimensions into three as she has done for a recent architectural commission for the entrance area of St. George's Hospital in south London (fig. 34). Vertical light beams, changing bands of light, and a delicately composed arrangement of bamboo panels create a spatial flow that appears curiously open and unfixed without being in any way destabilizing. Rather, the experience is akin to walking through one of Salter's paintings, with space (the architectural space of the entrance area) and time (the time it takes to negotiate the space) entering into a sophisticated dynamic.

As with all of Salter's work, materiality, albeit at its most subtle and understated, is at the forefront of this experience. Physical matter, by its very nature, is intrinsically linked to the problem of form and form-giving. "Making," then, is also a concept Salter repeatedly returns to in conversation. Salter cites the innate urge to make—as a way of being in the world and finding a measure of oneself vis-à-vis reality as it preexists us—as both the starting point

fig. 34: *Calligraphy of Light*, 2009
recycled glass and bamboo panel
main entrance, St. George's Hospital, London
(in collaboration with Gibberd, architects)

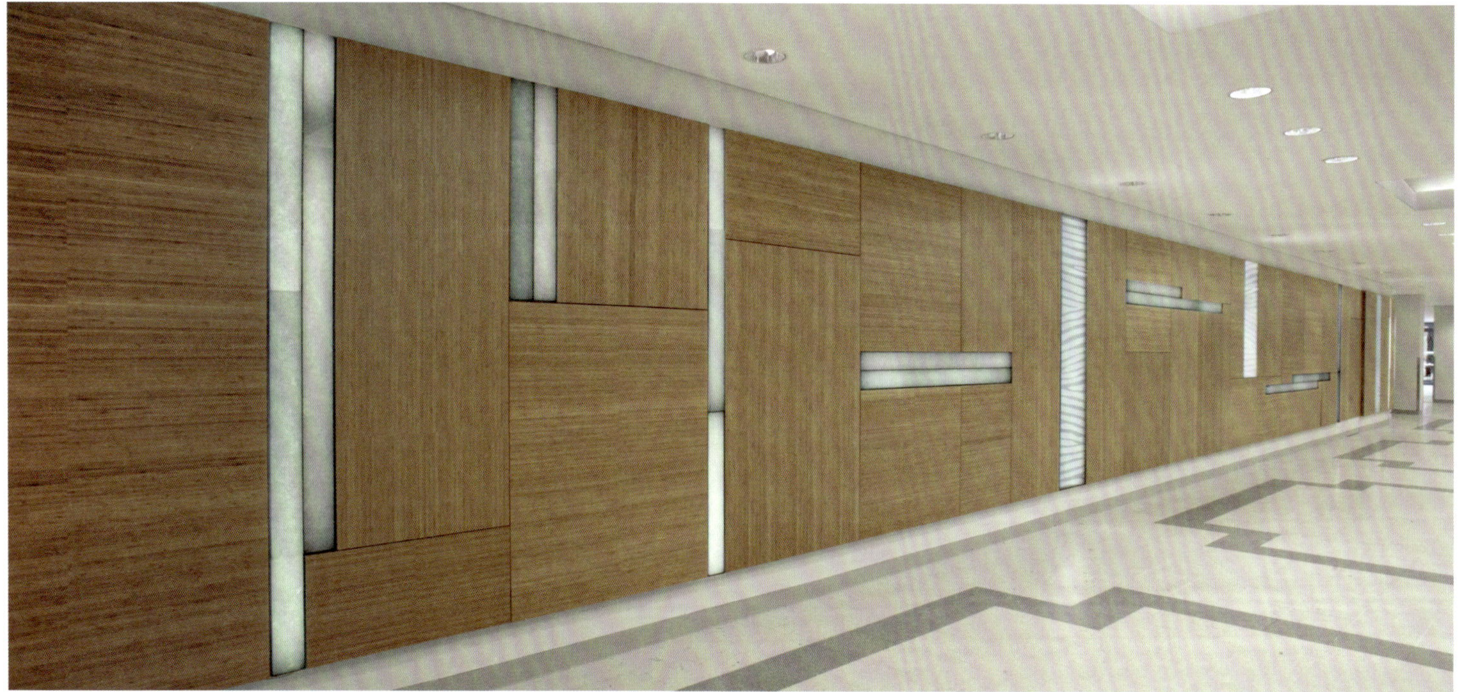

and the continuous motivation for her work. It is noticeable that though her work shows a strong sense of unity through the consistency of certain formal elements such as palette and surface treatment, it contains very few clearly conceived series as such. The series is an important concept in the history of modernist painting.[9] It replaces the notion of the masterpiece, the single definitive version, with the play of variations on a theme to express the sense of flux and ambiguity that is at the heart of the modern condition. A work within a series can realize the full potential of its meaning only in concert with its counterparts. Without these it is stunted and partially disabled. In contrast, each of Salter's paintings explores a unique set of material possibilities and patterns of behavior. This means that though they are related and though they share a common denominator, all of Salter's works survive perfectly on their own.

The issue of making in twentieth-century art at times has been a thorny one. On the one hand, the politics of production—how an object is made—in modernist thinking are essential to the meaning an artwork may be able to convey. On the other, there is a long tradition of denigrating the notion of making, especially in the sense of artists using manual skills, as bringing "high" art down to the level of "mere" craft. One of the most important legacies of Marcel Duchamp's revolutionary ready-mades is how they freed the artist from the constraints of craft and making, a legacy which is celebrated by some and repudiated by others. Consciously or subconsciously, there are also important ideas of gender at play here. The intellectual aspect of artistic production is associated with a superior, male way of inhabiting the world, while making and craft—skill-based and repetitive—are associated with the female sphere of enclosed domesticity. It is this very repetitive and mundane "doing it again, day after day" approach to painting, therefore, on which some of the most interesting female painters of the late twentieth century, such as Mary Heilmann (who incidentally also started out studying ceramics), have consciously based their practice in an attempt to subvert traditional gender conceptions about painting.

Historically, skepticism toward the material routine of painting, the banal aspects of dealing with "the stuff," dates back much further than the onset of modernity, to the very beginnings of painting. Prior to the invention of premixed paints in tubes it was left to studio apprentices to grind pigments and to mix these with a binding medium, leaving the intellectually superior faculties of the master artists unburdened with this task. This distinction between the physical and intellectual realms also informed the Renaissance dispute over the primacy of drawing (*disegno*) or color (*colore*) in painting, with the former winning for being more cerebral and less visceral.[10] Even twentieth-century painters such as Rothko would have assistants prepare their paints so that the emphasis in authorship lay in the act of conceptualization rather than making. In the wake of the long history of modern and modernist painting, Salter's insistence that making forms the basis of her practice is thus more subversive than at first may be apparent.

The centrality of the process of making to Salter's practice reveals itself to the viewer (who is excluded from witnessing this process firsthand) perhaps most easily in the slow perceptual process of deciphering the hues of color in her paintings, which often appear at first sight uniformly gray before subtle hints of other color values slowly come into play.

fig. 35: Jasper Johns
Two Flags, 1959, acrylic on canvas
79¼ × 58¼ in. (201.3 × 148 cm)
Museum Moderner Kunst,
Stiftung Ludwig, Vienna, on loan
from the Ludwig Collection, Aachen

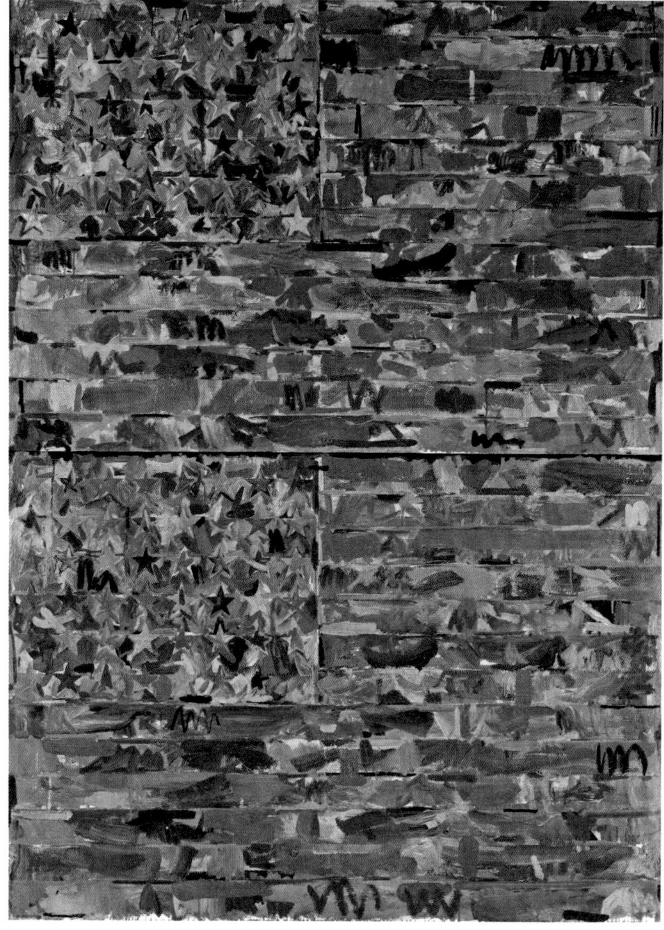

Salter recognizes erasure and subtraction, and the aggression involved therein, to be elements as valid to the creative process as the addition and construction of new layers. Many of her early paintings were started with layers of vibrant colors that were slowly covered, scraped back, and covered anew until the original chroma was barely visible. Effectively, this turned the surface into a palimpsest of earlier images known only to the artist. The slow recognition of barely perceptible hues of color in one of Salter's paintings is akin to a visual excavation of the protracted process of covering up what came before. Thus the muted grays of Salter's canvases are not dissimilar, yet they are very different from the programmatic use of gray, for instance, in the work of Jasper Johns or Gerhard Richter. Johns, in his early gray paintings from the 1950s, contrasted sharply defined geometric shapes such as the stars-and-stripes banner with a handling of paint that starkly emphasized its material, if not outright messy, properties (fig. 35). Over the following decades, he paradoxically deployed gray for an exploration of the perceptual qualities of color. This tension between color and non-color served to speculate about the relationship between the retinal and conceptual dimensions of painting. Gray came to stand for a state of irresolution between the two, this irresolution for Johns being one of painting's most vital qualities. Similarly, in his paintings from the mid-1970s, Richter deployed dull grays to manifest his distaste for any type of ideology and his unwillingness to being forced to take a position (be it on matters of aesthetics, politics or ideology) (fig. 36). In turn, Salter uses gray to tame what otherwise would be a force field of clashing visual cues and to generate a sense of precarious balance instead. Despite their otherwise extremely different approaches to painting, what these

fig. 36: Gerhard Richter
Gray, 1974
oil on canvas
98⅜ × 76¾ in. (250 × 195 cm)
Tate, London

Overleaf: Rebecca Salter, photographs of Japan, 1979–2010

artists share (and what is perhaps best expressed in their use of gray) is a fundamental belief in ambiguity as at the heart of the human condition. They uphold an acceptance of the limits of knowledge as being more truthful than the hunt for black-and-white orthodoxies. What the best of Salter's paintings tell us is that all we need to know—and all we possibly can know—is already in front of us. The only thing we have to do is to learn how to look and see.

notes: On the Surface of Things

1 Sigmund Freud, *Civilization and its Discontents* (London and Harmondsworth: Penguin Modern Classics, 2002, originally 1930).

2 "One thing that I am involved in about painting is that painting should give man a sense of place: that he knows he's there, so he's aware of himself." See *Barnett Newman, Selected Writings and Interviews*, ed. John P. O'Neill (New York: Knopf, 1990), 257–58, 289–90.

3 In the words of Brian O'Doherty, Rothko acted "as if he were the curator of a mystery which he could convey but to which he was not privy." See Brian O'Doherty, "Rothko's Endgame", *Mark Rothko: The Dark Paintings, 1969–70*, exh. cat. (New York: Pace Gallery, 1985), 9.

4 Conversation with the author in Salter's studio, 3 April 2010.

5 See *Clement Greenberg: The Collected Essays and Criticism: Modernism with a Vengeance, 1957–1969*, vol. 4 (Chicago and London: University of Chicago Press, 1993), particularly "After Abstract Expressionism," 121–34.

6 Rosalind Krauss,"Sculpture in the Expanded Field," *October*, vol. 8, Spring 1979, 30–44.

7 See Katy Siegel, ed., *High Times, Hard Times: New York Painting, 1967–1975*, exh. cat. (New York: Independent Curators International, 2006).

8 Conversation with the author in Salter's studio, 3 April 2010.

9 For an important early account of series and seriality, see John Coplans, *Serial Imagery*, exh. cat. (Pasadena, Calif.: Pasadena Art Museum, 1968).

10 See Ann Temkin, "Color Shift," in *Color Chart: Reinventing Color, 1950 to Today*, exh. cat. (New York: Museum of Modern Art, 2008), 16–27.

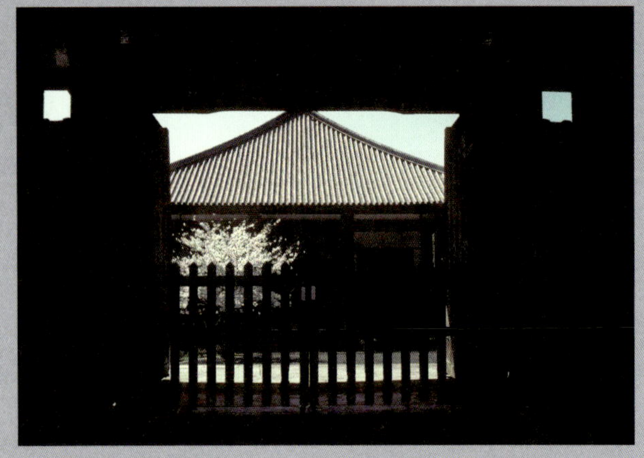

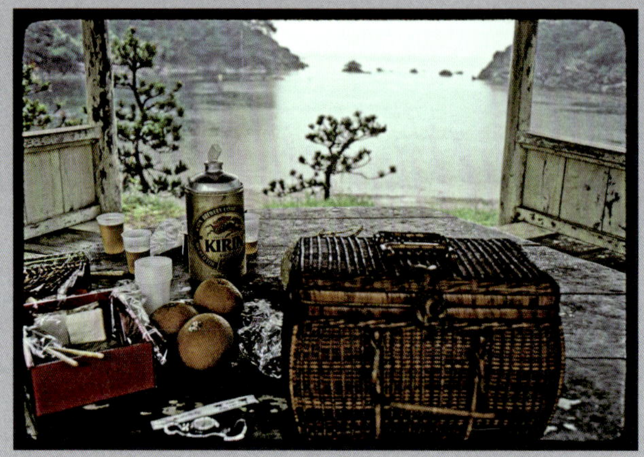

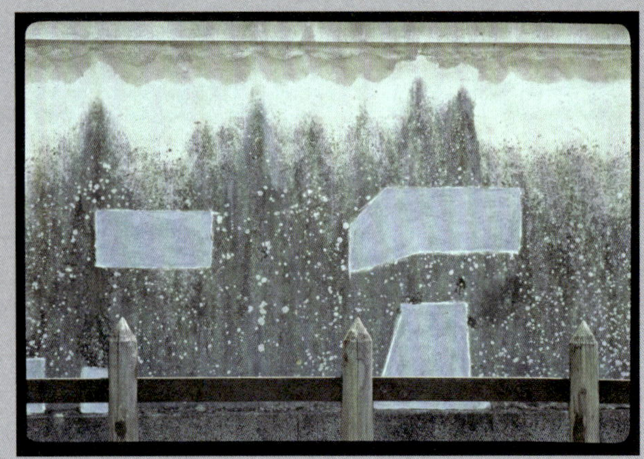

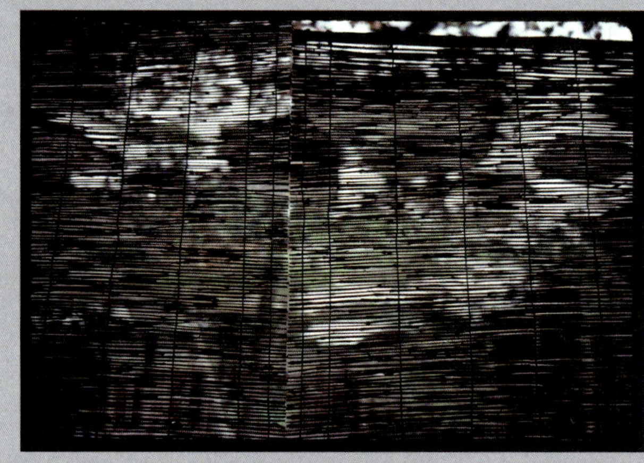

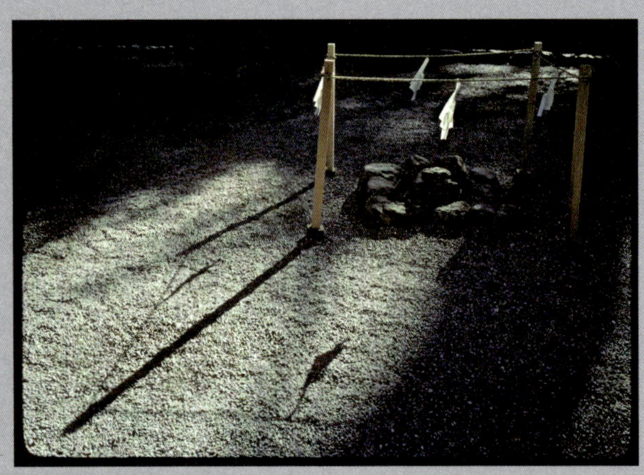

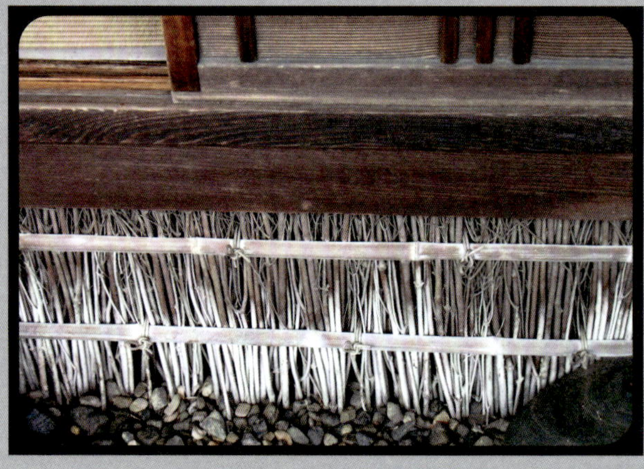

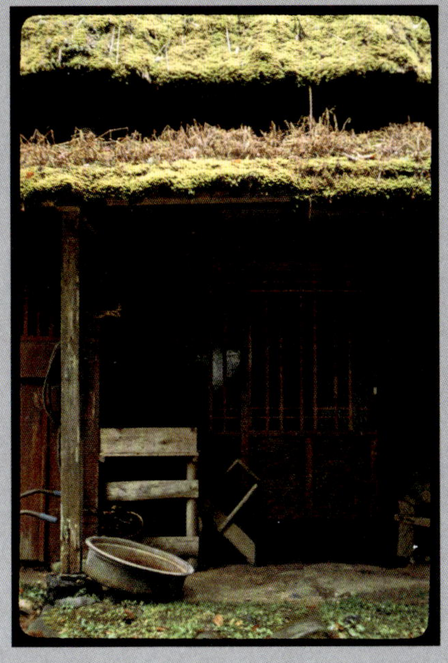
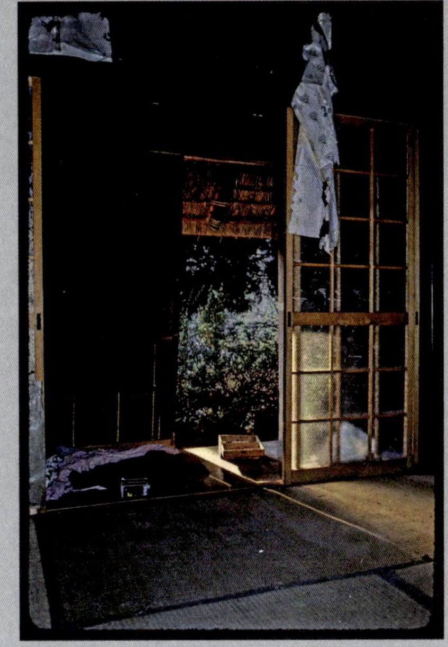
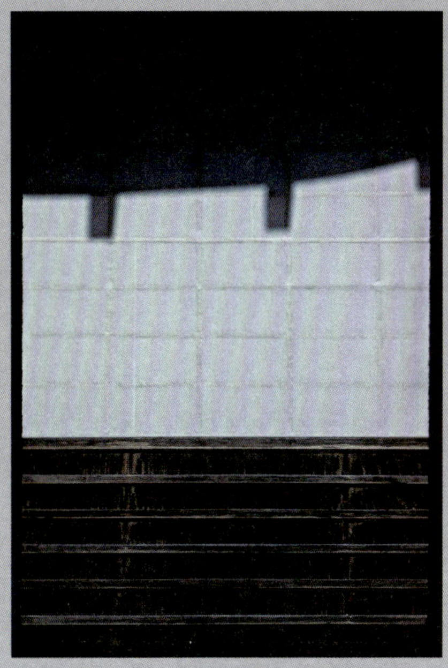
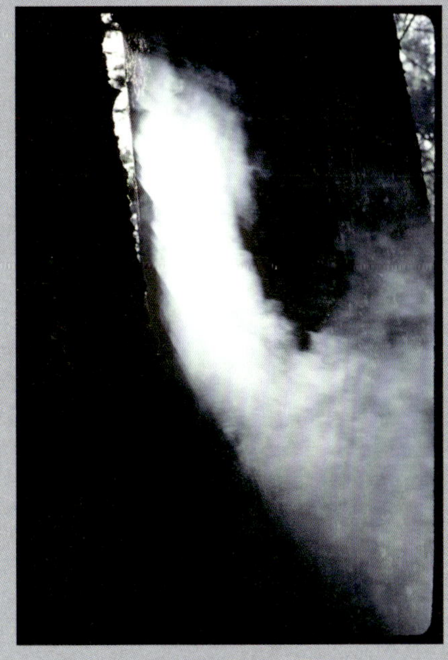
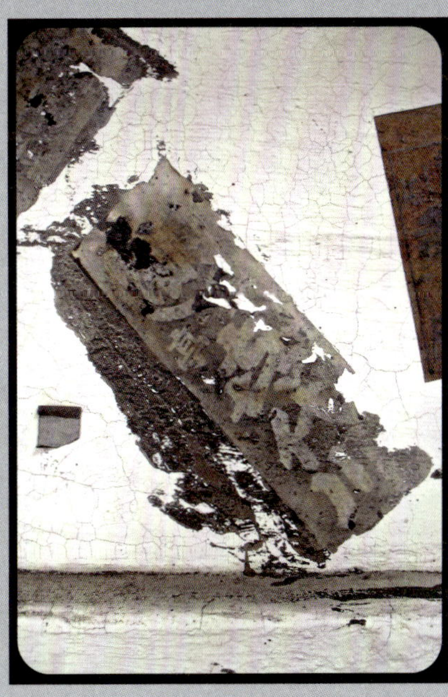
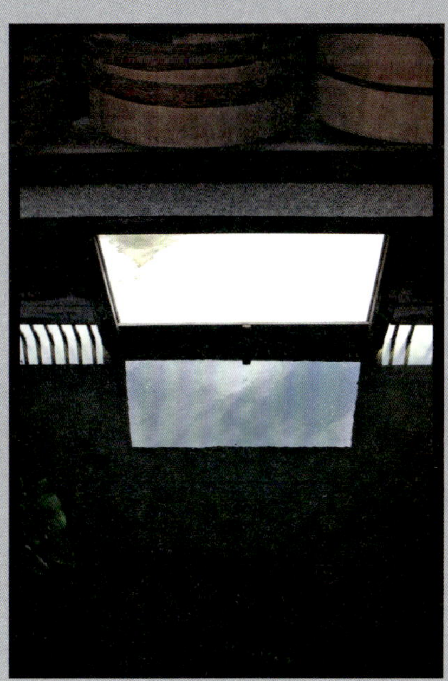

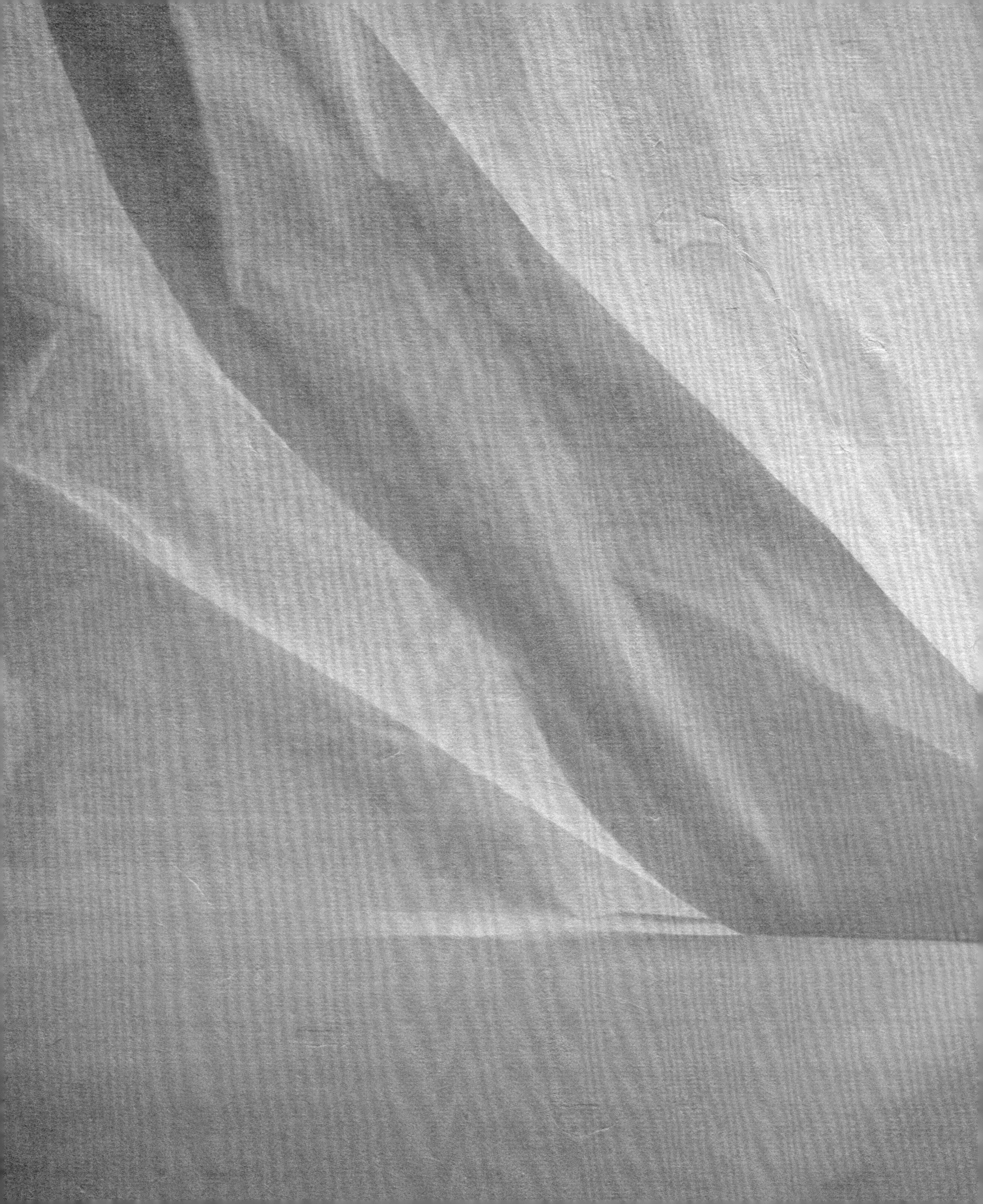

Rebecca Salter and Japan: Moments Layered in Time, Space, Color and Line

Sadako Ohki

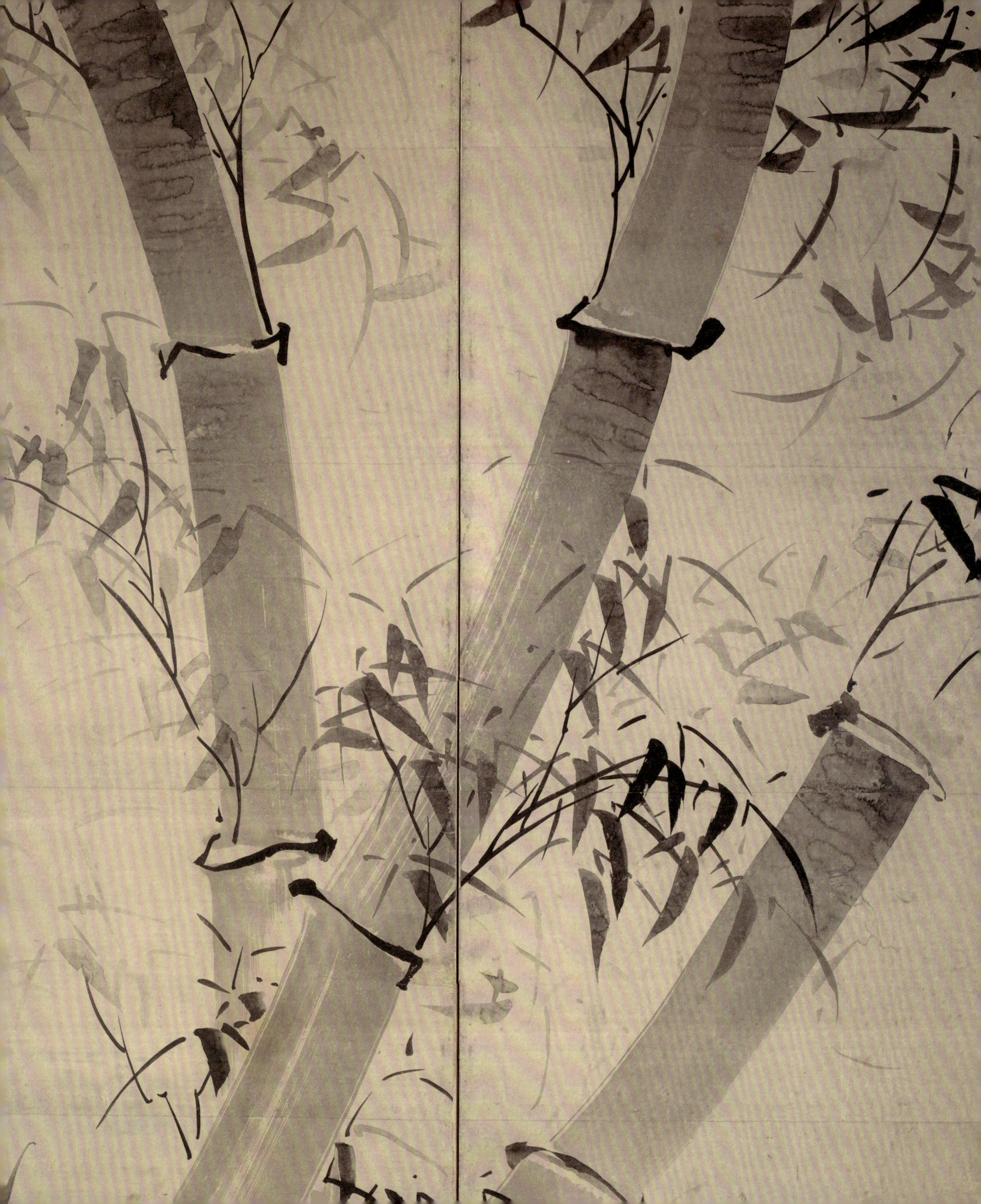

Rebecca Salter and Japan: Moments Layered in Time, Space, Color, and Line

Sadako Ohki

Formerly, we thought that the Orient had nothing to do with us; that we had no access to it. We know better now. We learned from Oriental thought that those divine influences are, in fact, the environment in which we are. A sober and quiet mind is one in which the ego does not obstruct the fluency of things that come in through our senses and up through our dreams. Our business in living is to become fluent with the life we are living, and art can help this.

John Cage, 1966

Rebecca Salter's move to Japan in 1979 proved to be a radical step that completely transformed the nature of her work and has had a lasting and deeply significant effect. Living in Japan was a liberating experience for Salter at this formative stage in her career, as the freedom accorded to a foreigner allowed her to experiment and develop her practice away from the constraints of the London art world.[1] Kyoto, where Salter made her base, was the old capital of Japan from 794 till 1868. The city encapsulates conflicting ideologies and images. Kyoto is extremely conservative, the home of imperial rulers and centuries-old family dynasties (in contrast to Tokyo, where a family who has lived there for three generations would be considered deeply entrenched), yet, paradoxically, radical left-wing movements in strong opposition to the establishment also flourished there in the mid-twentieth century, impacting even the notoriously conservative art world. In 1948 the ceramic sculptor Yagi Kazuo (1918–1979), with whom Salter originally had planned to study, founded *Sōdeisha* (Crawling in Mud Association) with Yamada Hikaru (1924–2001) and Suzuki Osamu (1926–2001). They created this organization with the radical objective of making abstract ceramic sculpture in response to Western avant-garde practices but within the matrix of Japanese art-making conventions (fig. 38). Yagi noted, "We knew that we wanted to develop in Japanese terms something that had not previously existed."[2]

Those who thrive in Kyoto's art world understand that tradition is not static; it has to renew itself by transformation through innovation. The Raku family of Kyoto, for example, has been making tea bowls by hand with simple tools since the time of Sen no Rikyū (1522–1591), the grand tea master of Japan. The current head of the Raku family, Raku Kichizaemon, is a fifteenth-generation descendant, who cut short his training in Western-style sculpture in Italy when his father died prematurely, in order to return to Kyoto and assume his role as head of the family. His tea bowls often exhibit sculptural qualities that run counter to traditional Japanese ceramic forms, a characteristic of the works made by the influential *Sōdeisha* artists. During Salter's time in Kyoto in the early 1980s, *shokunin*, or craftspeople, working in such traditional practices as textile, paper-, and printmaking, as well as the

fig. 37: Ike Taiga
Moonlight Bamboo, ca. 1758–60
ink on paper
detail: see fig. 41

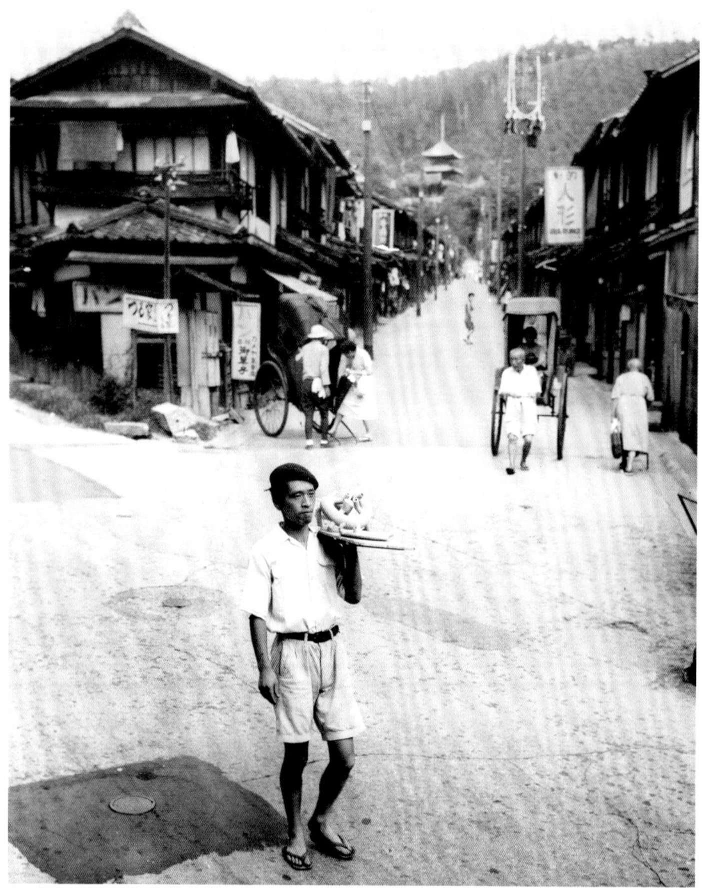

fig. 38: Yagi Kazuo carrying his unfinished avant-garde ceramic sculptures down Gojōzaka slope to the common kiln in Kyoto, 1954 Courtesy of Yagi Akira

creation of iron vessels for the tea ceremony, were also exploring more avant-garde forms. This climate of experimentation provided a highly stimulating environment in which Salter could develop her practice and form enduring friendships with fellow practitioners. She began by making tea bowls and soon moved on to making a series of thin, screen-like, ceramic objects (see figs. 2 and 29). These avant-garde, non-functional pieces have not survived, but photographs taken of the works when they were exhibited show that they had base colors of earth tones with added abstract lines and planes that hint at Salter's earlier interest in the spatial paradigms of Japanese *emakimono*, or picture scrolls.[3]

Salter's experimental ceramics were transitional works; soon after creating them she abandoned clay altogether, moving on to make two-dimensional abstract works—initially drawings and prints, and, after her return to Britain, paintings. This logical and positive shift in practice was influenced in part by practical considerations, as Salter did not have ready access to a wood kiln to fire her ceramics after leaving the Kyoto City University of Arts, but the more significant factor was her desire to exert greater control over the outcome of her labors. It is impossible to determine the final appearance of ceramics fired in a wood kiln; this element of chance is a critical aspect of the potter's practice, but the unpredictability was unsatisfactory to Salter, and working with other media gave her a greater degree of control. She is constantly aware of this powerful desire to control, however, and strives to permit the accidental to play its part in the creation of her works. Salter's background as a ceramicist is in other respects important for our understanding of her current working methods and aesthetic. Her laborious process of repeatedly layering on pigment and scraping the surfaces of her paintings is a vestige of her ceramic work, but of more significance is her recasting of her paintings as three-dimensional works, a distinctive and radical aspect of her practice examined by Achim Borchardt-Hume in his essay in the present publication. Salter's incorporation of the temporal framework of ceramic making into the creation of her paintings is also an important legacy of her earlier practice.

The artist's engagement with Japan is always discernible in her work and has been stimulated through regular visits to that country since her return to Britain in 1985, but it is more clearly manifested at some moments in her career than others. In this essay, I trace the centrality of Japanese artistic practices, aesthetics, and architectural concepts in Salter's art making, focusing on the four fundamental themes of time, space, line, and color. I also offer some comparisons between Salter's assimilation of Japanese models and that of twentieth-century American painters and suggest that her practice is markedly distinct from that of artists such as John Cage, Brice Marden, and Jackson Pollock.[4]

TIME AND SPACE

Without a powerful compositional centre, the viewer is free to wander an apparently endless space. The complexity of the surface allows the painting to become a labyrinthine narrative to explore. I consider that they [the paintings] are also about time—the time they take to make and the time they take to arrest the viewer. The density of the surface dictates the pace of this engagement.

Rebecca Salter, unpublished artist's statement (2010)

Salter's first significant encounter with Japan was reading *The Tale of Genji* and seeing reproductions of the most famous set of Japanese narrative picture scrolls when she was an undergraduate. She became fascinated by two key elements of the work: the visual scheme used for the presentation of space and the physical experience of viewing the horizontal scroll, which is unfolded from right to left, section by section, at about the width between one's two arms comfortably extended from one's body.[5] The viewing is very controlled because one cannot flip pages, as one can with conventional books, to take a random look. As Salter notes, "The action of the story is both reflected in and anticipated by the pictorial element. By only seeing a part of the narrative at a time, the 'moment' becomes very powerful."[6] In *The Tale of Genji* and other Japanese picture scrolls, the story is communicated visually in two different forms: first, by the calligraphy, which essentially provides the storyline (though the flowing calligraphic writing can also be aesthetically admired); and, second, by the actual images. Through the physical action of unrolling and rolling the scroll, the viewer anticipates what is coming and reflects on what has been already seen. Salter also notes that the notion of the "moment" is a crucial element of Japanese culture, as evidenced, for example, by the preference for single-petal cherry blossoms, which bloom for one day and then scatter the next day, over double-petal cherry blossoms, which last for longer periods. The single-petal blossoms' transitory existence is treasured, as are the calligraphic lines in picture scrolls, drawn with spirited concentration and without hesitation.

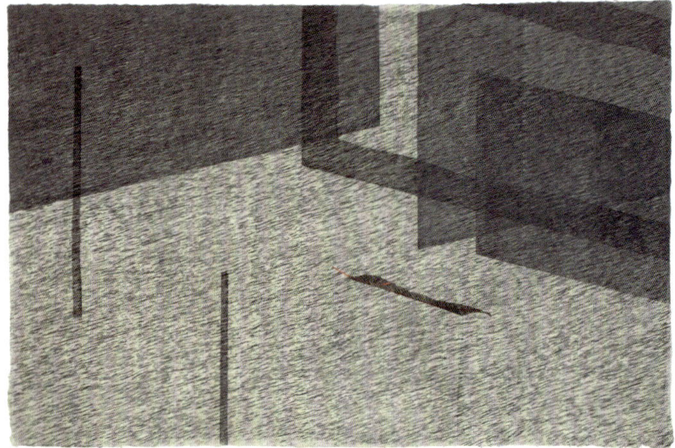

fig. 39: *Untitled A21*, 1982
mixed media on Japanese paper
25¼ × 37¾ in. (64 × 96 cm)

fig. 40: *Untitled B34*, 1984
mixed media on Japanese paper
33⅛ × 39⅜ in. (84 × 100 cm)

As Arata Isozaki has noted, in Japanese the concepts of space and time are expressed by a single word, *ma*, which has been glossed, variously, as "the natural distance between two or more things existing in a continuity"; "a space delineated by posts and screens (rooms)"; and "the natural pause or interval between two or more phenomena occurring continuously."[7] The concept runs counter to the Western notion of time and space being discrete, linear, and sequential entities. In Japan, *ma* is embedded in the structures of everyday life, underpinning architecture, the visual arts, music, drama, literary forms, and domestic rituals.[8] This radically different conception of the spatial and temporal made a significant impact on Salter after her arrival in Japan, and her experience there of secular and religious architecture and formal gardens provided as much stimulation for her practice as did the visual arts. Her rapid mastery of the language enabled her to read seminal texts not readily available in English, and her knowledge deepened when she was invited by Tada Michitarō, a scholar at Kyoto University, to translate texts which investigated Japanese ideas of space, in particular the boundaries between the sacred and profane.[9] The concept of time and space being inherent in the creation of works of art became a central element of Salter's practice, first with her ceramics and later with her two-dimensional works.

Untitled A21 (fig. 39), a woodblock print on paper Salter made in 1982 in Japan soon after her transition from ceramics to two-dimensional works, seems to reference a pictorial convention of Japanese scrolls present in the *The Tale of Genji*: the blown-away ceiling (*fukinuki yatai*), which allows viewers to look down into an architectural structure from a forty-five degree angle. Using a combination of straight and diagonal lines that seem to suggest architectural structures as well as the edges of *tatami* straw mats, Salter creates what appears to be a three-dimensional space where narratives can potentially unfold. Salter also used the traditional Japanese medium of the woodblock print to explore her preoccupation with *ma*, making use of the narrative possibilities offered by the print portfolio format to work out complex spatial relationships (see plates 152, 153).

A series of works on paper Salter made two years later includes *Untitled B34* (fig. 40), *Untitled B119* (plate 19), and *Untitled B120* (plate 20), each consisting of two rectangles inscribed with horizontal lines that subtly denote spatial intervals. These works embody the Japanese love for empty space, or *śūnyatā*, often associated with the Zen philosophy of meditative "nothingness." The Zen belief that each person has innate power, which can be cultivated through sustained meditation to achieve a state of no desire, and, ultimately, enlightenment, freeing one from karma and the vicious cycle of rebirth, had a significant impact on American artists, writers, and composers in the 1960s, notably John Cage. This influence was a direct outcome of the activities of the Kyoto School of modern Japanese philosophy, which, as Alexandra Munroe notes, "created a secular, aestheticized, and universal theory of Zen that elided its historical, cultural, and doctrinal complexities as an organized religion."[10] As Munroe argues, Western artists, for the most part, appropriated essentialized notions of Zen for their practices without taking on board its deeper philosophical concerns.[11] Salter's practice, however, embodies a profound engagement with Japanese philosophy, which was cultivated by her sustained efforts to master traditional artistic and craft practices. Her experience of learning calligraphy, which requires the artist to experience the unique "moment" of making each brushstroke, a process analogous to meditation in Zen, was particularly influential, as was her interest in ink painting, an art form practiced by Chinese Ch'an and Japanese Zen monks and, later, literati artists in Japan.[12]

fig. 41: Ike Taiga
Moonlight Bamboo, ca. 1758–60
ink on paper, six-panel folding screen, without mounting
61 × 141⅛ in. (154.8 × 358.4 cm)
Yale University Art Gallery, Leonard C. Hanna, Jr., B.A. 1913, Fund

The structures of *Untitled B34*, *Untitled B119*, and *Untitled B120* recall *Moonlight Bamboo* (figs. 41 and 37), a six-panel folding screen by Ike Taiga (1723–1776), one of the most celebrated eighteenth-century Japanese literati artists, who had a strong interest in Zen. The screen is a superb example of Japanese ink painting, particularly striking for its depiction of empty space. Accompanied by a few lines of calligraphy at the top right, three gigantic

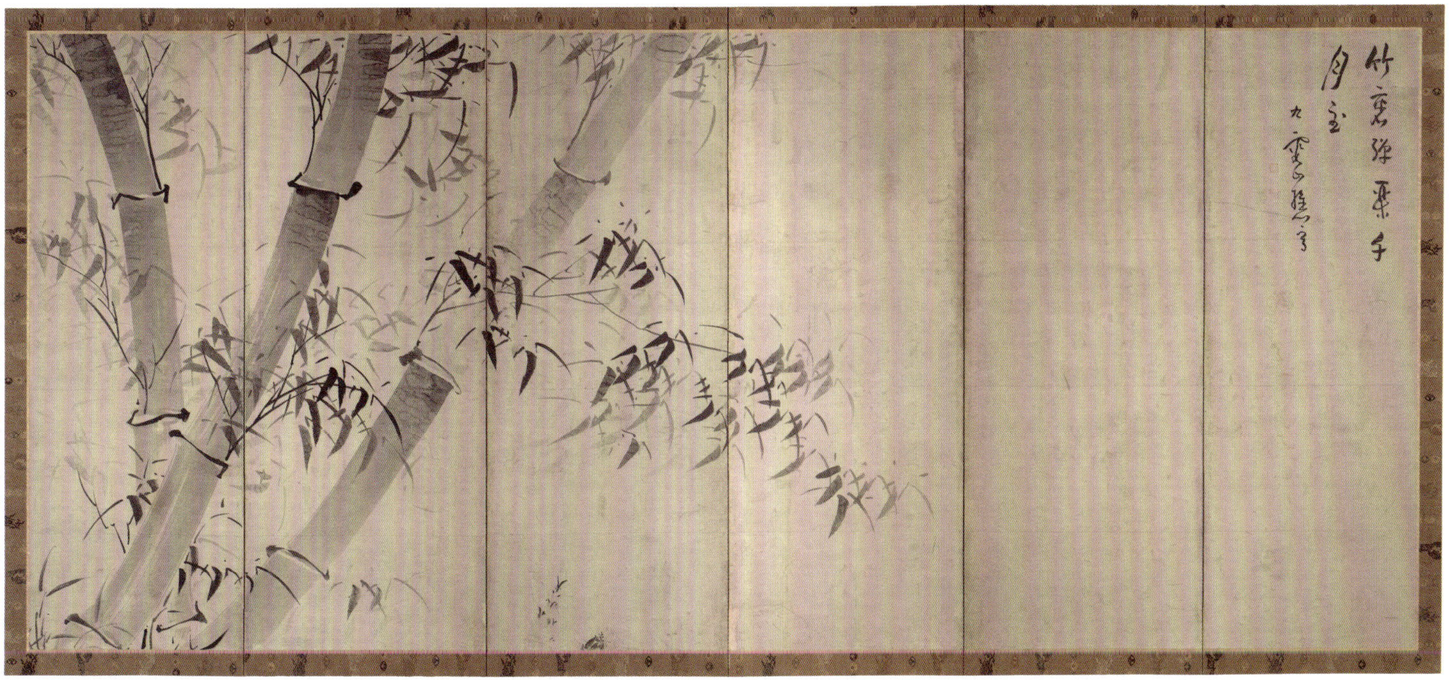

bamboo trunks spread out to the night sky like a Japanese folding fan. From the bamboo joints, sparse clusters of leaves extend and blow in a light breeze. Although there is neither moon nor directional light (in keeping with traditional Japanese painting practice), viewers can perceive the moonlit bamboo drawn in a light ink, while the darkly inked bamboo leaves suggest a shaded area. As Junichirō Tanizaki noted in his celebrated essay on Japanese aesthetics, *In Praise of Shadows*, a highly influential text for Salter, "We find beauty not in the thing itself but in the pattern of shadows, the light and the darkness, that one thing against another creates."[13] The drawn bamboo, which was given living breath by Taiga's brush, is just like a living, meditating human. The profound serenity surrounding the bamboo is reinforced by the void on the right-hand side of the screen, which functions as a space where stillness reverberates, an analogue to the state induced by Zen meditation. The calligraphy on the top right says, "Playing the *koto* in a bamboo grove, the moon came from a thousand miles away (to shine upon me)." The line is believed to have been composed by Taiga himself, and alludes to a poem about a bamboo grove composed by Wang Wei (ca. 701–761),[14] a great Chinese poet of the Tang dynasty, who is considered the fountainhead of the Literati School of painting from which the Japanese literati movement derived. The literary allusion hovers over the screen, breaching the limitations of historical time and stimulating the viewer to use the empty space to conjure up imaginary events, such as Taiga meeting with Wang Wei in the bamboo grove.

Salter's work demonstrates her understanding of the significance of empty space in works such as Taiga's. The scarcely decorated rectangles in *Untitled B34* (fig. 40) as well as *Untitled B119* and *Untitled B120* can be read as the blank pages of a notebook and even as two blank panels of a folding screen. Although there is no calligraphic poem or figurative bamboo in Salter's abstract drawing, the void space and the distressed surface prompt us to let our imaginations soar.

The temporal and spatial are also critical elements of Salter's complex working methods, which have evolved over the course of her career. For some of her early Japanese two-dimensional works, such as *Untitled B34*, she printed the creamy-white base color on the backs of sheets of paper using oil-based inks; brushed the printed area over with *sumi*, which dyed the unprinted areas and seeped through the printed areas to provide a sense of depth; and then laminated the sheets onto backing papers with rice glue. After she started to make paintings in the late 1980s, she began to create works through a laborious process of construction and deconstruction, alternately layering pigments on the surface of the support (which could be *washi* paper, canvas, linen, calico, or another material) and removing them by abrading and scraping, often using a sharp instrument such as a razor.

Salter's layering technique and overall surface design recall *Meditative Series #5* (fig. 42), a watercolor made in 1954 by the American artist Mark Tobey, whose practice was strongly influenced by his experience of Asian art, aesthetics, and religious practice.[15] In that work, Tobey layered subtle pale blue and dark pink washes over a warm tan-colored paper and, manipulating chalky white and pearl gray color with the tip of a flat brush, made infinitesimal squares, which cover the entire surface like a mosaic. Of all the abstract expressionists who were influenced by Asian art, Tobey developed the techniques most closely aligned

fig. 42: Mark Tobey
Meditative Series #5, 1954
watercolor on paper
16¾ × 10¼ in. (42.5 × 25.9 cm)
Yale University Art Gallery, Gift of George Hopper Fitch, B.A. 1932

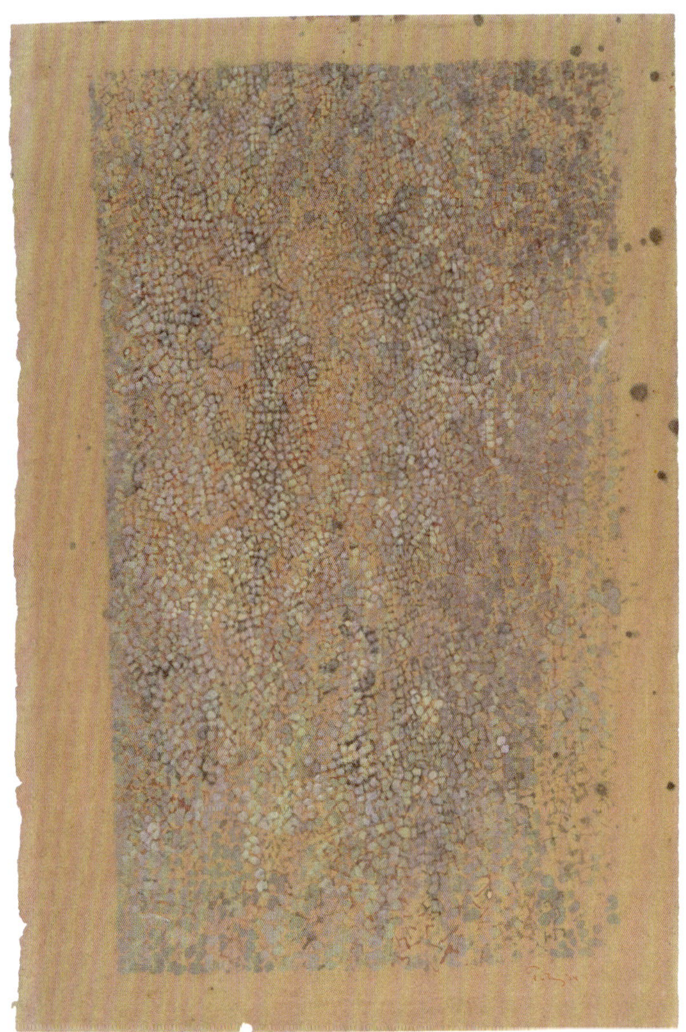

with those used by Salter, and the two artists also share a sophisticated comprehension of Asian philosophy and aesthetics.[16]

The surfaces of the rectangles of printed color in *Untitled B34* are anything but neat, and their edges in particular have a worn appearance, which speaks to another aspect of the temporal in Salter's work inspired by Japanese aesthetics: the philosophy of *wabi*. Derived from the Japanese word *wabishi*, meaning "in want" (as in the feeling of longing for someone or something, but used in a positive sense in which this condition is pleasurable), *wabi* celebrates the visible manifestations of the aging process. This quality is present in a Negoro tray dating from the fourteenth or fifteenth century (fig. 43), which was made for monks at the Negoro-ji temple in the present-day Wakayama Prefecture, south of Kyoto, and was used for serving meals. The tray is worn in the areas that have been touched repeatedly by human hands, and the red lacquer applied over black pigment has weathered to reveal an attractive horizontal pattern of lines along the wood grain underneath. Beginning in the sixteenth century, such aged utensils were rediscovered and given new life by Japanese aficionados of the tea ceremony, and they have become prized objects.[17]

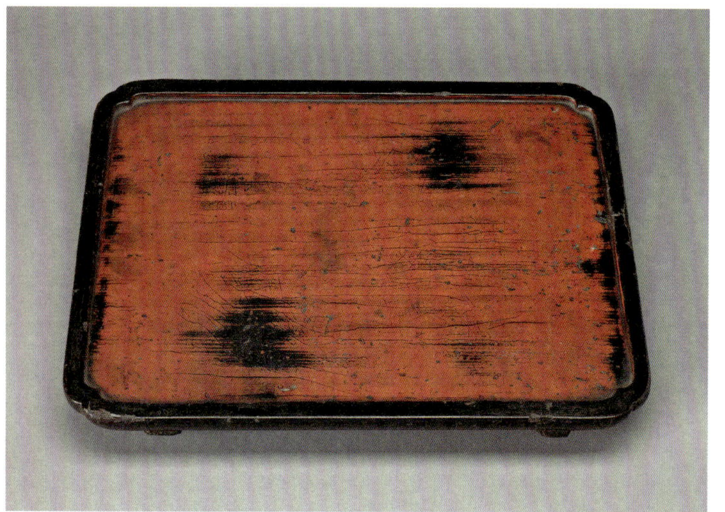

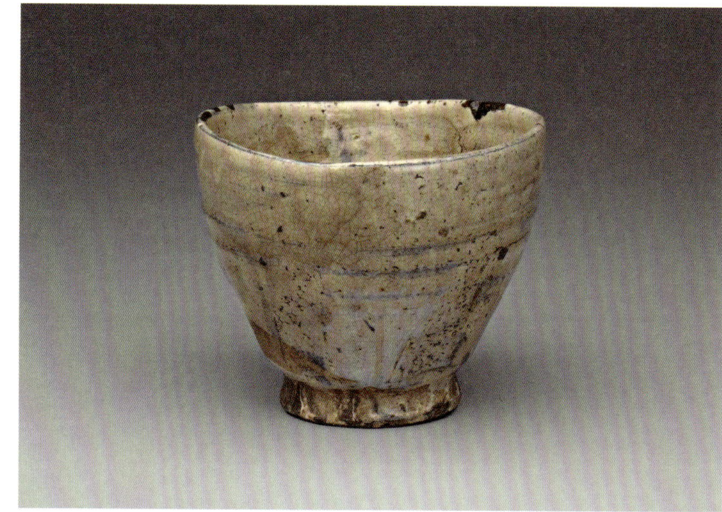

fig. 43: Kasuga Tray, Negoro ware,
14th–15th century
lacquer on wood, mother-of-pearl
1⅝ × 11⅛ × 14⅞ in. (4 × 28.2 × 37.8 cm)
Collection of Peggy and
Richard M. Danziger, LL.B. 1963

fig. 44: Tea Bowl, *Wind in the Pines*,
Hagi ware, early 17th century
stoneware with slip
3¾ × 5⅜ in. (9.5 × 13.7 cm)
Promised gift of Peggy and
Richard M. Danziger, LL.B. 1963
to Yale University Art Gallery

The layering of pigments that characterizes Salter's paintings also recalls the creation and the surfaces of Japanese tea bowls, exemplified by a Hagi bowl named *Wind in the Pines* (fig. 44). The Hagi kiln in the present-day Yamaguchi Prefecture in southwestern Japan was built by displaced Korean potters who were brought to Japan by Toyotomi Hideyoshi, a military general, at the time of his infamous failed invasion of the Korean Peninsula at the end of the sixteenth century. The gray clay body of the tea bowl is covered with a slip of reddish brown-gray, followed by a creamy-white glaze, and finally a thin bluish-gray glaze. The aesthetics of *wabi* can be observed in the subdued *shibui* colors of the Hagi bowl and as well as in the honoring of natural blemishes on the body that may have occurred during the firing as well those imperfections caused by wear.[18]

In 2006 Salter was commissioned to redesign the entrance hall and reception area of St. George's Hospital in Tooting, a residential neighborhood in south London (fig. 45). The terms of the commission required her to use the existing architectural structures; this was in some respects a considerable limitation, but the project provided her with an ideal opportunity to apply her decades-long preoccupation with *ma* to an actual space rather than a two-dimensional surface. The entrance hall as originally conceived was unwelcoming and very challenging to navigate, and Salter's primary objective was to create a sympathetic and calm space that mediated between the bustling outside world and the heart of the hospital, and, as she put it, "to introduce an intuitive way of navigating using light and texture."[19] Salter chose, whenever possible, sustainable and recycled materials. The walls of the reception area and hallway are now covered with bamboo panels; at rhythmic intervals vertical and horizontal columns of recycled glass emanate light. Behind the reception desk, there is a large horizontal panel of glass backlit with colored light,[20] while predominantly vertical lights are embedded into the tan-colored bamboo wall, which curves gently and leads the visitor down to a waiting room on the left. The bamboo wall unfolds like a calligraphic hand scroll, opening from right to left; the vertical changing lights of varied length and height appear like the poetic lines of calligraphy. At the project's inception, Salter determined where to place the light panels on the wall by kneeling in front of a long, horizontal piece of paper with eyes closed drawing lines in ink with an Asian brush to evoke the rhythm of writing on a Japanese poem scroll.

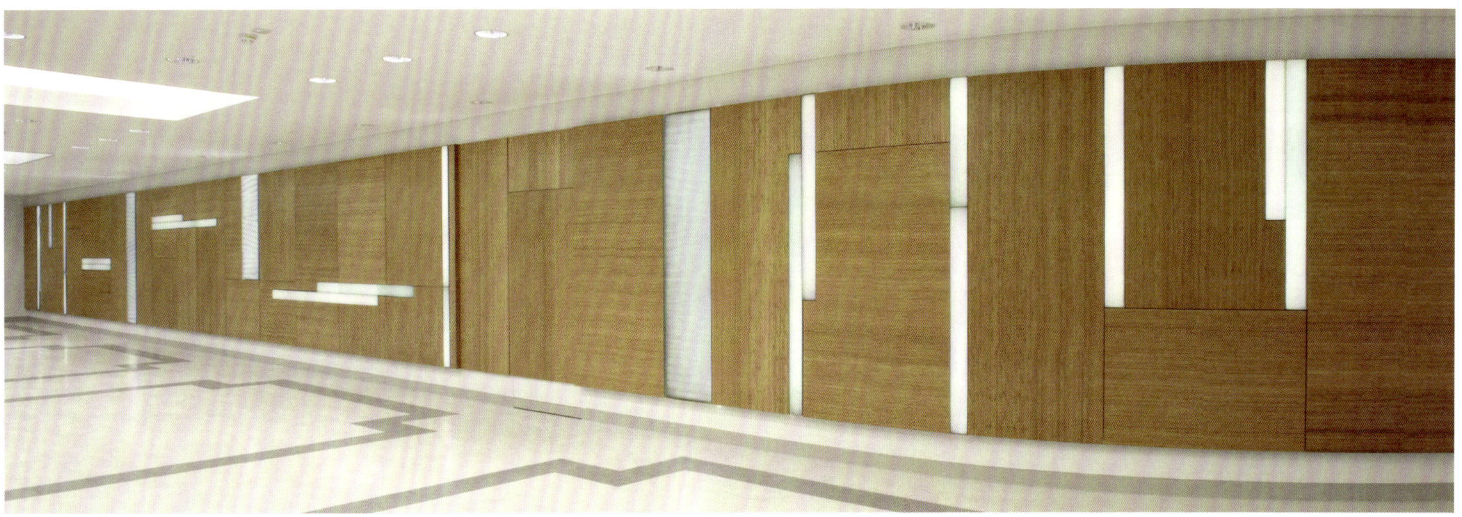

fig. 45: *Calligraphy of Light*, 2009
recycled glass and bamboo panel
main entrance, St. George's Hospital, London
(in collaboration with Gibberd, architects)

Salter has named the work *Calligraphy of Light*. This project can be understood as her creation of a sacred space for hospital patients, as if to follow the model of Japanese Shinto priests and priestesses, who demarcated sacred space for the descent of divinities. The empty space they would leave in a central area was to be empowered or blessed by a descending divine spirit. In present-day Japanese culture, people still invite Shinto priests to cleanse a space before building even non-traditional concrete buildings. From that perspective, Salter's decision to close her eyes to avoid exerting excessive control in making her blueprint perhaps suggests the invocation of a divine spirit to bless the hospital space, or at least her willingness to allow chance to operate in the creative process.[21]

COLOR AND LINE

In Japanese art there is no directional light, therefore no shadows. Space is defined and separated by colour and texture. My colour is concealed and revealed only when the eyes "settle." The colours are so close, the differences are hard to perceive. ... The lines are not for reading or explaining—they are entities in a dialogue with the surface and the colour. They are bold or tentative or dynamic or intimate but the challenge is always to keep them alive and autonomous—to stop them becoming predictable or formulaic. And that requires a state of skilled "unknowing."

Rebecca Salter, unpublished artist's statement (2010)

Snow White, Chalk, Pearl Gray, Silver Gray, Lead Gray, Ash Gray, Mouse Gray, Celadon Gray (*Rikyū nezumi*),[22] Olive Gray, Dove Gray, Blackberry, and *sumi* ink; these are just a dozen out of 250 names of colors ranging from Snow White to Ivory Black listed in *Nihon no haishoku* (Traditional Japanese Color Palette).[23] They also describe the color range one encounters in Salter's practice, and her restricted palette recalls that of Japanese calligraphy, ink painting, and textiles—subdued yet pungent, a concept that the Japanese term *shibui*. If a Japanese person had to name a quintessentially British color, gray would be the most likely choice, evoking as it does the drizzling rain which in popular perception typifies Britain's climate.

fig. 46: *Untitled RR21*, 2009
mixed media on linen
detail: see plate 148

To use the matrix of the *Traditional Japanese Color Palette*, it might be a color between Pearl Gray and Celadon Gray. It could have a tint of blue, which would make it Smoke Blue #2 (*masuhana iro* 185); if it tends toward brown, it is close to Aqua Gray (*fukagawa nezumi* or *minato nezumi* 174); if it has both light blue and light gray, it is close to Light Saxe Blue (*sabi asagi* 175). Salter's preference for a restrained palette seems to be a synthesis of her British background and the Japanese color aesthetic.[24] She often juxtaposes colors with very subtle differences, as in the aforementioned *Untitled B34*. This practice could be likened to the ancient Japanese practice of *kasane*, or layering combinations of colors, particularly in evidence during the Heian period (794–1185), when court ladies would often wear twelve different layers of silk garments.

Salter's study of calligraphy in Japan profoundly impacted her artistic practice in ways beyond its influence on her palette. When the term "calligraphic" is used in English, it typically describes line movement on the surface of a plane, but in Asian calligraphy the drawn lines are analyzed in three-dimensional terms. Paper is not two-dimensional, as is obvious when one looks at it with a microscope, so one might ask, for example, how much pressure from the brush is translated into the sheet of paper.[25] Watching a calligrapher at work, one notices that the drawn lines are merely a part of a bigger, dynamic movement that takes place for the most part in the air. At what angles the brush engages the paper, at what speed, where it makes a sudden stop, and how one-tenth-of-a-second stops pick up velocity again are all factors that determine the character of the lines and dots on the paper. The art of calligraphy in Japan is also part of Zen practice. Grinding ink with water on a stone is a time for meditation, and when a calligrapher picks up the brush and starts writing, the act is spontaneous, and he or she "disappears." An artist cannot make a masterpiece while he or she is still "willing" and "intending" to make one. This state of arriving at spontaneous engagement was termed "flow" by the psychologist Mihály Csíkszentmihályi.[26]

Asian calligraphers are expected to devote years to learning their craft, but Salter wisely avoided that professional path. Its rigors would have required a commitment she was not prepared to make but, more importantly, considering the scope of her technical variety, focusing exclusively on calligraphy would have severely limited her practice. Nevertheless, both the technical and philosophical aspects of calligraphy are critical to Salter's working practices; line is always an essential structural component (though through her processes of layering and erasure, it often lies under the surface of her finished paintings), and she speaks of her works as places into which she can "disappear" while she is making them.

Viewers look at Salter's works by "reading" them as they would the calligraphic lines of an ink landscape painting. The artist knows where she began and where she ended, but, because her works are often devoid of a central or focal point and seem, at least initially, labyrinthine, viewers need to move their eyes around before they can begin reading or understanding the works. It is at the point when he or she discovers the concealed colors in the layering and the eye finally settles that the viewer can wander in the narrative space Salter has created.

Untitled RR21 is suggestive of a painting of a lily pond by Monet, though one where thin killifish-like creatures swim in place of water lilies (fig. 46). The work is not chromatically

fig. 47: Calligraphic practice sheets, 2006
sumi ink on paper
20 × 29 in. (50.8 × 73.7 cm)

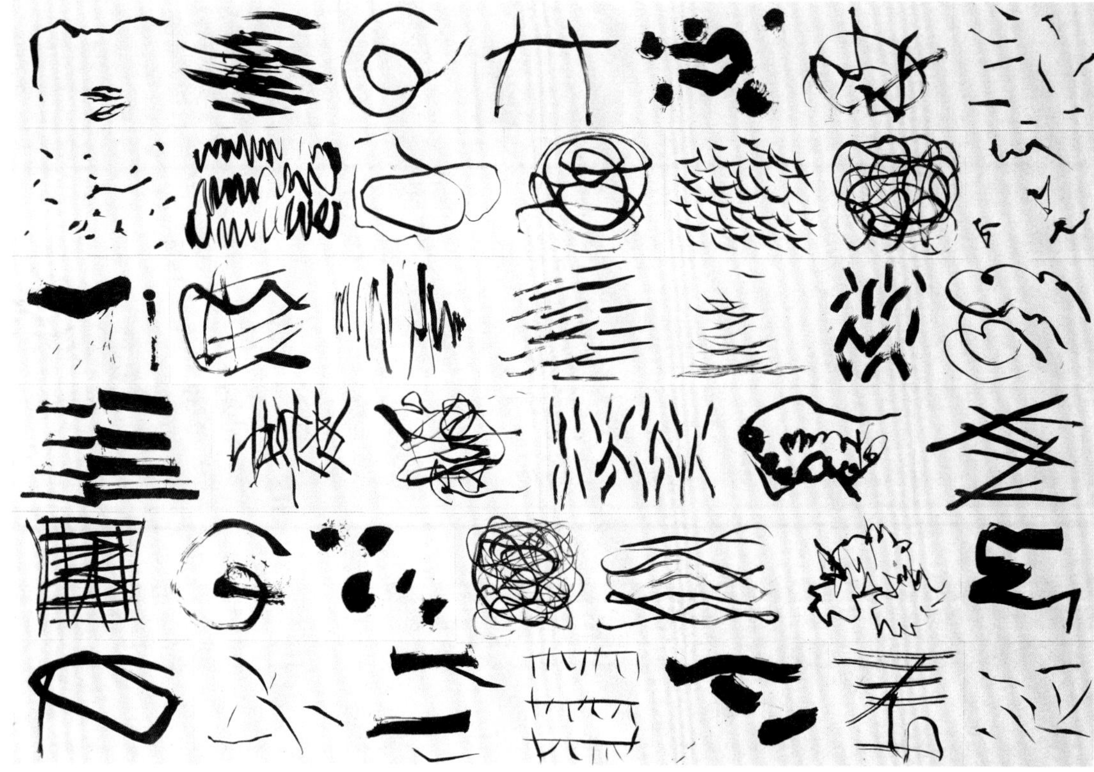

exuberant; Salter painted the linen first with bluish gray, then with a darker purplish gray and a wash of thin white, before vigorously (one might almost say violently) abrading the surface to create an aurora-like, shimmering effect suggestive of a curtain. The tiny, fish-like forms, drawn with sepia ink, which has a purplish tone, resemble the sinograph *ichi*, known in calligraphic practice as the "number one" stroke.[27] For a Japanese calligrapher (like myself), Salter's *ichi* in ink are particularly impressive, drawn using strong and spontaneous lines with varied gestures and well-thought-out placements.[28] The lines maintain the basic horizontality required for the *ichi* stroke, with angles of no more than forty degrees up or down, but demonstrate gentle variations of line thickness and curvature created by varying the pressure and lifting the brush. Many of the lines display the pressure at the beginning and the end of the strokes that characterize *ichi* made by a Japanese calligrapher using a soft Asian brush. Salter's familiarity with this technique can be observed in some of her practice sheets of brush drawing in ink on *washi* paper (fig. 47). The lines in *Untitled RR21* are sharp and strong without being mechanically straight; they are drawn with a dip pen rather than a brush, and with resilience, while the moist spread of ink gives them softness and diversity. The lively horizontal lines have "breath" (*ki* in Japanese; *qi* in Chinese) in each stroke, and such vitality is further emphasized by the ample space around each line, allowing it to take a deep breath.

The strokes on Salter's practice sheet recall a section from the two left-hand panels of Ike Taiga's *Moonlight Bamboo* (fig. 5), though Taiga's brush handling is more deliberate because he has a specific plant's physicality in mind rather than drawing unspecified lines.

fig. 48: *Untitled 2007–18*, 2007
mixed media on paper
14⅛ × 18⅞ in. (36 × 48 cm)

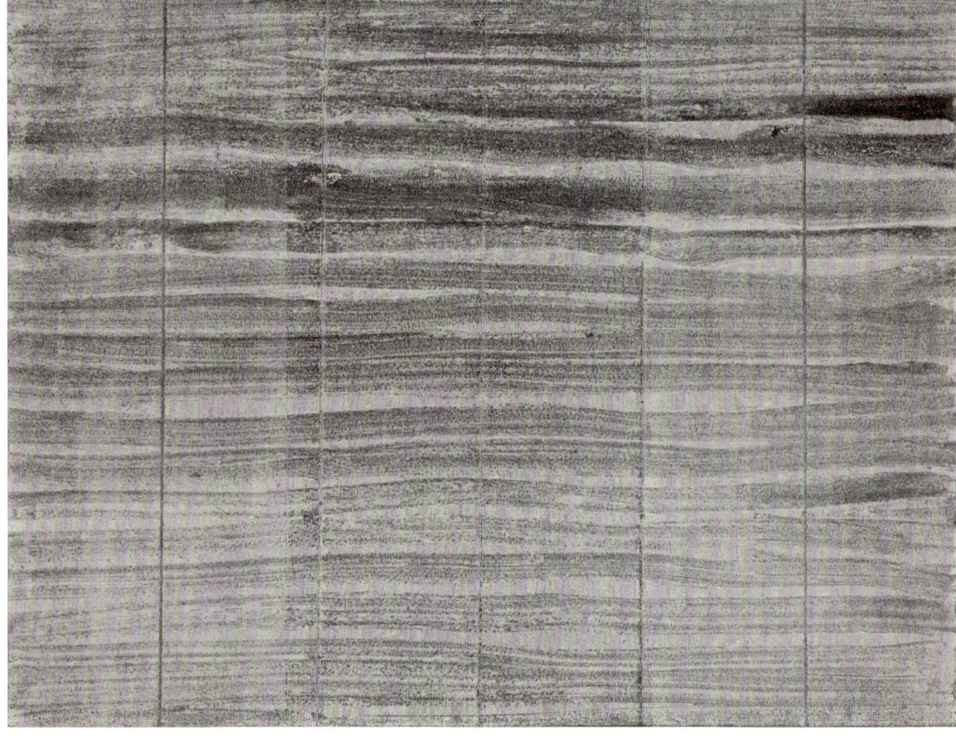

fig. 49: Tōshūsai Sharaku
The Actor Arashi Ryūzō as the Money-lender Inshibe no Kinkichi, from the play The Iris Soga of the Bunroku Era, 1794
polychrome woodblock print
15 × 10 in. (38 × 25.5 cm)
Yale University Art Gallery, The Hobart and Edward Small Moore Memorial Collection, Gift of Mrs. William H. Moore

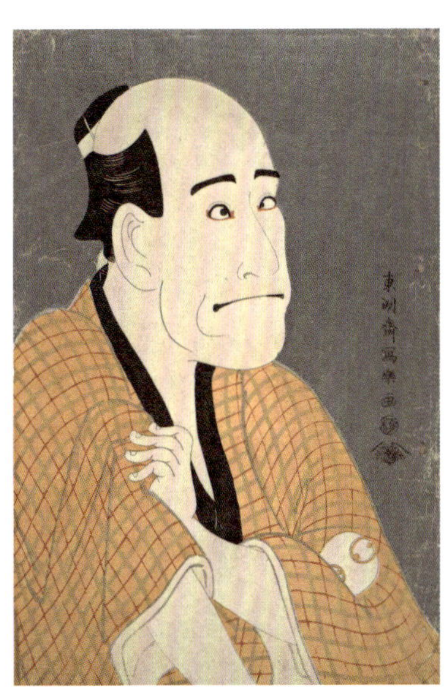

Large, aged bamboo trunks with surface dryness and possible cracks are indicated by broken and split traces called "flying white," created by a coarse-hair brush.[29] Such straight, wide marks with broken brush traces are made with a single stroke and are not touched up by drawing over them. Salter has used a similar broken and split stroke technique in several of her drawings, including *Untitled 2007-18* (fig. 48). Salter's variation of "*V*" shapes on her practice sheet (fig. 47) resemble some of Taiga's thin bamboo stems and leaves drawn in thick black ink that appear to grow out of the joints vertically and diagonally. Taiga's bamboo leaves, with their various shapes, sizes, and ink tonality, may be compared with Salter's strokes: both indicate that there was initial pressure and then a sharp ending of the stroke as the brush came off the paper. Thick, black, ropey lines signaling the location of the bamboo joints in Taiga's painting may be compared with Salter's example drawing in the righthand column, second from the bottom in fig. 47. Taiga applied the darkest of black inks in rendering these joints, which are the pivotal finishing points of any painting of bamboo. (In the same way, in a painting of a dragon the eyes are inked in last to make it come alive.) The black bamboo joints are calligraphic variations of the *ichi*, written very quickly and with great pressure, and reminiscent of a Kabuki player's tightly closed mouth that suggests pent-up emotion, as often depicted in *ukiyo-e* prints (fig. 49). The stroke in the lower section of Salter's practice piece is of the same quality—it has the strength of a cord that can withstand great pressure at the knot.

Ichi can also be found in Salter's *Untitled RR26* (fig. 50), drawn in ink pen on lines of wet egg-white. The *ichi* do not have a horizontal orientation but instead were drawn in all directions like a cascade. Salter's painting *Untitled MM33* consists of Japanese *ichi* of diverse

fig. 50: *Untitled RR26*, 2009
mixed media on paper
detail; see plate 136

fig. 51: *Untitled MM33*, 2008
mixed media on linen
41⅜ × 43¼ in. (105 × 110 cm)
Private collection, London

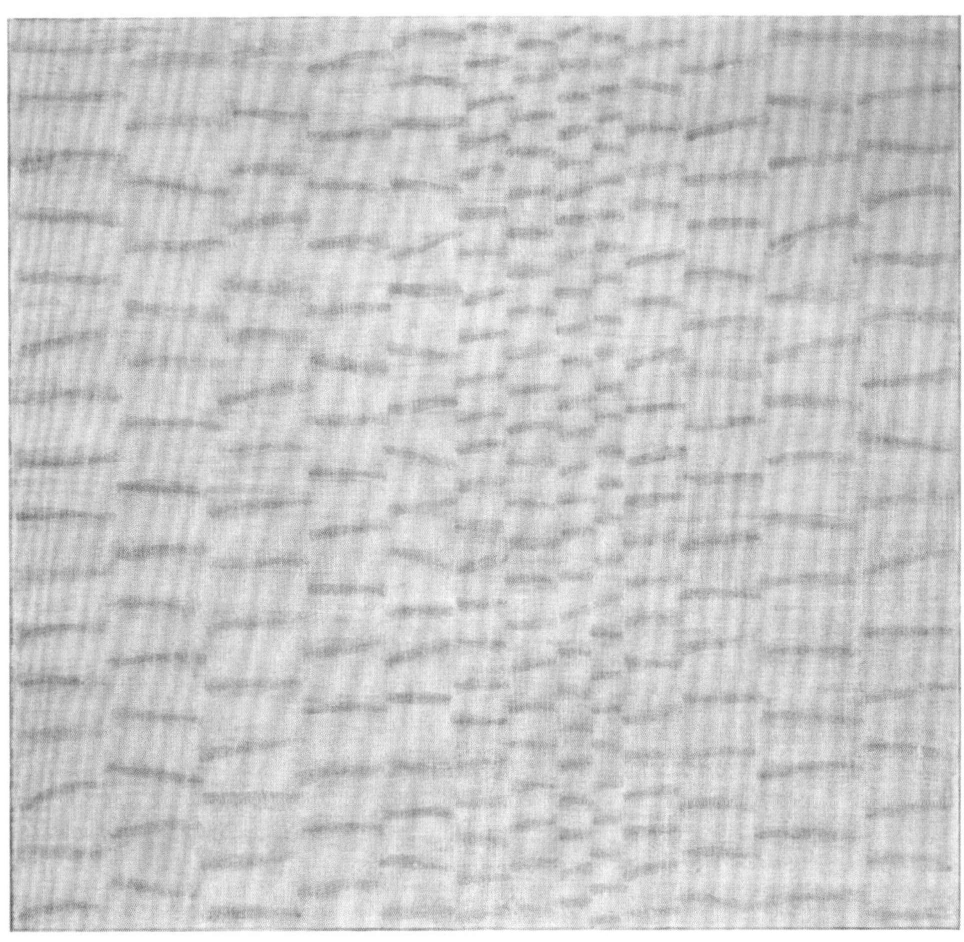

widths, rendered with a fuzzy blurriness and arranged in thirteen columns (fig. 51). In the center, four narrow columns create an optical illusion of a space that draws viewers toward it, inviting them to wander the small passages created by the stick-like *ichi*, which upon close view look like humorous caterpillars. Humor is another distinctive characteristic in Japanese art. Even Taiga's serene *Moonlight Bamboo* shows two perky bamboo shoots peering up from the bottom of the screen to enliven the scene.

Salter's twentieth-century Western predecessors, for the most part, have been less rigorous in their use of calligraphic methods and less willing to acknowledge the influence of Asian drawing on their practices. The reluctance of the abstract expressionists in particular to engage fully with the aesthetics and processes of Asian art has been attributed by scholars to the perception during their heyday of the East as "feminized," and therefore problematic to those participating in an artistic milieu where masculinity was regarded as a hallmark of creative excellence.[30]

With his orchestration of ink drips in numerous consistencies, colors, and combinations on a grand scale, Pollock mastered the circular movement of calligraphy, yet he did not take much interest in the possibilities an Asian brush could offer. An Asian brush has the ability to convey minute variations in strength of lines and tonality. The way in which an artist's hand holds the bamboo trunk of the brush determines the amount of pressure that is transmitted on the brush hairs; this in turn determines the gestural quality of the brush-work as well as the tonal effect created on the paper. This "pressure" is crucial to a fuller understanding of Asian calligraphy. A sensitive Asian brush is capable of revealing the artist's emotional state as well as the level of the artist's calligraphic skill. But Pollock often eliminated the brush—a sensitive medium—and aggressively poured or flung paint directly at the canvas. American action painters like Pollock had little interest in either conveying a subtle sensitivity or revealing his state of mind through his brush strokes. Franz Kline often denied his connection with Asian calligraphy, claiming he was more interested in mass than in lines; yet it is evident that he used a pointed Asian brush for his drawing. The sharp endings of several of the lines in an untitled work of 1951 (fig. 52) were produced by lifting the tip of the brush from the paper in the middle of a rapid stroke. Western flat brushes cannot make such marks; such a movement requires an understanding of how much pressure to apply on the brush hair and when to release it as it comes off the paper in the split of a second. We know that Kline did practice the art of pressure in this drawing. While most of the lines in Kline's better known paintings are straight rather than circular, in this drawing he experimented with the brush to create semicircles, drips, and broken traces.

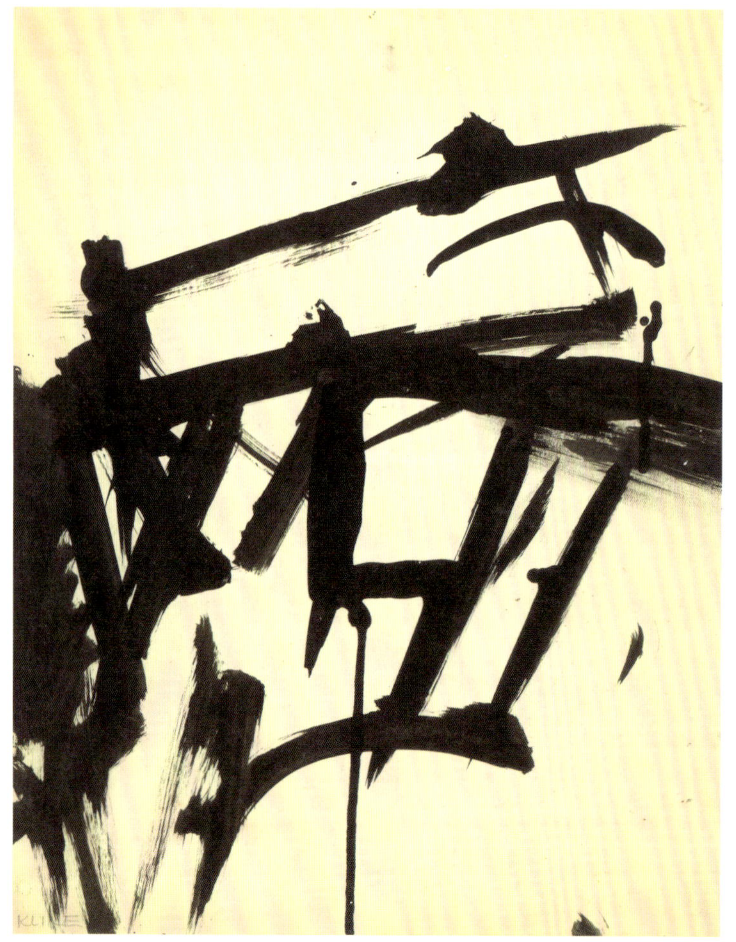

fig. 52: Franz Kline
Untitled, 1951
ink on paper
13⅞ × 11 in. (35.2 × 28 cm)
Yale University Art Gallery,
Richard Brown Baker, B.A. 1935,
Collection

fig 53: Brice Marden
Tu Fu Dog I, 1986–91
black ink over embossed platemark on paper
9⅜ × 7⅛ in. (23.7 × 18.1 cm)
Yale University Art Gallery, Richard Brown Baker, B.A. 1935, Collection

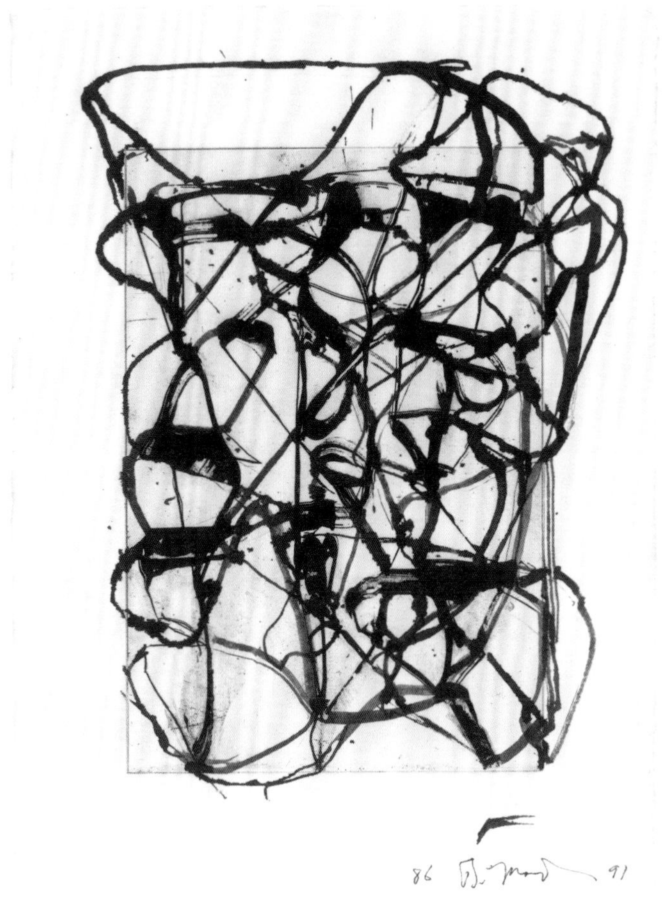

Even the way in which the background paper shows through reminds us of "flying white" in Asian calligraphy.

Brice Marden's *Tu Fu Dog I* is a calligraphic piece with modulated ink tones and split traces made not by a brush, but with a long stick (fig. 53). He may not have felt comfortable with the propensity of the sensitive Asian brush to reveal the hidden depths of human character and emotions. More than calligraphy, Marden seems to be interested in the interaction between the drawn line and the space around it; he appears less interested in the lively and singularly drawn line itself. Time and again, he retraces and retouches his lines in pursuit of a certain effect; he clearly has no use for the golden rule of Asian calligraphy: do not retouch what you have already put down on the paper. Salter also uses a variety of methods for applying paint, sometimes painting with her left hand or using unusual brushes, like the sign writers' brush with long hair, to let accidental effects reveal themselves on her working surface. For both artists, these tools—the long stick for Marden and a variety of instruments for Salter—are a means of letting spontaneity occur and of avoiding too much intentional control.[31]

As the artist Yvonne Audette notes, Mark Tobey found "the basis of universality in calligraphy," but his engagement with Asian aesthetics and philosophy seems to have impacted negatively on his reputation.[32] Tobey's style was considered too "Oriental" and subtle; the critic Thomas Hess observed of one of Tobey's works of 1946, "The beauty of *Structure* is almost as minor as its parts, perfectly articulated though it and they may be. The flaw here is, perhaps, not so much with Tobey as it is with the Oriental models to which he is so attached. Understatement to the point of preciosity and restraint to the degree where statement is innocuous—both flaws which often mar Oriental painting—are evident in this modest tempera."[33] Feeling marginalized in the American art world, Tobey permanently moved from New York to Europe in 1960.

Rebecca Salter has lived both in the East and in the West, and she has many friends who are *shokunin*, or professional artisans, in Japan.[34] Like their work, Salter's art belongs to the sensitive realm of art where one finds disciplined technical execution. The masters among *shokunin* are called *meijin* in Japanese—those who have achieved mastery with divine spirits, the "flow." We have observed Salter's extended efforts to invite these divine influences into her art and to strive for immediacy without clinging to self-expression. Salter is an artist who insists on a tenacious penetration of pigments and stain into the surface of the paper, canvas, linen, or whatever material she uses for a ground, in layer upon layer; she then carves and removes the paint to reveal the remnant and disclose what is left. She has produced an unprecedented body of work of wide-ranging ink tones and subdued *shibui* colors, executed with great delicacy and the perseverance of *wabi* aesthetic. With variations of gesticulated calligraphic lines and juxtaposing textured colors, she demonstrates a profound understanding of space, including empty space and time. Salter has tried to free her art from intention so that the viewers may experience the natural "flow." We must remember that behind her subtlety and subdued colors there are great efforts concealed, just as years of laborious training remain unseen in the work of the *shokunin* artisans of Japan.

As viewers, we must experience her art in person to see with our own eyes the art of persistence that offers serenity. As Salter notes of her work, "The lines are not for reading or explaining—they are entities in a dialogue with the surface and the color. They are bold or tentative or dynamic or intimate but the challenge is always to keep them alive and autonomous—to stop them becoming predictable or formulaic. And that requires a state of skilled 'unknowing.'"[35] This statement summarizes her endeavor and conveys how much of it overlaps with the Japanese practice of making calligraphic lines come alive to reach the point of "flow," or non-intentional stage of achievement. Her quiet and yet persistent assemblage of constant, unyielding drawing, incising, and scratching displays another kind of strength. It does not possess an obvious or masculine beauty; it is subtle yet tough, delicate yet persevering, small but intricate, at a glance soft-looking but with a depth of layering and even violent defacing, without the intention of clean perfection but using abrasion as a call for empathy—much like *Wind in the Pines*, the Hagi tea bowl of *wabi* aesthetic that invites you to hold it in your hand.

notes: Rebecca Salter and Japan

This essay is based on a series of conversations with Rebecca Salter, which took place in New Haven between 2003 and 2010 and in London at Salter's studio in 2009. Unless otherwise noted, all translations are my own. Many thanks go to Rebecca Salter and Gillian Forrester who patiently guided me to sound evaluation of modern western art history that is beyond my normal scope. The epigrah is from John Cage, "Memoir," in Richard Kostelanetz, ed., *John Cage* (London: Praeger, 1970), 77.

1 See Gillian Forrester's introductory essay above (pp. 5–27) for a fuller account of Salter's career.

2 Yagi Kazuo, "Watashi no jijoden (My autobiography)," in *Yagi Kazuo, Kokkoku no hono-o* (Steady flame) (Kyoto: Shinshindo shuppan, 1981), pp.15–16. For *Sōdeisha*, see Louise Allison Cort and Bert Winther-Tamaki, *Isamu Noguchi and Modern Japanese Ceramics: a Close Embrace of the Earth*, exh. cat (Washington, D.C.: Arthur M. Sackler Gallery, Smithsonian Institution; in association with Berkeley: Univ. of California Press, 2003), 156–86.

3 Rebecca Salter, "Emakimono: *The Tale of Genji* Scrolls" (BA thesis, Bristol Polytechnic [now the University of the West of England], 1977).

4 The topic of the relationship between Asian and American artistic practice in the twentieth and twenty-first centuries was explored fruitfully in *The Third Mind: American Artists Contemplate Asia, 1860–1989*, an exhibition at the Guggenheim Museum, New York, in 2009, and in its accompanying publication of the same title, to which this essay is indebted.

5 This type of scroll is therefore called a "hand scroll," to differentiate it from the vertical hanging scrolls that can be observed at once in their entirety. *The Tale of Genji* hand scrolls have been cut into several pieces and are now housed in the Tokugawa Art Museum in Nagoya and the Gotō Museum in Tokyo.

6 Artist's statement, 18 April 2010 (unpublished), sent to the author.

7 Arata Isozaki, *Space-Time in Japan: MA*, exh. cat., (New York: Cooper-Hewitt Museum, 1979), 12. Isozaki's account of *ma* has been a key reference for this essay. See also Gunter Nitschke, "Ma: The Japanese Sense of Place," *Architectural Design* 36 (March 1966): 108–56.

8 See Isozaki, *Space-Time in Japan*.

9 The texts translated by Rebecca Salter can be found in Tada Michitarō, "Sacred and Profane: The Division of a Japanese Space," *Zinbun: Memoirs of the Research Institute for Humanistic Studies*, Kyoto University, 17 (1981).

10 Alexandra Munroe, "The Third Mind: An Introduction," in Munroe, ed., *Third Mind*, 27. On the Kyoto School, see James W. Heisig, *Philosophers of Nothingness: An Essay on the Kyoto School* (Honolulu: Univ. of Hawai'i Press, 2001).

11 Munroe notes that "this iteration of secondary Orientalism nurtured American artists, writers, and critics whose uses and interpretations of Zen reflect the *processes of mediation* rather than the historical reality of Zen itself." Munroe, "Introduction," in Munroe, ed., *Third Mind*; see also Munroe, "Buddhism and the Neo-Avant-Garde: Cage Zen, Beat Zen, and Zen," in the same publication, 198–215.

12 The so-called literati artists originated with the Tang dynasty poet and painter Wang Wei (ca. 701–761), who is considered the fountainhead of literati painting. This style of painting later became known as the Southern School of painting in contrast to the Northern School of painting, which was largely practiced by professional artists. Literati painting gained further credence when Chinese scholar-statesmen began practicing ink painting in defiance of Mongol rules during the Yuan dynasty (1271–1368). The painting style was brought to Japan by Chinese Ch'an (Zen) monks during the seventeenth century and became one of the major styles during the Edo period (1615–1868). Unlike in China, a majority of the Japanese literati artists were professional artists; however, some, like Ike Taiga, took a strong interest in Zen.

13 Junichirō Tanizaki, *In Praise of Shadows*, trans. Thomas Harper and Edward G. Steidensticker (London: Vintage, 2001), 47.

14 Wang's poem reads, "Seated alone in a bamboo grove / Playing the *qin* [zither] and singing along / In the deep woods no-one knew where I was / But the bright moon came and shone on me." The poem is entitled "Zhuliguan" (Bamboo grove pavilion) in vol. 13 of the commentary on the collection of the works by Wang Youcheng found in the electronic version of Siku Quanshu (Wengnange Edition)/Collection group/ Appended section/from Han to Five Dynasties/ Wang Youcheng ji jianzhu. Unpaginated.

15 Tobey, who traveled widely, was exposed to Asian calligraphy when he was a student in Seattle. Later in his career he lived in Kyoto. Bert Winther-Tamaki, "The Asian Dimensions of Postwar Abstract Art: Calligraphy and Metaphysics" in Munroe, ed., *Third Mind*, 147–48.

16 Salter's works with numerous surface designs appear to recall some of Tobey's works. See, for example, Salter's *Untitled 2009-27* (plate 138), *Untitled MM2* (plate 131), *Untitled 2006-28* (plate 112), *Untitled KK40* (plate 125), and *Untitled JJ11* (plate 109).

17 See Yuriko Saito, "The Japanese Aesthetics of Imperfection and Insufficiency," *Journal of Aesthetics and Art Criticism*, 55, no. 4 (Fall 1997): 377–85. In this essay, Saito stresses the significance of the Zen Buddhist aesthetic of appreciating the worn and the imperfect and discusses its political implications.

18 The earliest mention of the term *shibui* can be found in an early Heian period (794–1185) record describing the pungent taste of an unripe persimmon (Kōjien electronic dictionary SR-M4000). Today *shibui* is a popular term used to describe a gustatory, aesthetic, or visual quality that is subtle, astringent, and understated. See Donald Keene, *Landscapes and Portraits: Appreciations of Japanese Culture* (Tokyo and Palo Alto: Kodansha International Ltd., 1971), 11. There was a popular *shibui* boom in the 1960s in the United States. For details, see John Whitney Hall, "On the Future History of Tea," in *Tea in Japan: Essays on the History of Chanoyu*, ed. Paul Varley and Kumakura Isao (Honolulu: Univ. of Hawai'i Press, 1989), 250–51.

19 Rebecca Salter, "A Short Reflection on a Long Project," unpublished essay, 2009 (copy held at the Yale Center for British Art).

20 A range of colors is available, and the receptionist selects a color for his or her shift.

21 See Richard Cork's essay in the present publication.

22 "Rikyū" refers to Sen no Rikyū (1522–1591), the most celebrated grand tea master of Japan, who firmly established the subdued aesthetics of the so-called *wabi* tradition of *matcha* tea.

23 Sano Takahiko and Hamada Nobuyoshi, *Nihon no haishoku* (Traditional Japanese Color Palette) (Tokyo: PIE, 2009), 166–82. In my essay text, colors that I have identified in Salter's art works are indicated by English translation of the traditional Japanese color name, followed by the Japanese name and the color numbering used in the above book in parentheses.

24 For Salter's use of gray in relation to that of Jasper Johns and Gerhard Richter, see above, pp. 44–45.

25 Kazuaki Ōsawa's publication, *Sho wo kagakusuru* (Scientific analysis of calligraphy) (Tokyo: Mokujisha, 1974), is one of the earliest attempts to examine calligraphy scientifically.

26 Mihály Csíkszentmihályi, *Flow: The Psychology of Optimal Experience* (New York: Harper Perennial, 1991).

27 Calligraphy schools and teachers in Japan often develop their own methodology for study based on Chinese historical records. In addition, the Japanese Ministry of Education (*Monbushō*) regulates textbooks for elementary school children on rendering sinographs and Japanese syllabaries with a brush or a pen. For English-language works on basic calligraphy strokes, see Chiang Yee, *Chinese Calligraphy: An Introduction to Its Aesthetic and Technique*, 3rd ed. (Cambridge, Mass.: Harvard Univ. Press, 1973), 153–65.

28 We see some variations based on the *ichi* in other paintings by Salter; for example, *Bethany Squares* (see fig. 23). See also *Untitled BB32* (plate 79), *Untitled MM33* (fig. 51), and *Untitled RR35* (fig. 25).

29 The Japanese term *hihaku* ("flying white"), derived from the Chinese term *feibai*, is often used to describe the favorable traces made on rough paper by a brush when its ink is running out. These strokes reveal the texture of the paper support, and they are admired by calligraphers, though the style of the stroke and the degree of dryness shown on the paper can differ according to the artist, his or her historical context, and his or her country of origin.

30 On the "feminized Orient," see Bert Winther-Tamaki, "The Asian Dimensions of Postwar Abstract Art: Calligraphy and Metaphysics," in Munroe, ed., *Third Mind*, 152; and Bert Winther-Tamaki, *Art in the Encounter of Nations: Japanese and American Artists in the Early Postwar Years* (Honolulu: Univ. of Hawai'i Press, 2001), especially 54–55 in the section "Mark Tobey: A Janus-Faced America."

31 Marden also wanted to view the drawn surface from a distance, particularly when he was producing large works, and the long stick gave him that perspective.

32 Inscribed in the pages of Audette's sketchbook dated 1955. See Kirsty Grant, *Yvonne Audette: Different Directions, 1954–1966*, exh. cat. (Melbourne: National Gallery of Victoria, 2007), quoted by Winther-Tamaki, "The Asian Dimensions" in Munroe, ed., *Third Mind*, 153 n. 67.

33 Thomas Hess, *Abstract Painting: Background and American Phase* (New York: Viking, 1951), 121. "Structure" is reproduced on p. 127.

34 Salter has a number of printmaker friends in Japan. She has also filmed and conducted interviews with Japanese *shokunin* artisans in an effort to preserve their arts. On the subject of Japanese artisans, see Rokusuke Ei, *Shokunin*, Iwanami shinsho 464 (Tokyo: Iwanami shoten, 1996).

35 Artist's statement, 6 April 2010 (unpublished), sent to the author.

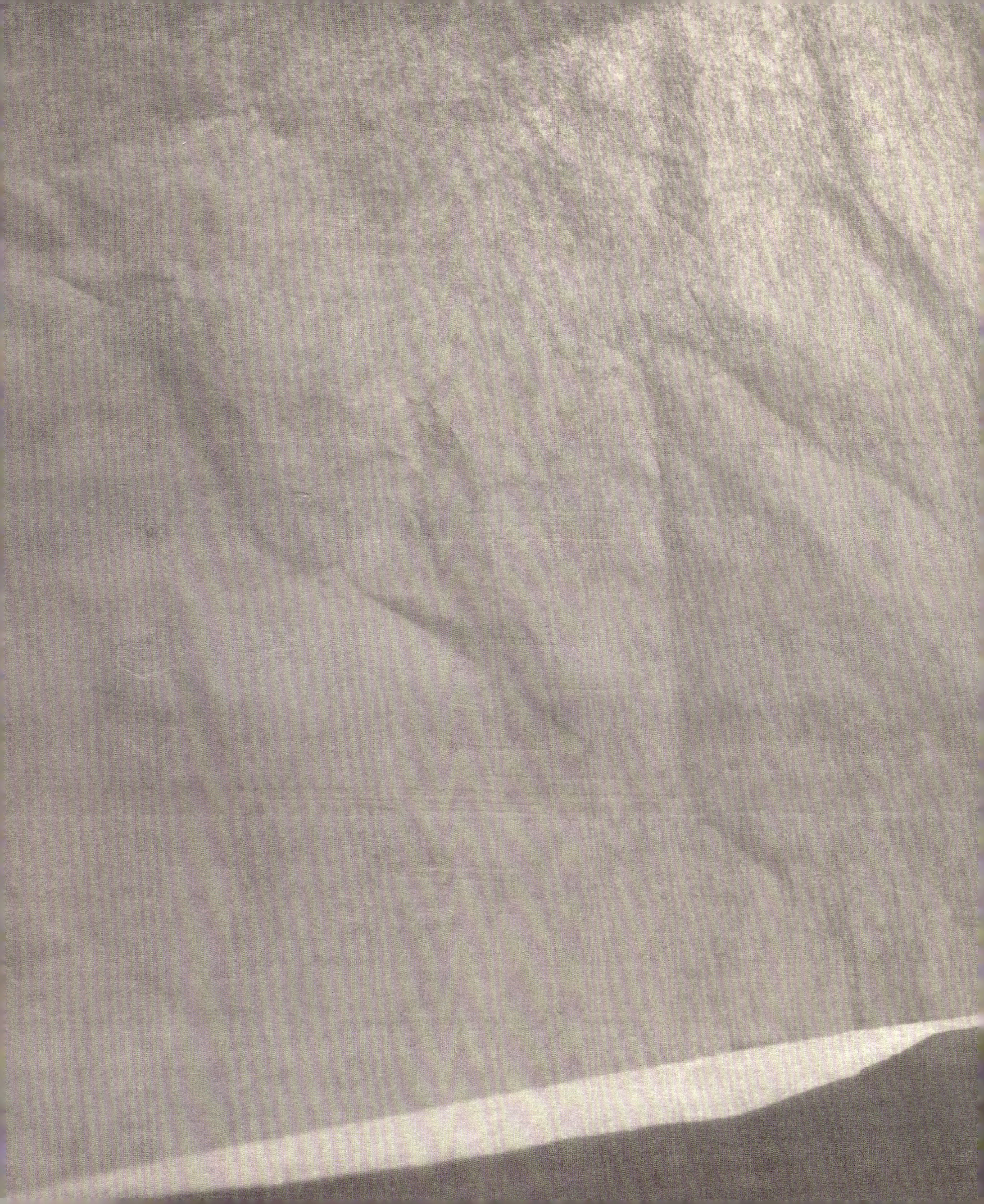

Calligraphy of Light

Richard Cork

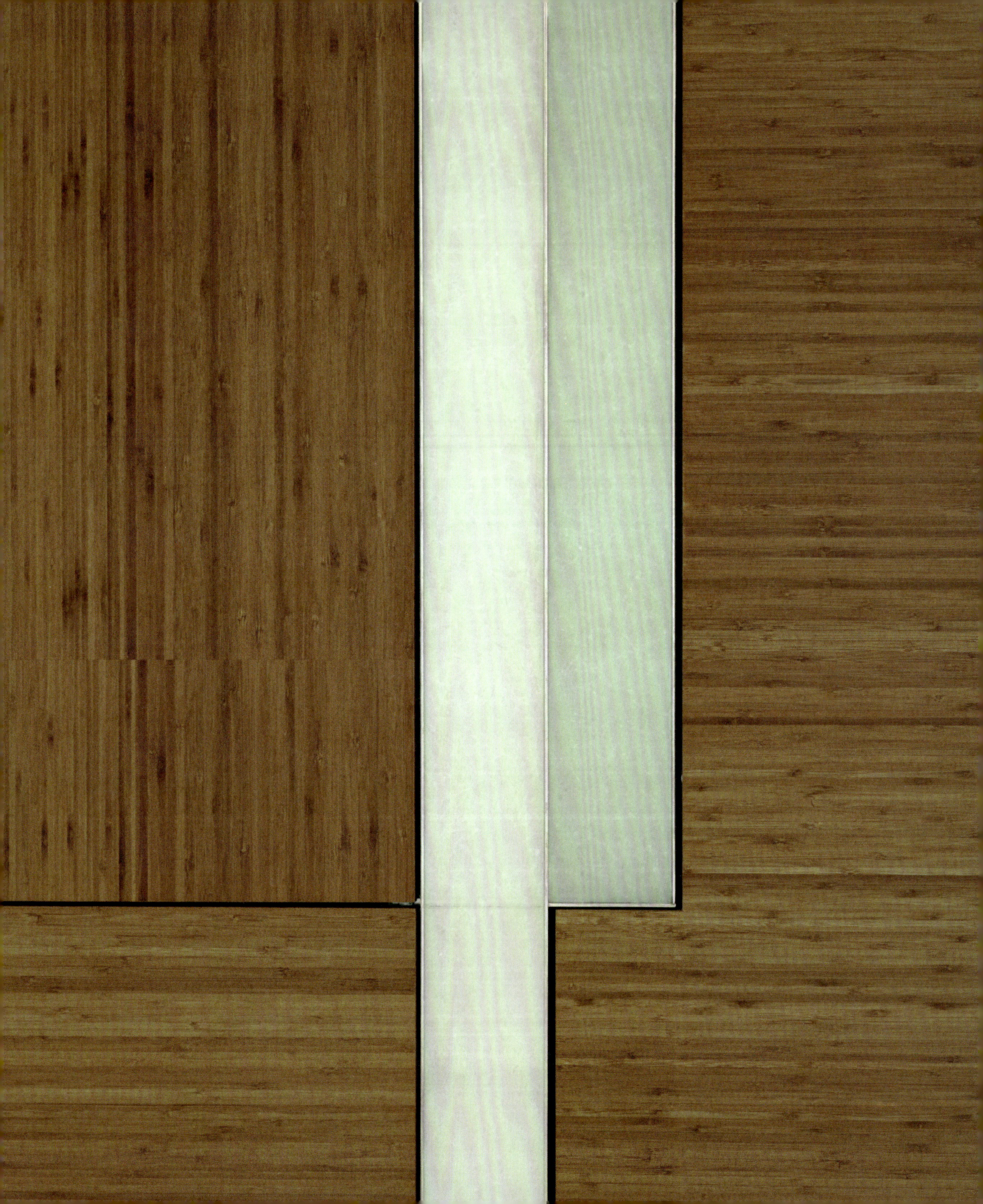

Calligraphy of Light

Richard Cork

All too often, people find that approaching and entering a hospital can generate acute feelings of anxiety. Patients and their relatives, already worried about illness, find themselves growing even more agitated. Hospital facades invariably look forbidding, while their entrances combine clinical bleakness with a blizzard of bewildering signs and instructions.

But the truth is that they need not be so grim. For many years I have been writing a major history of Western art in hospitals, including works from the Renaissance right through to the twentieth century.[1] I became fascinated by the extraordinary richness of a subject that involved so many of the greatest masterpieces in Western art. Starting with the works of Piero della Francesca in San Sepolcro, Rogier van der Weyden at Beaune, Hans Memling in Bruges, and Matthias Grünewald at Isenheim, the book journeys through to those of El Greco at Toledo, William Hogarth in London (fig. 55), Francisco de Goya in Madrid, Vincent van Gogh at Saint-Rémy, and Frida Kahlo and Diego Rivera in Mexico.

There is an increasing awareness that art can do an immense amount to humanize our hospitals, alleviate their clinical harshness, and offer genuine, lasting pleasure to patients,

fig. 54: *Calligraphy of Light*, 2009
recycled glass and bamboo panel
St. George's Hospital, London
detail; see plate 162

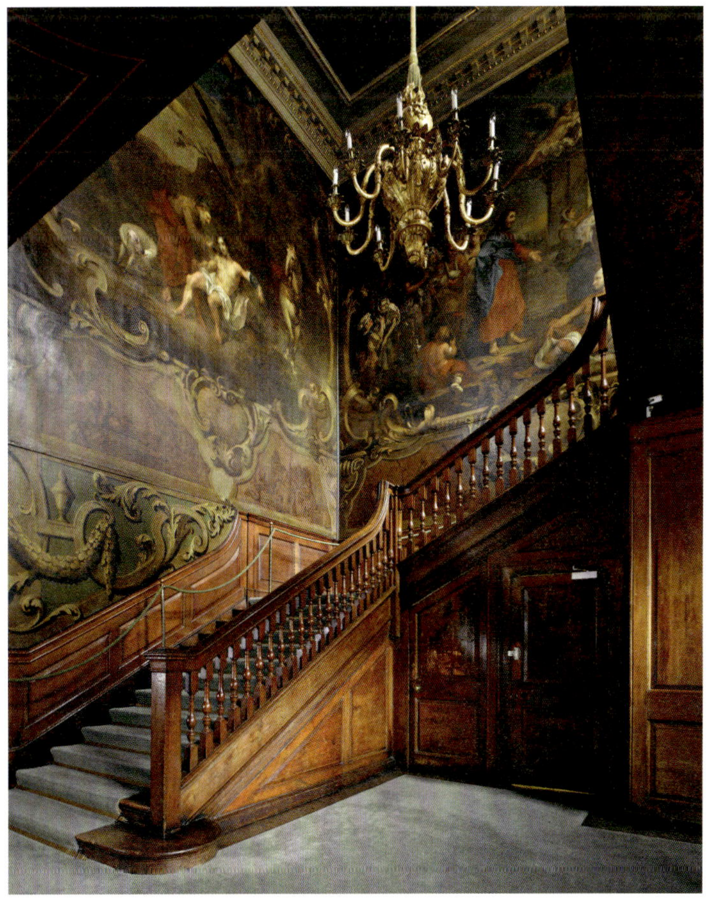

fig. 55: View of staircase at St. Bartholomew's Hospital, London, showing William Hogarth, *The Pool of Bethesda*, 1736 and *The Good Samaritan*, 1737

staff, and visitors alike. Hospitals can undoubtedly do more to look after the whole person, not just the patient's bodily ailment. Many critical moments in our lives occur there, from birth to death, and they deserve to take place in surroundings that honor their true significance.

From the Renaissance onward, artists have experimented with subjects ranging from dramatic confrontations with suffering to the most sublime, airborne celebrations of resurrection and heavenly ecstasy. Some, like Leonardo da Vinci's incisive drawings, are based on uncompromising, firsthand study of hospital patients. Others explore a redemptive world where Christ is born, orphans are rescued, and plague victims are given shelter. Western culture is at its most potent when focused on the healing power of art.

And now, in the twenty-first century, one of the oldest London hospitals has managed to renew this outstanding tradition by transforming its crucial entrance area. A few years ago, St. George's Hospital in Tooting decided that the ten thousand patients, visitors, and staff passing through its portals each day deserved to be greeted by a more welcoming, pleasurable, and vibrant environment. Buses used to drive right up to the hospital's Grosvenor Wing entrance, but the traffic-ridden road and roundabout have now been replaced by a quiet, specially designed garden. Beech hedges in curving segments, accompanied by a soothing profusion of lavender and rosemary, are growing in this new landscape designed by Charles Funke Associates, where people can rest on benches before making their way to the hospital itself.

Inside, Rebecca Salter has ensured that they are consoled at once by a soothing sense of quietness. The confusion and stress engendered by so many hospital entrances are replaced here by luminosity and poise. Salter calls her installation *Calligraphy of Light*, and we immediately become conscious of an interior where the artist has been able subtly to echo the garden (fig. 56). Aiming at a potent distillation of color and light, Salter succeeds in bringing about a mood of beneficence. Without resorting to dubious theatricality, she has made the whole entrance undergo a dramatic metamorphosis. Her profound understanding of purged, minimal elements ensures that *Calligraphy of Light* sweeps everyone through the space with serenity and grace.

The spirit of simplicity at work here should not deceive anyone into imagining that it was easy to achieve. The project, which was coordinated by Belinda Harward, the adventurous arts officer of St. George's, was the result of a complex collaboration between the in-house design team, the architectural practice Gibberds, the St. George's Hospital Charity, and the hospital's arts committee. Although Salter has devoted most of her career to studio-based work for gallery exhibitions, Harward felt instinctively that she would be a good choice. Salter's work reminded her of that of Agnes Martin, whose exhibition at the Palazzo Grassi in Venice had impressed Harward on a profound level. "I felt that Rebecca's work had exactly the right feel and sensitivity," she explains. "It had for me a quality which was absolutely right for the space. When we met, I also wanted to know that she was the right kind of person."[2] The project evolved through a lengthy but ultimately extremely productive series of conversations, and, as Harward notes, throughout the process Salter was persuasive and tenacious.

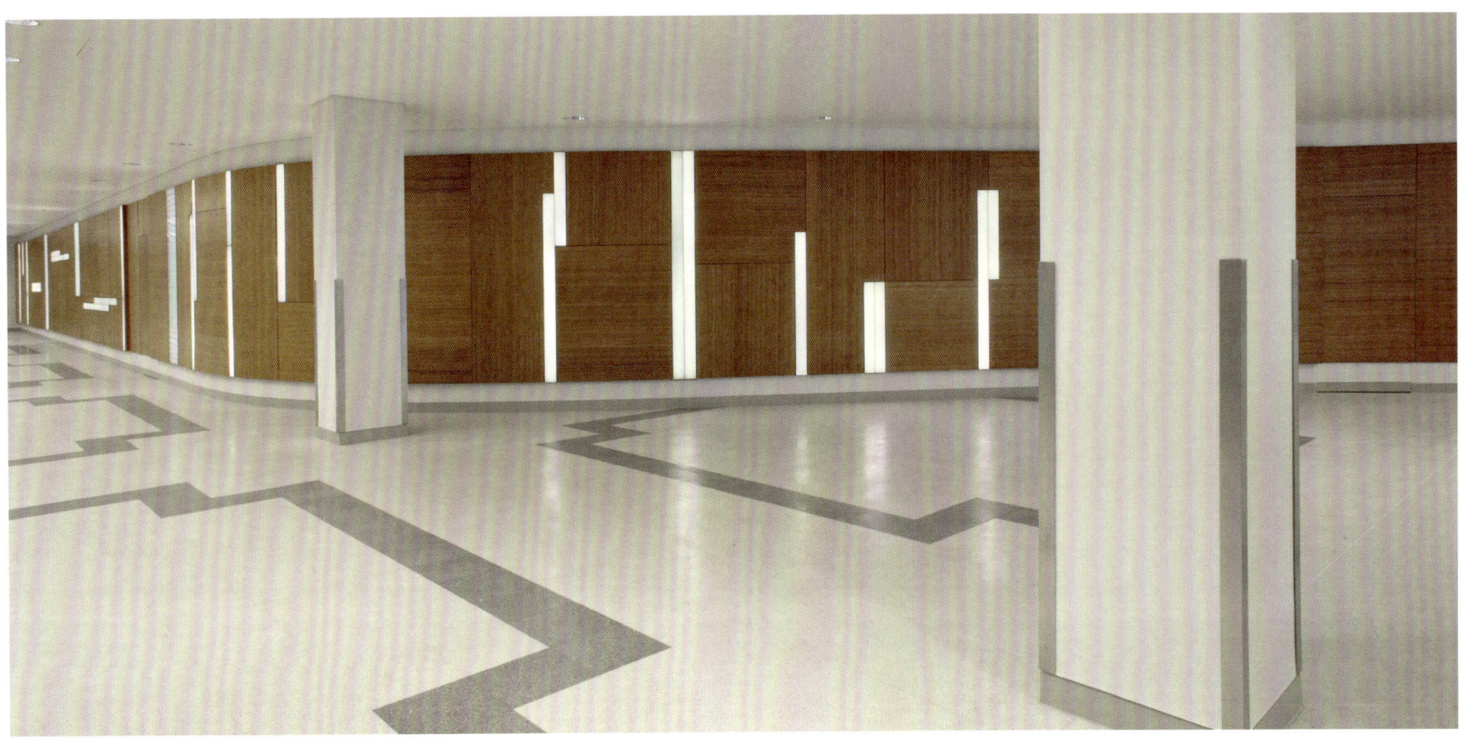

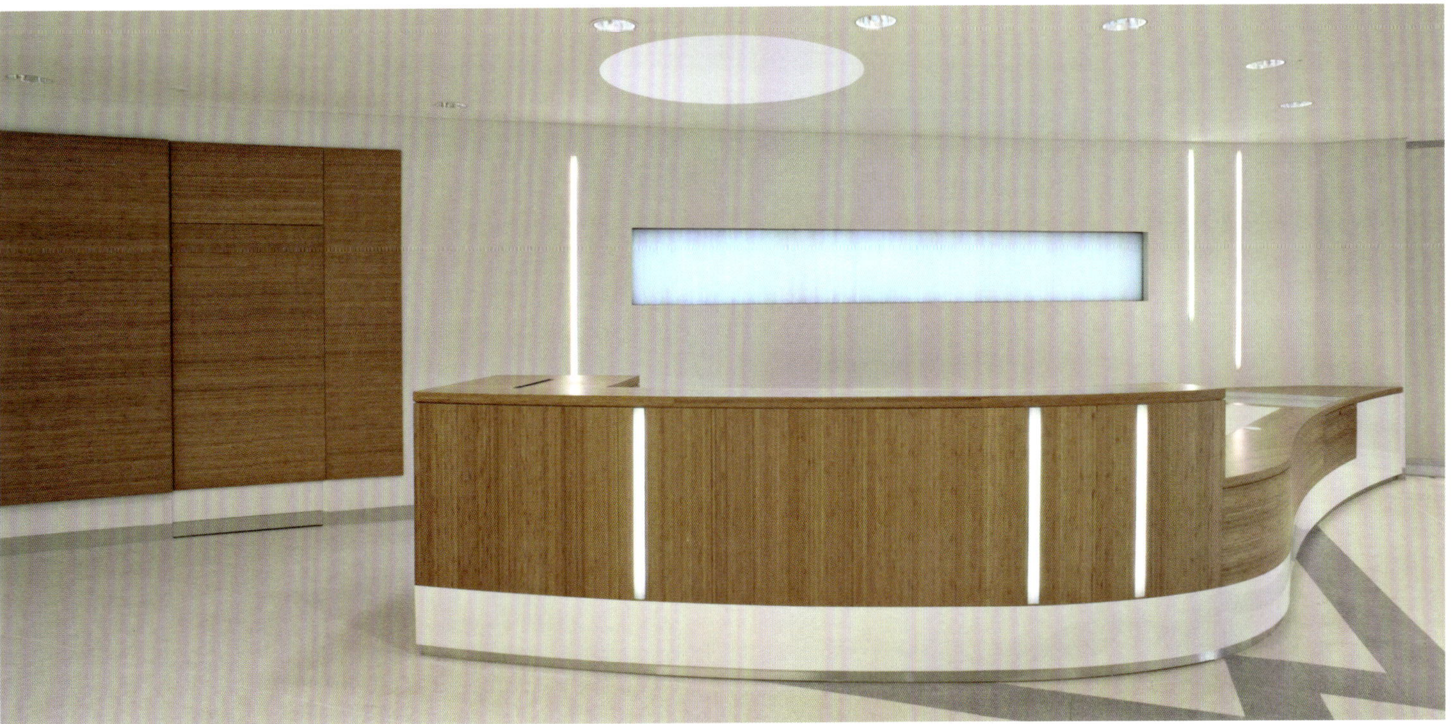

fig. 56: *Calligraphy of Light*, 2009
recycled glass and bamboo panel
main entrance, St. George's Hospital, London
(in collaboration with Gibberd, architects)

fig. 57: *Calligraphy of Light*, 2009
recycled glass and bamboo panel
reception area, St. George's Hospital, London
(in collaboration with Gibberd, architects)

fig. 58: View of the main entrance, St. George's Hospital, London, showing *Calligraphy of Light*, 2009 (in collaboration with Gibberd, architects), with Edward Bawden's painting of the former St. George's Hospital, Hyde Park

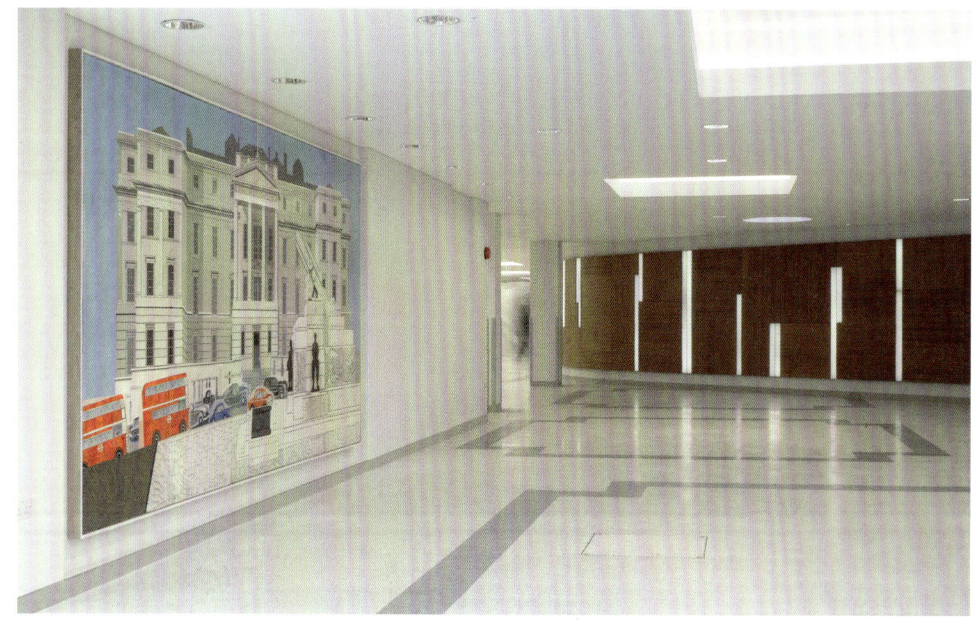

No sense of struggle can be detected in *Calligraphy of Light*. Completed in 2009, it greets visitors with a softly glowing horizontal panel built into the curved wall behind the reception desk (fig. 57). Made with strips of recycled glass illuminated from behind with LED lights, the panel emits an ever-changing and inexhaustible sequence of colors ranging from orange and purple to pale blue. Even at their most arresting, they are not at all aggressive or ostentatious. But they do possess a shimmering, sensuous allure which immediately offers an inviting alternative to the starkness of hospital interiors. This seductive glass panel is composed of around 250 vertical segments. When viewed up close they reveal an organic patterning reminiscent of tree bark. The panel's overall horizontality is contrasted with slim, elegant vertical slits on either side. Made of glass without color, they seem to be suspended weightlessly in space rather than attached to the wall. They are reinforced by the vertical slits in the desk's curved front and contrast with the discreet circularity of the sun-pipes lodged in the ceiling.

Certain elements of the original entrance hall had to be retained—most notably the floor and Edward Bawden's monumental painting of the old St. George's Hospital

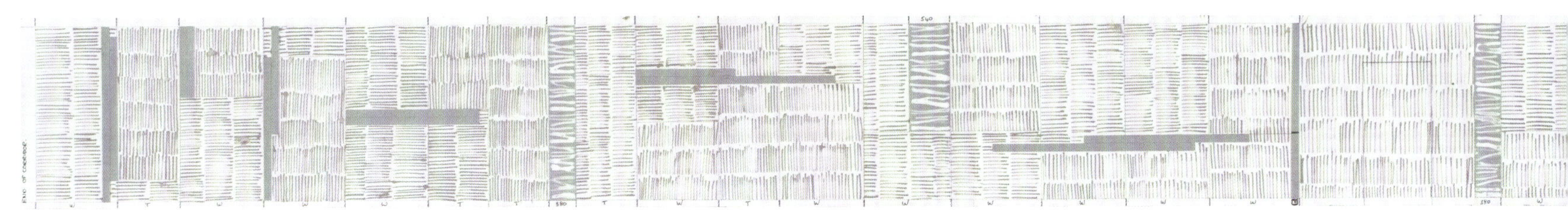

fig. 59: *Calligraphy of Light*
preliminary ink drawings, 2007
sumi ink on Japanese paper
each drawing: 9½ × 13 in. (24.1 × 33 cm)

fig. 60: *Calligraphy of Light*
final layout drawing, 2008
ink and gouache on paper
4¾ × 83⅜ in. (12.1 × 211.8 cm)

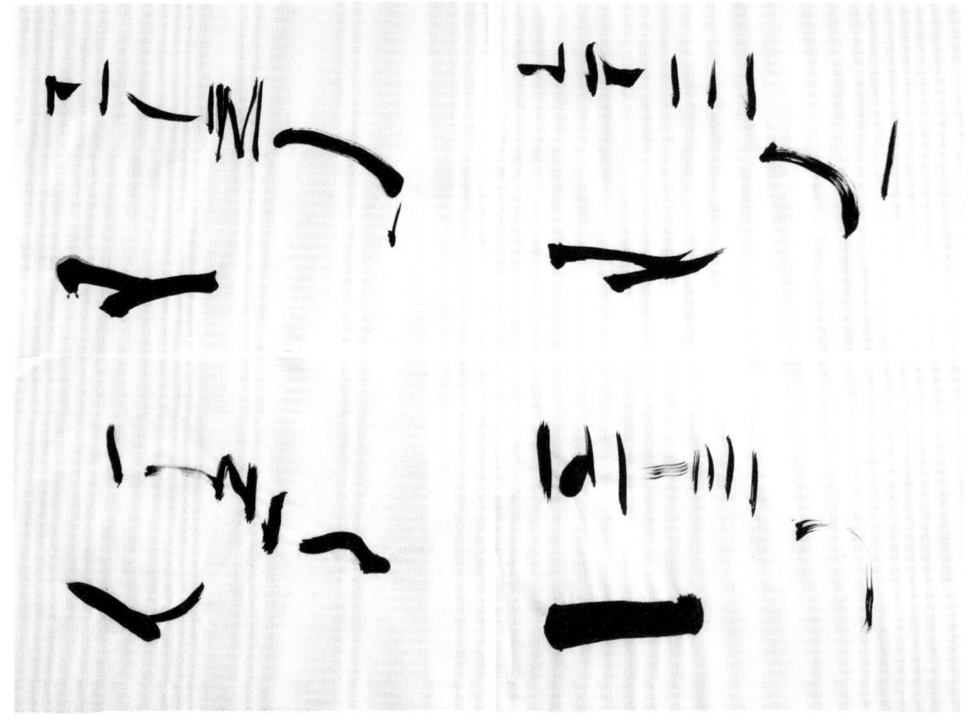

overlooking Charles Sargeant Jagger's Royal Artillery Memorial at Hyde Park Corner (fig. 58). But Salter knew precisely how to integrate them with her work. As a result, even Bawden's enormous painting does nothing to interrupt the flow of the wide corridor curving into the hospital beyond the reception desk. Salter's transformation of this space is at once understated and remarkably serene. People walking through it now are able to locate their destinations intuitively, without being bombarded by arrows and instructions.

How did she set about attaining such an admirable result? Salter was under no illusions about the complex nature of the task confronting her at St. George's. She knew, from the outset, what kind of artwork was needed. But she also realized that in such an undertaking "the challenge is to maintain the clarity of the original concept as it meets and has to take account of problems on the way. This process is complicated enough if the building in question is what might be called a 'neutral' space such as an office. It is a far

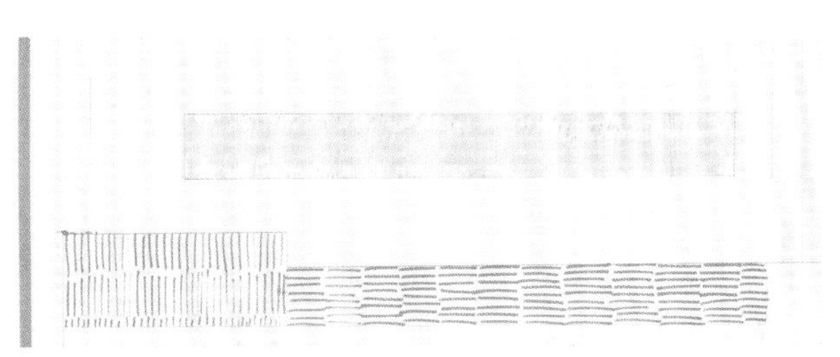

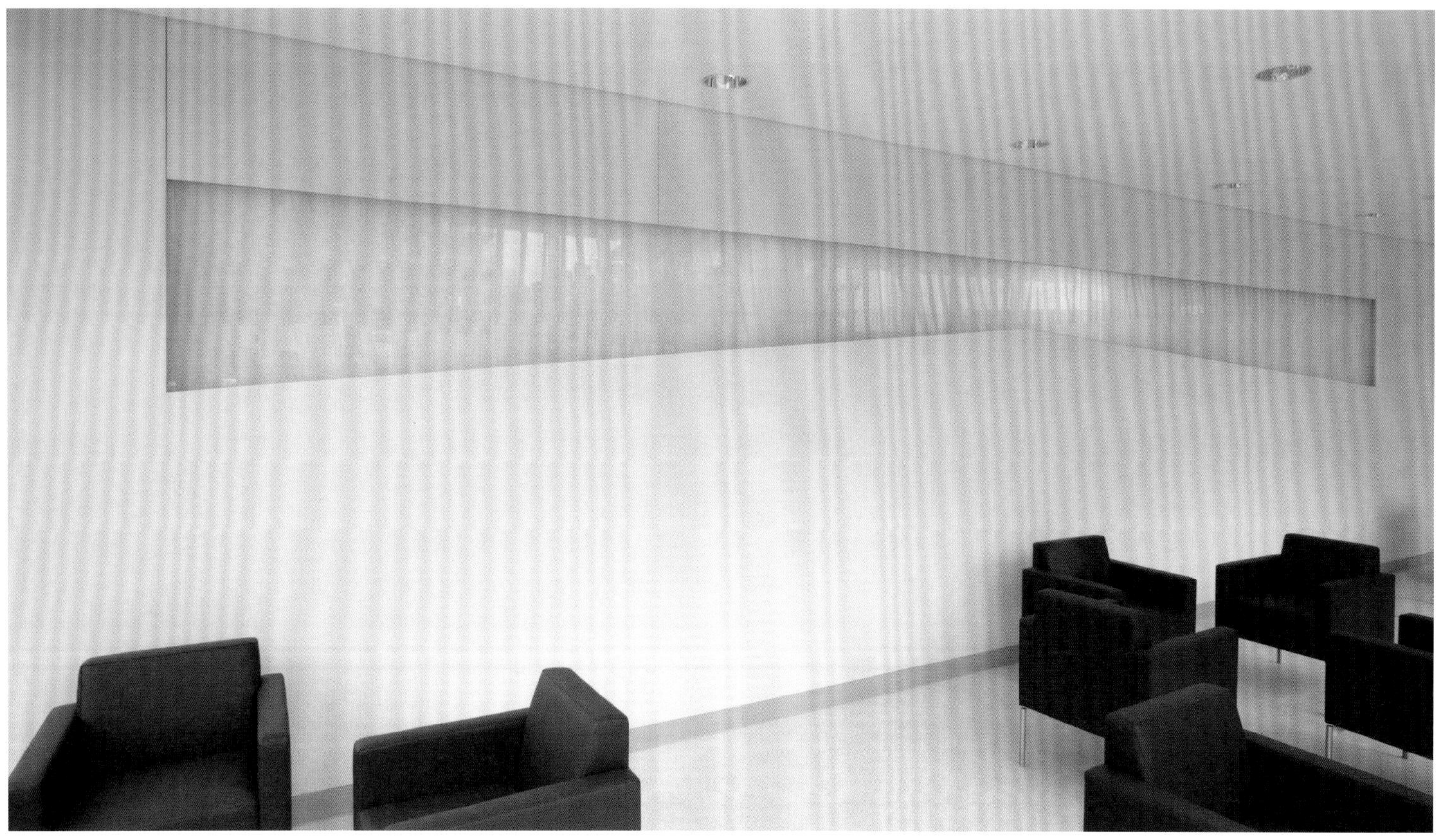

fig. 61: *Delta*, 2009
sandblasted glass with LED backlighting
waiting area, St. George's Hospital, London
(in collaboration with Gibberd, architects)

more daunting task when the space is loaded with meaning. Hospitals originally had close links with religious organizations, and, like places of worship, they are not just ordinary spaces. They are 'spaces' which are 'places' with meaning and they both bear witness to some of life's most important events."[3]

The depth of Salter's understanding, combined with the strength of her resolve, ensured that she gave this entrance the fundamental calm it deserves. Determined to bring the outside inside, she turned back for her primary inspiration to aspects of Japanese art and architecture encountered earlier in her career. As a research student on a Leverhulme scholarship between 1979 and 1981, Salter was based at the Kyoto City University of Arts. She spent an immense amount of time looking at and living in Kyoto's gardens, discovering above all how the Japanese make space feel larger than it actually is.

Salter has never forgotten this profound source of enlightenment, and it can be felt at St. George's in the way she composed the strips of backlit glass set in the corridor's bamboo paneling as it curves toward the hospital interior. Working with a Japanese ink brush on Japanese paper and with her eyes closed, she made series of drawings in her studio to define the almost musical rhythm of the glass pieces as they move along the wall (fig. 59). Then she produced a pen and ink drawing of the bamboo paneling itself, which echoes the embracing warmth of the beech trees in the hospital's garden (fig. 60). Although Salter is essentially an abstract artist, she is capable of evoking nature—especially in the corridor's bigger, sand-blasted glass panels, which reminded the viewer of water's rhythm as it runs freely along the earth.

Opposite Salter's bamboo wall there are visual elements of a very different kind, most notably of the hospital's frenetic shop. But nothing can impair her uncanny ability to push

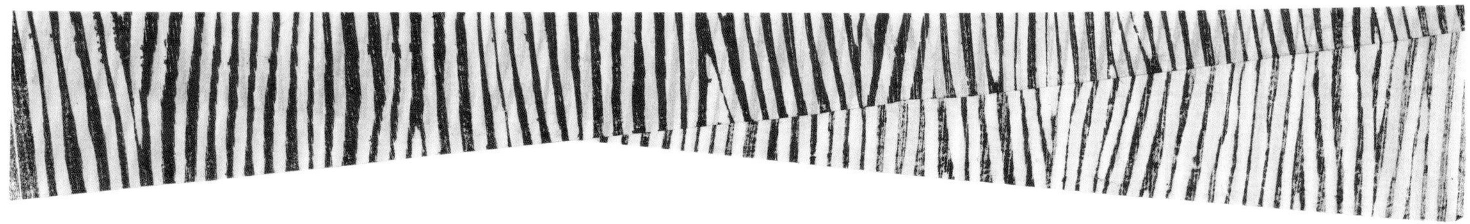

fig. 62: *Delta*
original design, 2008
woodblock on Japanese paper
2⅛ × 15 in. (5.4 × 38.1 cm)

fig. 63: *Delta*
color scheme, 2008
collage on Japanese paper
3⅛ × 10¼ in. (7.9 × 26 cm)

Overleaf: *Calligraphy of Light*, 2009
recycled glass and bamboo panel
reception area, St. George's Hospital, London
(in collaboration with Gibberd, architects)

back (in visual terms) the expanse of bamboo wall to create more space. This corridor used to be like a gloomy funnel, generating stress and growing darker as people walked deeper into the hospital. Yet now, enhanced by the white and pale green glass set with such felicitous judgment in the carefully orchestrated bamboo paneling, it reassures visitors with an overall sense of tranquility.

The dialogue with the garden reaches its climax in the last room to be completed: a seating area at the front of the building. Salter created a contemplative interior dominated by a long, delta-shaped work in backlit glass (fig. 61). Stretching eight meters along one wall, it originated in a woodblock, which she printed on Japanese paper. Salter twisted the image (fig. 62) and then made a collage to work out the color scheme for the final glass version (fig. 63) that quietly echoes the forms and colors of the planting visible in the garden beyond the window. The result is immensely restful.

Anyone moving out of this room to leave the building can look back at the rest of the entrance and relish its serenity. Always busy, St. George's is one of Britain's largest teaching hospitals. But Salter has created an environment where intimidating bewilderment has no place. Balancing certitude with sensitivity, she has been able to counter anxiety and, on these lucent walls, arrive at a redemptive stillness.

notes: Calligraphy of Light

1 Richard Cork, *The Healing Presence of Art: A History of Western Art in Hospitals* (New Haven and London: Yale Univ. Press, forthcoming, 2011).

2 Belinda Harward, conversation with the author at St. George's Hospital, 14 May 2010.

3 Rebecca Salter, "A Short Reflection on a Long Project," unpublished essay, 2009 (copy held at the Yale Center for British Art).

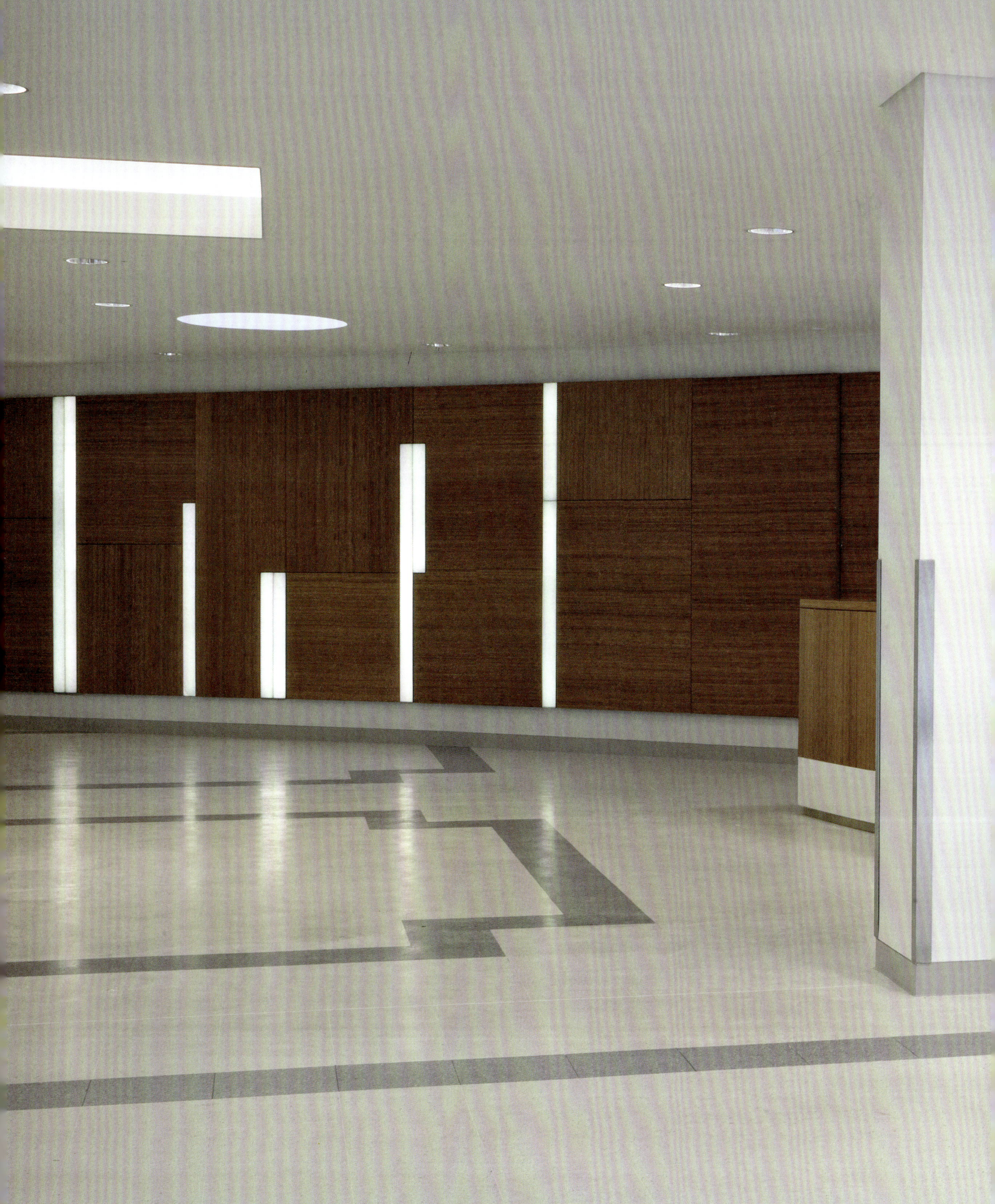

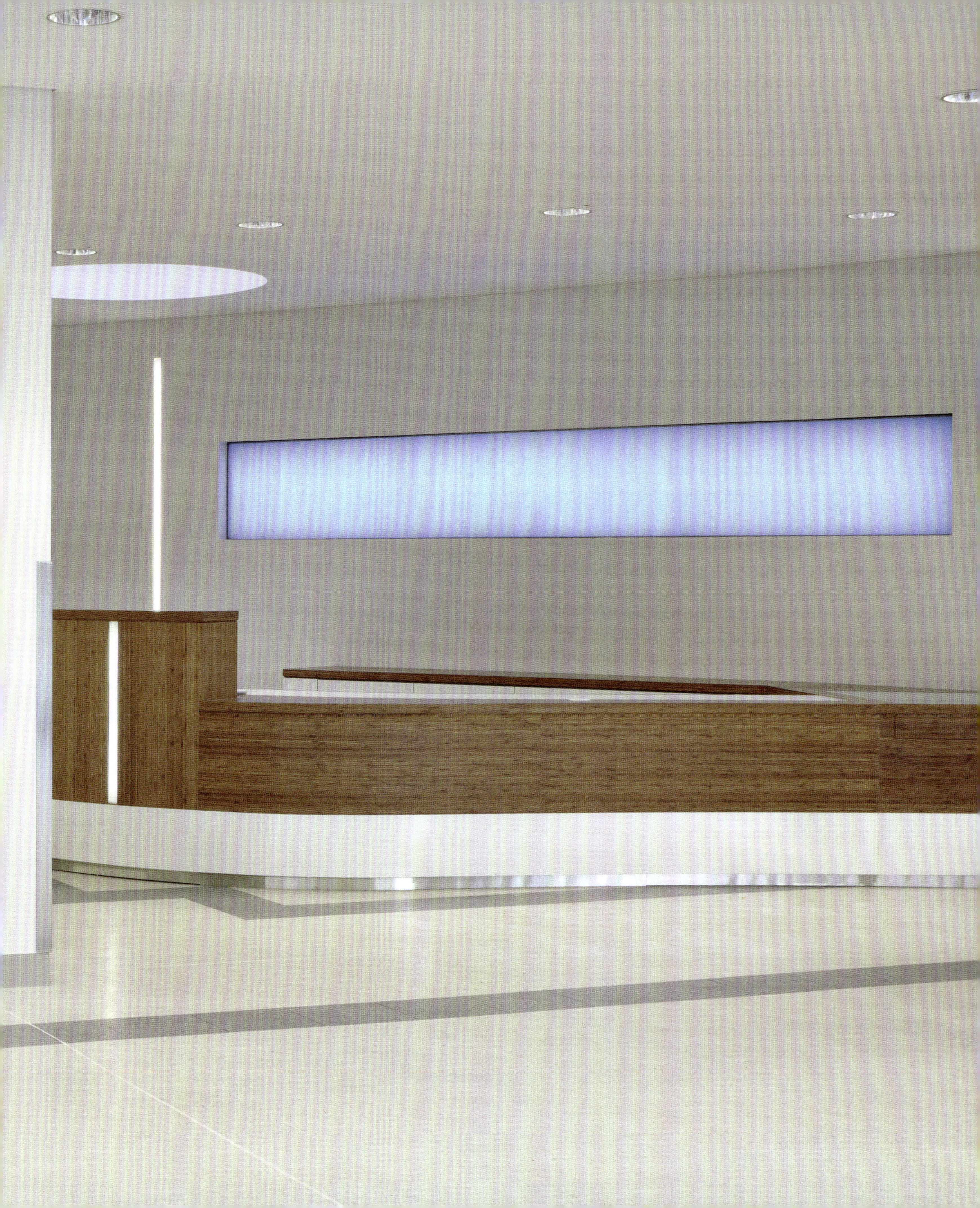

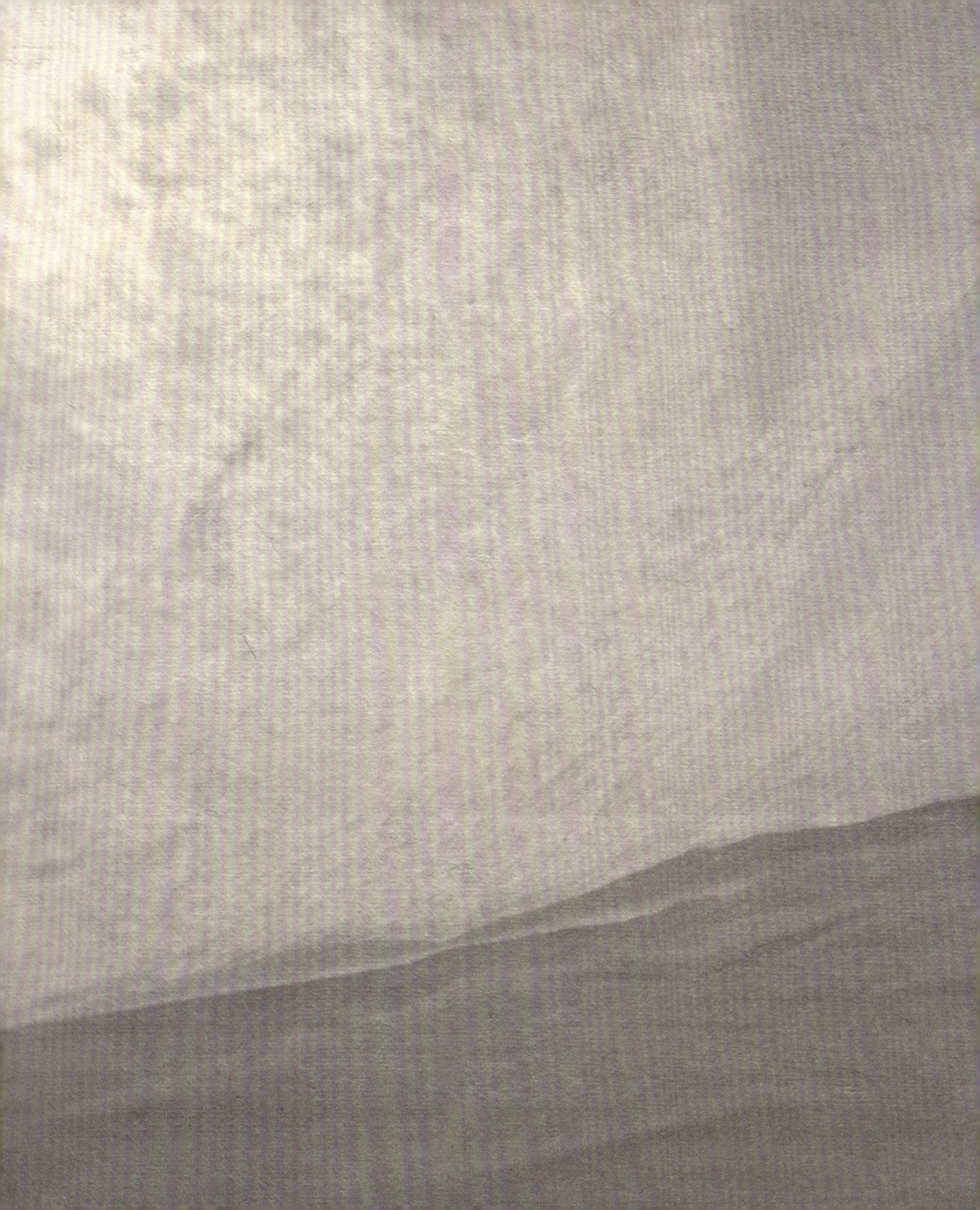

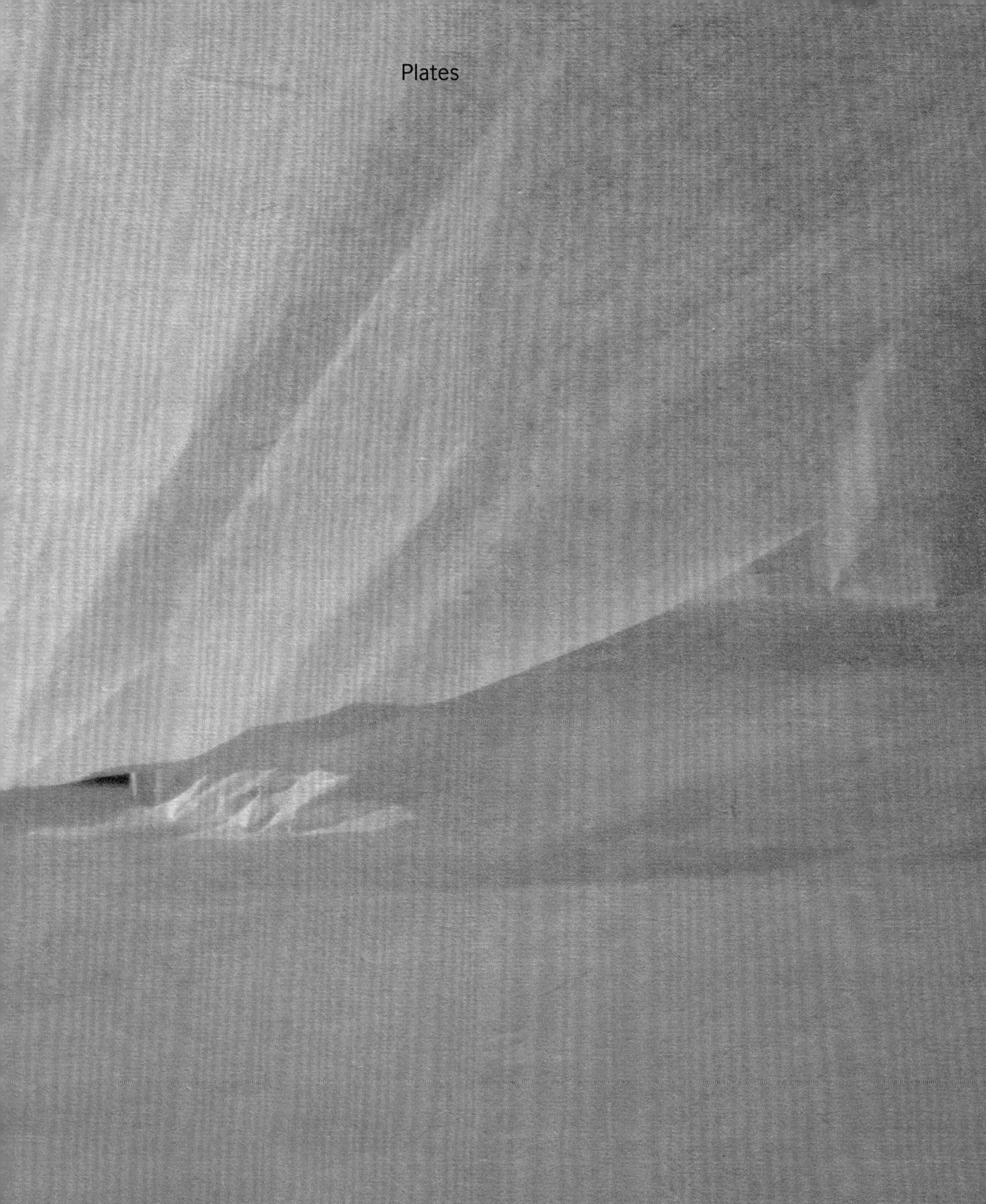

Plates

1 *Untitled B116*, 1982
mixed media on Japanese paper
16⅞ × 12⅝ in. (43 × 32 cm)

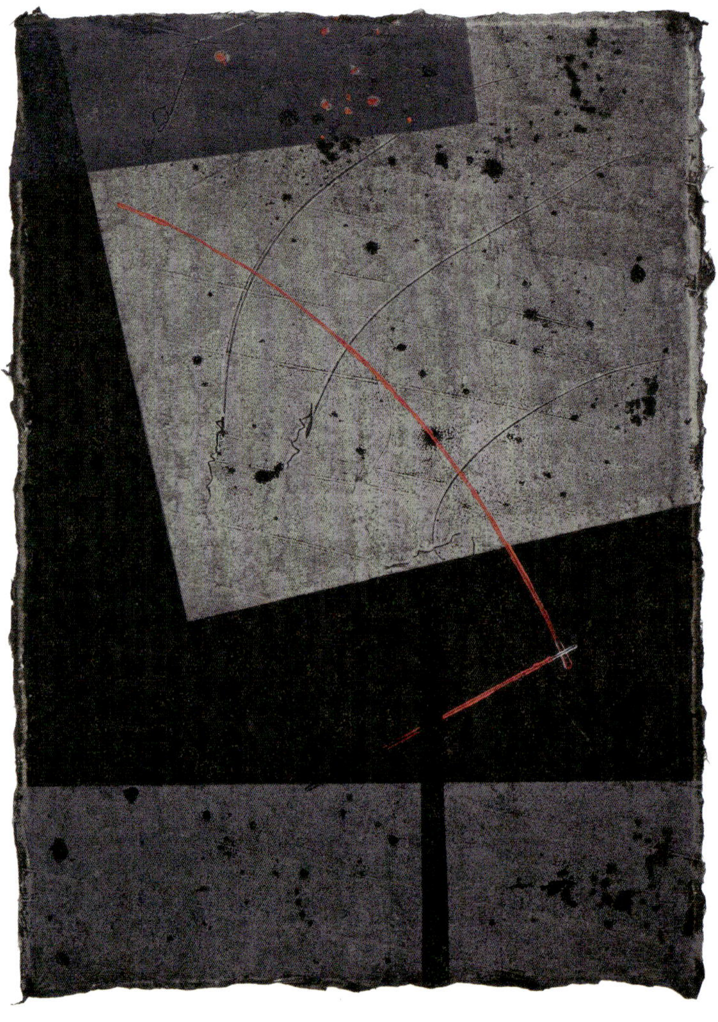

2 *Untitled B168*, 1982
mixed media on Japanese paper
16⅞ × 12⅝ in. (43 × 32 cm)
Klingensmith Collection, London

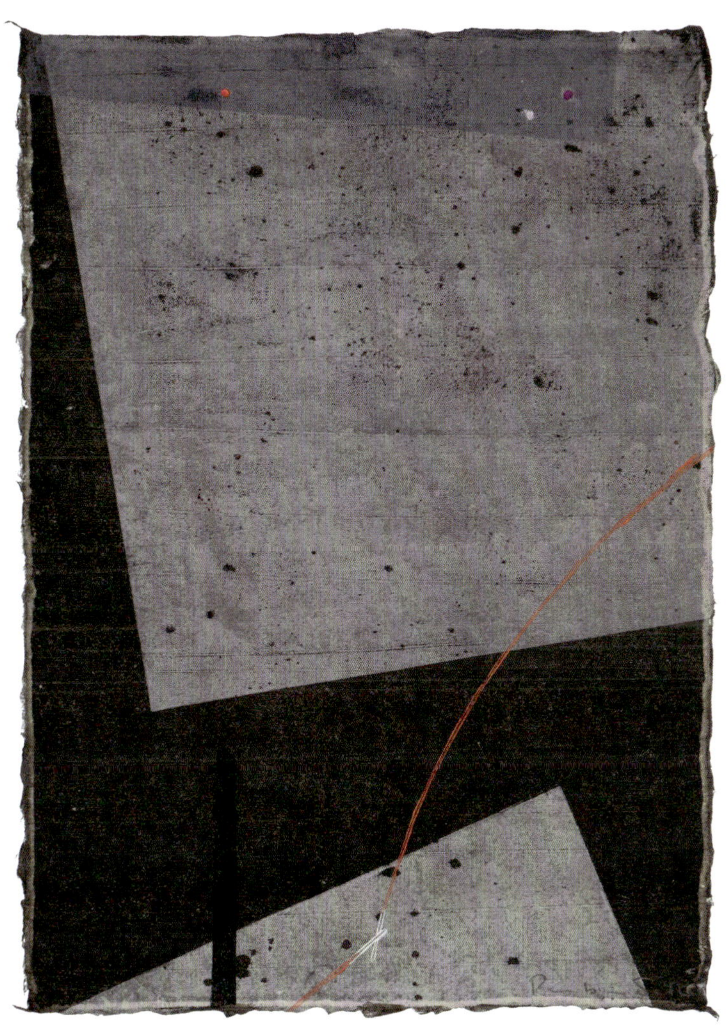

3 *Untitled A34*, 1982
mixed media on Japanese paper
25¼ × 37¾ in. (64 × 96 cm)

4 *Untitled A21*, 1982
mixed media on Japanese paper
25¼ × 37¾ in. (64 × 96 cm)

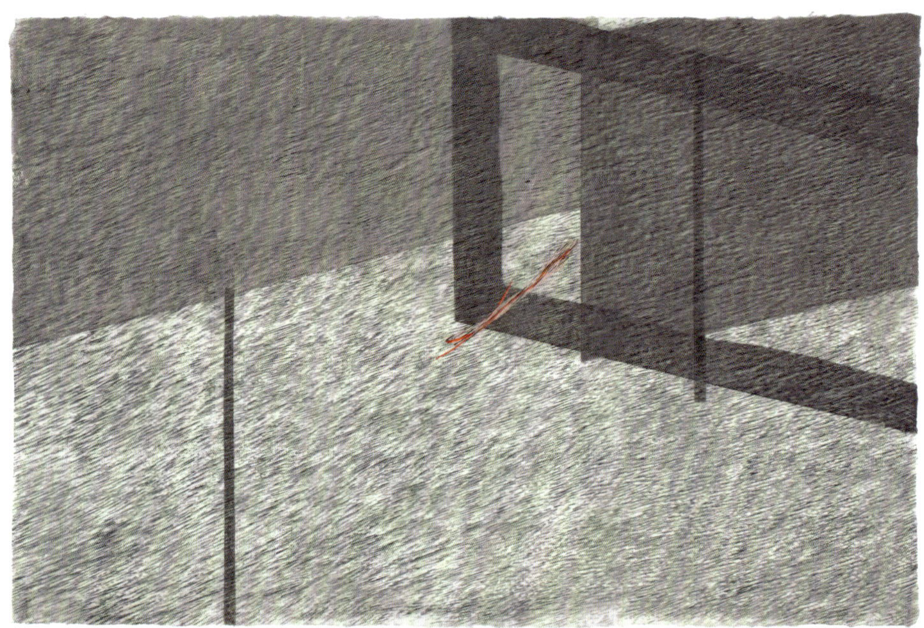

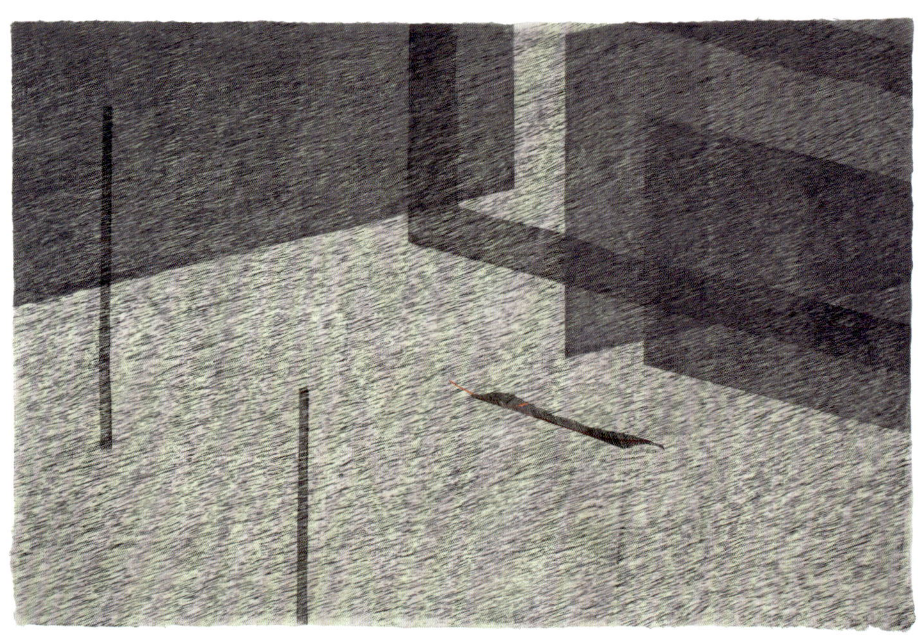

5 *Untitled A57*, 1982
mixed media on Japanese paper
25¼ × 37¾ in. (64 × 96 cm)

6 *Untitled A56*, 1982
mixed media on Japanese paper
25¼ × 37¾ in. (64 × 96 cm)

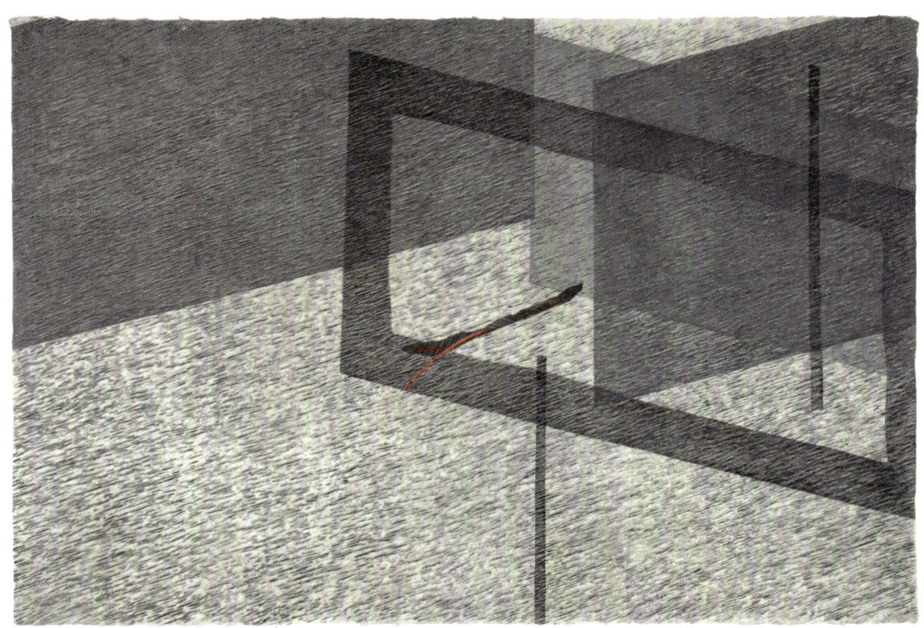

7 *Japan I*, 1982
woodblock on Japanese paper
7⅛ × 9 in. (18 × 23 cm)

8 *Japan II*, 1982
woodblock on Japanese paper
7⅛ × 9 in. (18 × 23 cm)

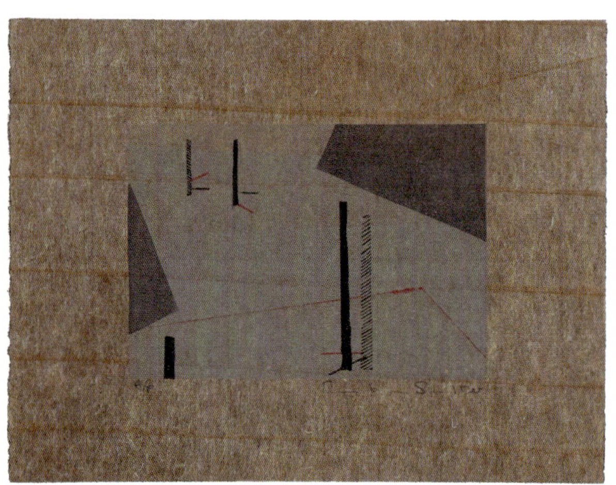

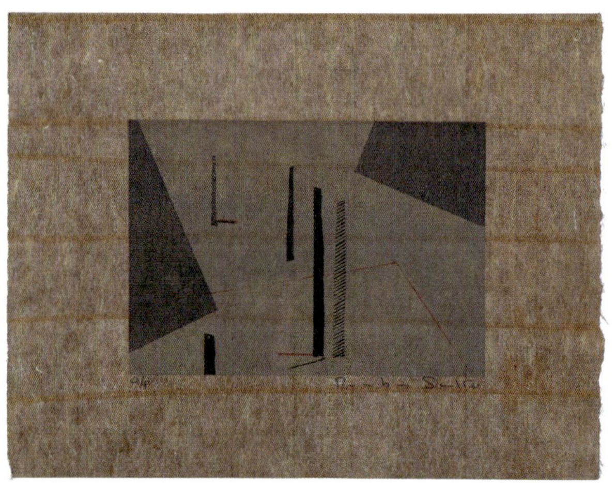

9 *Japan III*, 1982
woodblock on Japanese paper
7⅛ × 9 in. (18 × 23 cm)

10 *Japan IV*, 1982
woodblock on Japanese paper
7⅛ × 9 in. (18 × 23 cm)

11 *Japan V*, 1982
woodblock on Japanese paper
7⅛ × 9 in. (18 × 23 cm)

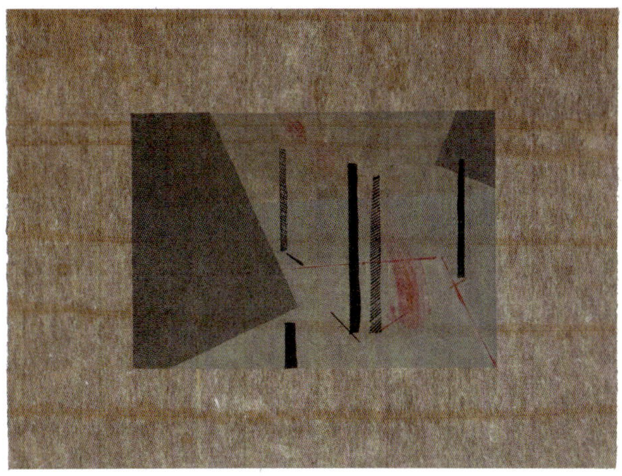

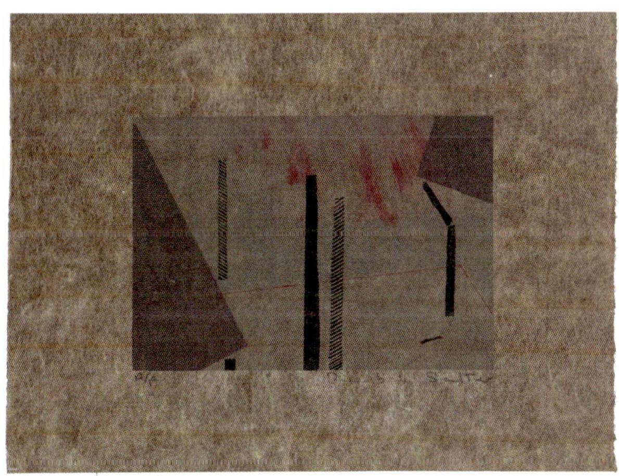

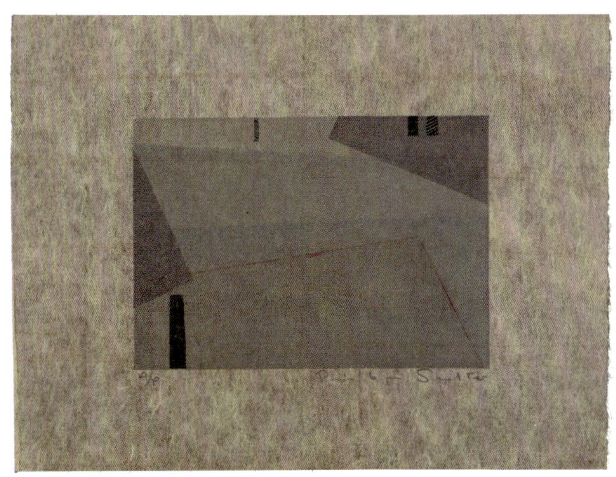

12 *Untitled A72*, 1983
mixed media on Japanese paper
25¼ × 37¾ in. (64 × 96 cm)

13 *Untitled A39*, 1983
mixed media on Japanese paper
26⅜ × 38¼ in. (67 × 97 cm)

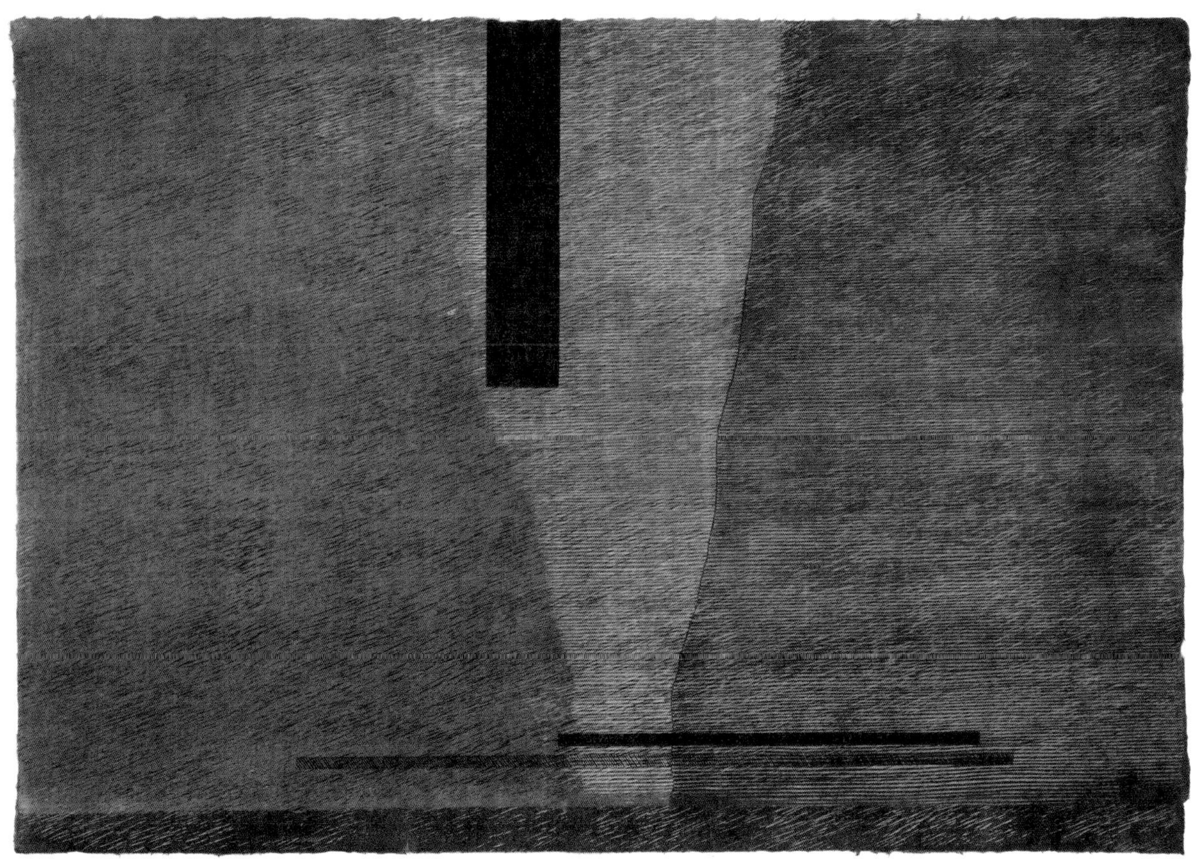

14 *Untitled A50*, 1982
mixed media on Japanese paper
25¼ × 37¾ in. (64 × 96 cm)

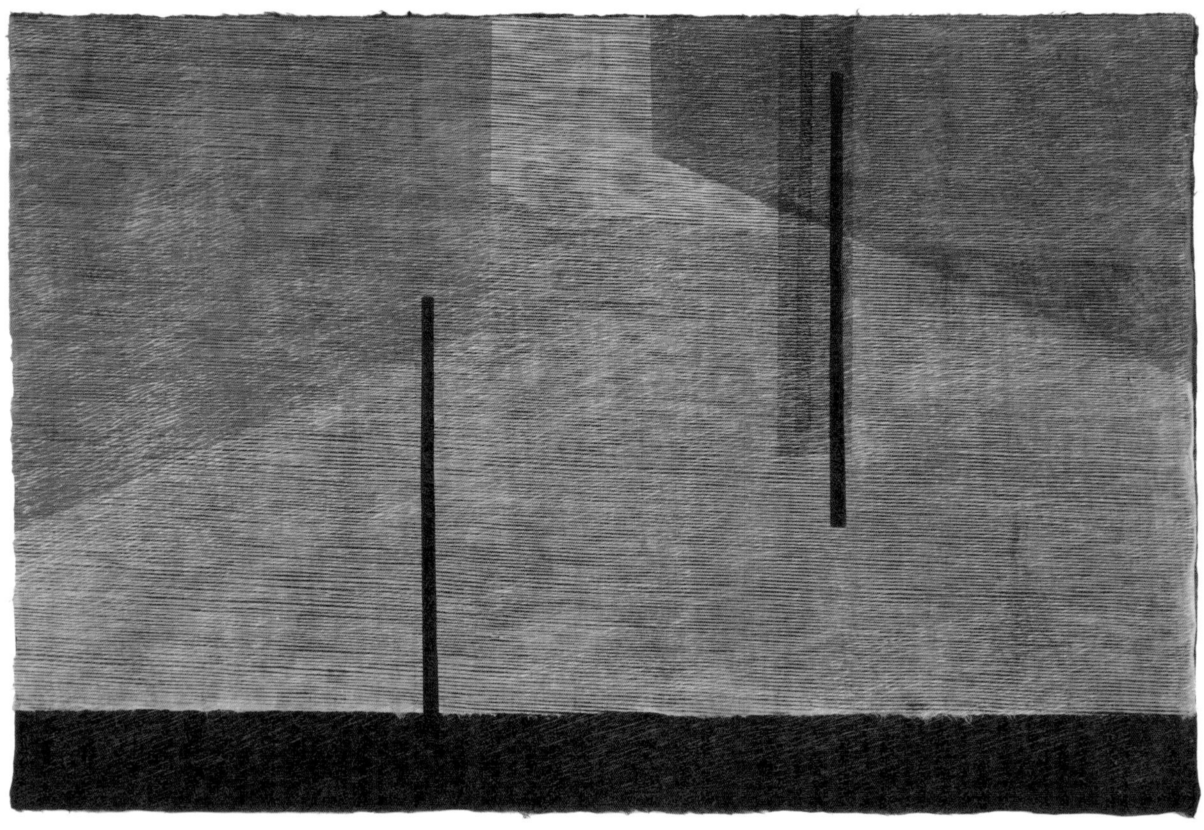

15 *Untitled A65*, 1983
mixed media on Japanese paper
25 × 38 in. (63.5 × 96.5 cm)

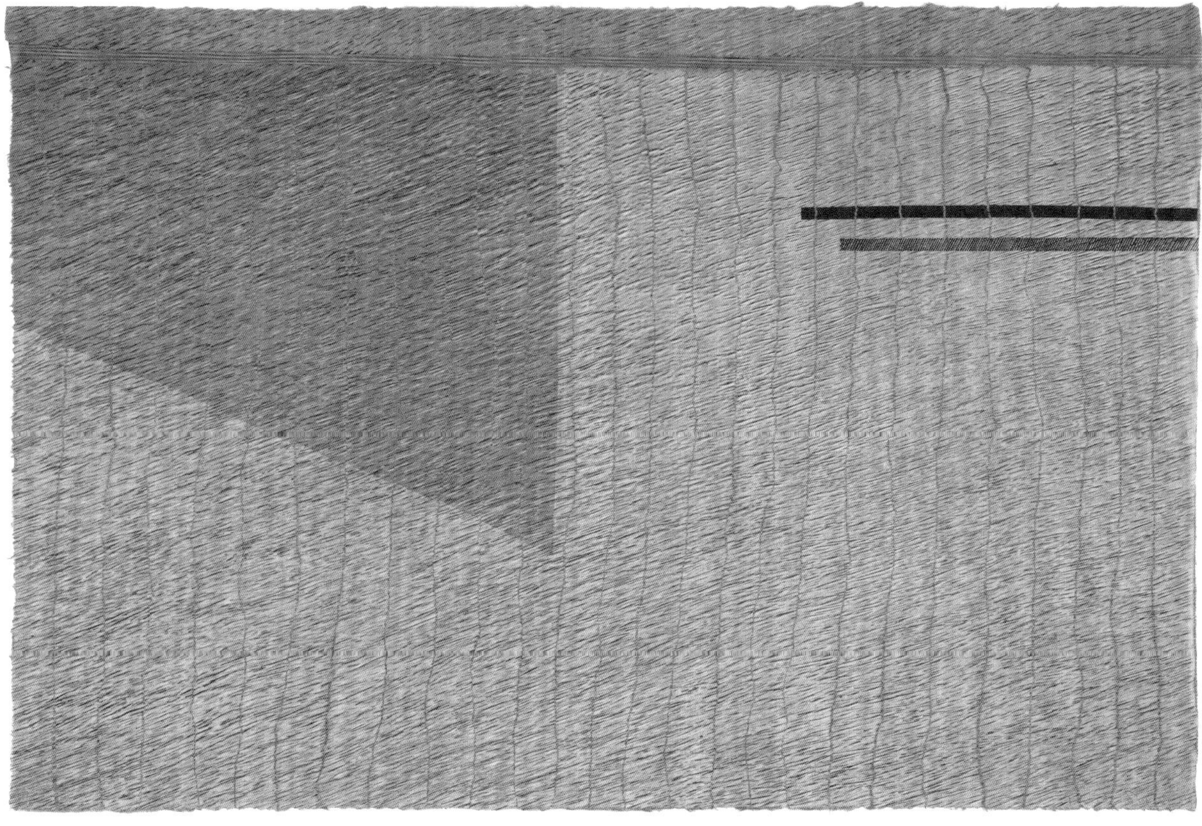

16 *Untitled B96*, 1983
mixed media on Japanese paper
39⅜ × 25¼ in. (100 × 64 cm)

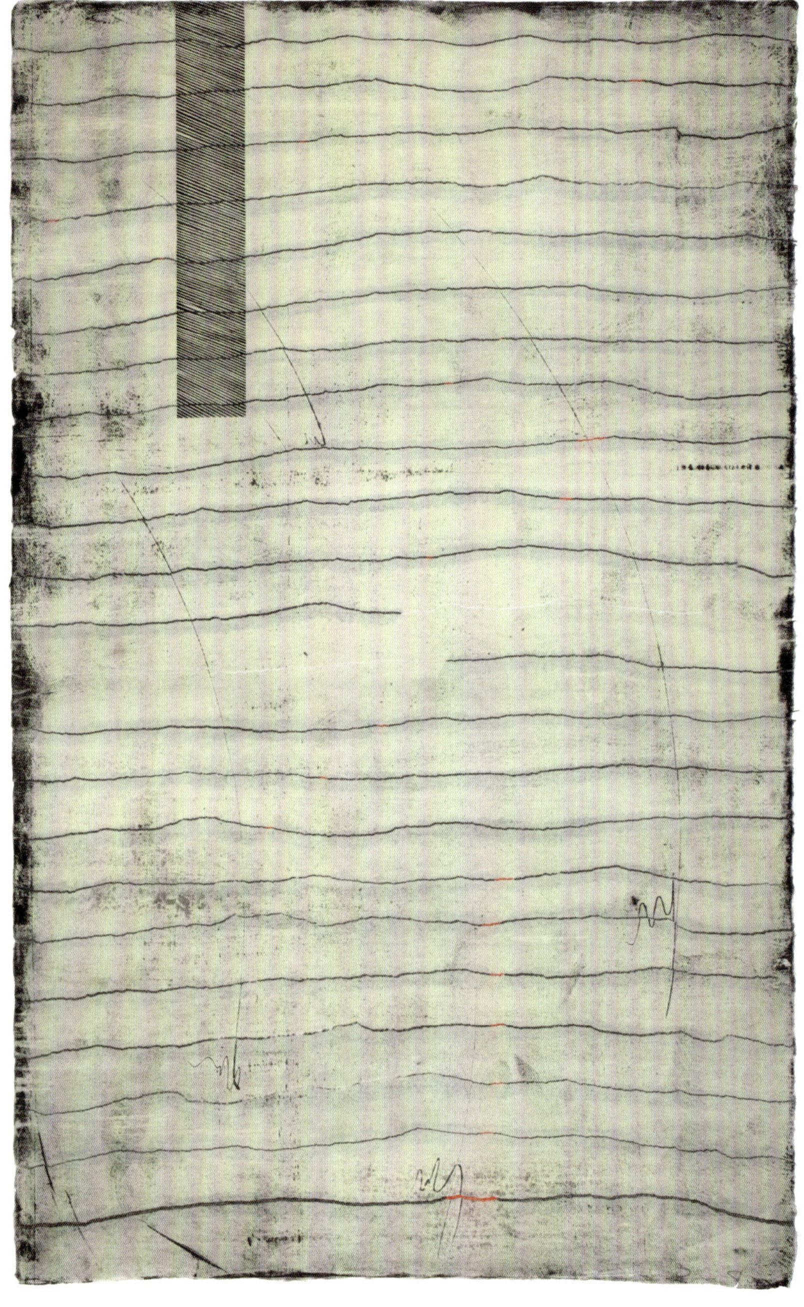

17 *Untitled B117*, 1984
mixed media on Japanese paper
38⅛ × 25¼ in. (97 × 64 cm)

18 *Untitled B34*, 1984
mixed media on Japanese paper
33⅛ × 39⅜ in. (84 × 100 cm)

19 *Untitled B119*, 1984
mixed media on Japanese paper
25¼ × 37 in. (64 × 94 cm)
Promised gift, Nina Collection, Kyoto

20 *Untitled B120*, 1984
mixed media on Japanese paper
25¼ × 37 in. (64 × 94 cm))
Promised gift, Nina Collection, Kyoto

21 *Untitled B161*, 1984
mixed media on Japanese paper
25¼ × 63 in. (64 × 160 cm)

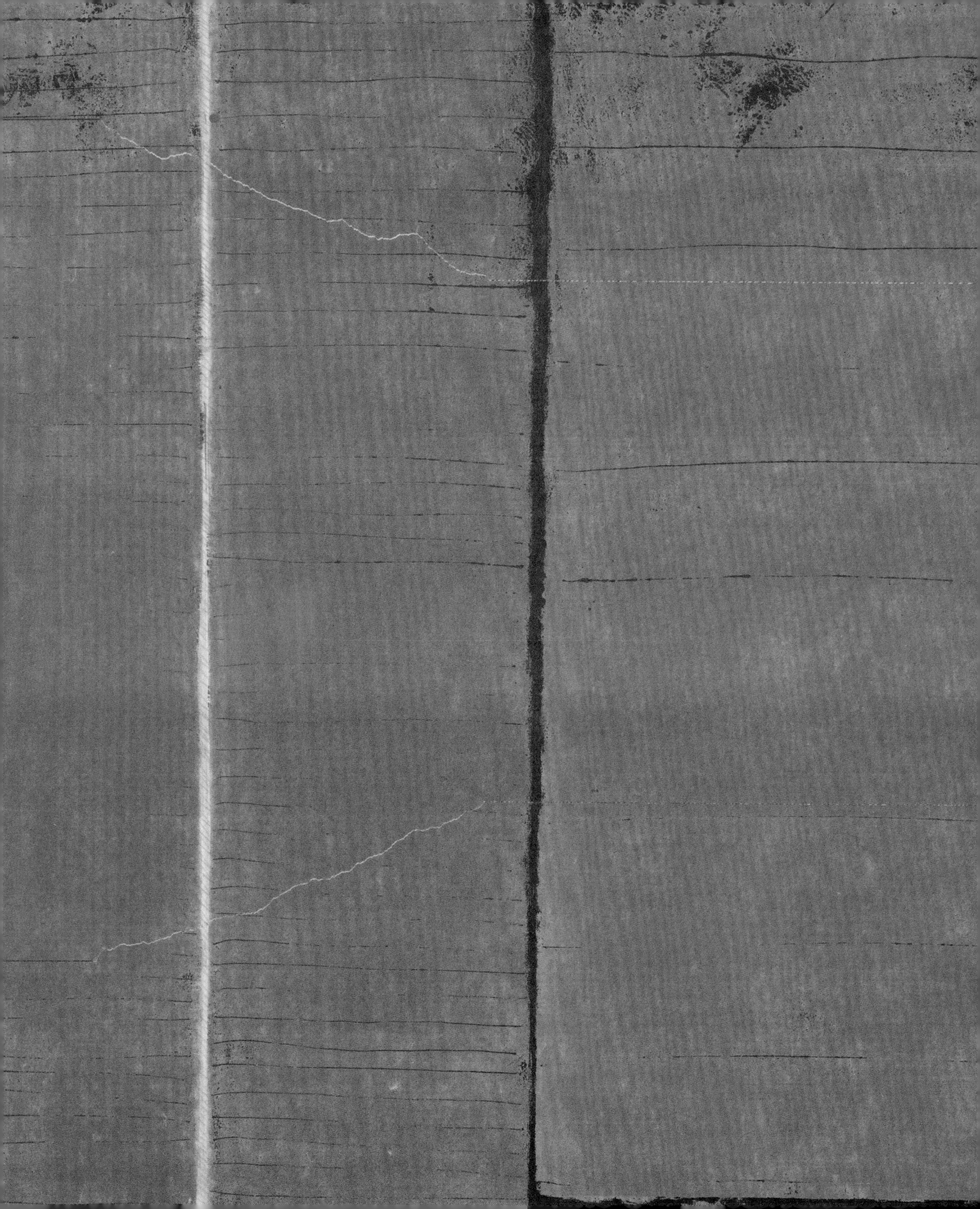

22 *Untitled B140*, 1983
mixed media on Japanese paper
36¼ × 58⅝ in. (92 × 149 cm)

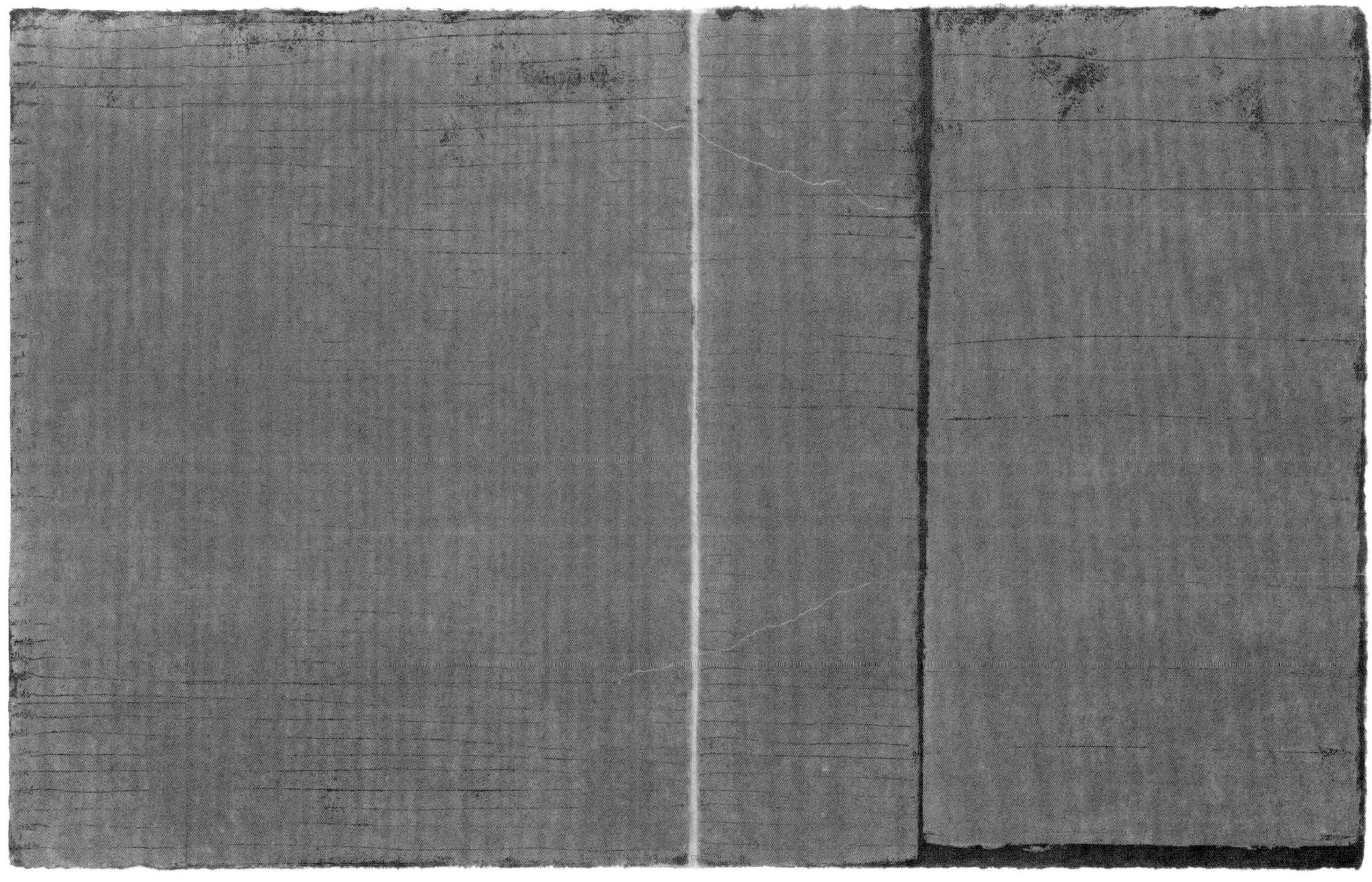

23 Narumi Hiroyuki, *Boku Sorekkiri* *(Just Me)*
ぼく、それっきり
鳴海裕行
designed by Rebecca Salter
printed book, 1983
in plastic or paper slipcase
cover: 7¾ × 6 in. (19.8 × 15.2 cm)

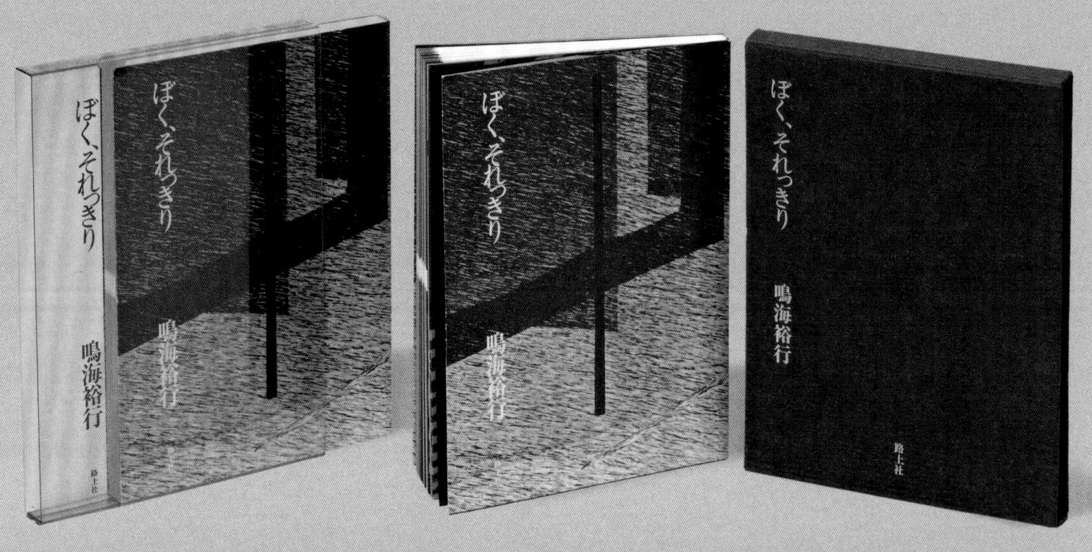

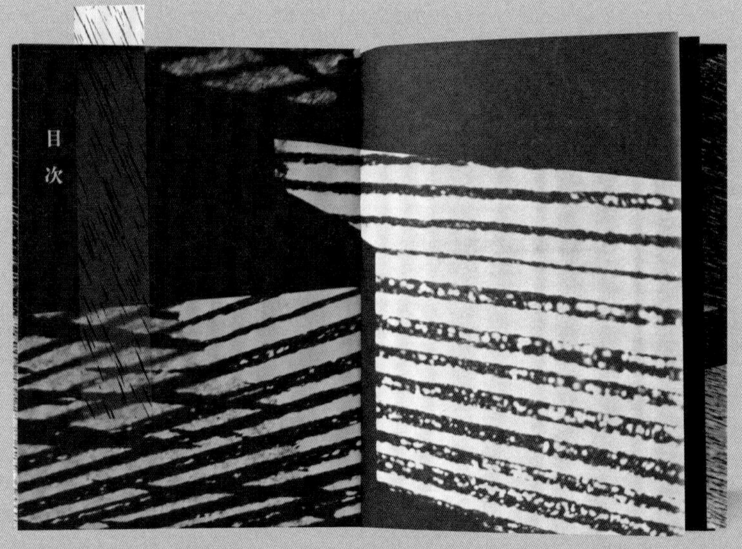

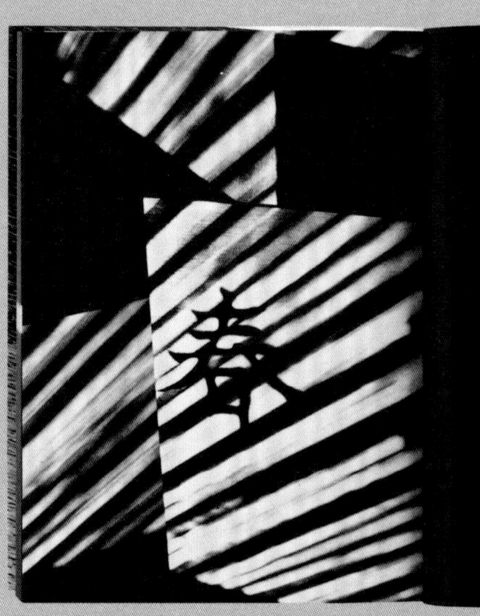

24 Elias Canetti, *The Human Province,*
cover design incorporating a work on
paper by Rebecca Salter
printed book, 1986
cover: 7¾ × 5⅛ in. (19.7 × 12.9 cm)

25 J. Marvin Spiegelman, *Active Imagination,*
J.M.シュピーゲルマン
能動的想像法—内なる魂との対話
cover design by Nishioka Tsutomu
incorporating drawing by
Rebecca Salter
printed book, 1994
cover: 7⅝ × 5⅜ in. (19.4 × 13.7 cm)

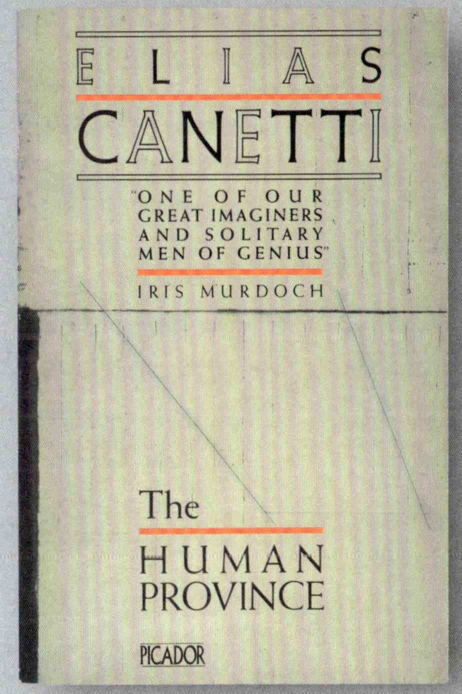

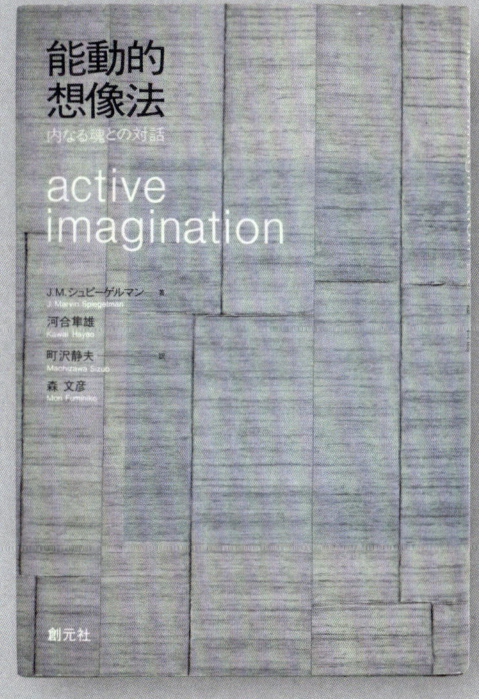

26 *Untitled F48*, 1990
mixed media on paper
49¼ × 51⅛ in. (125 × 130 cm)

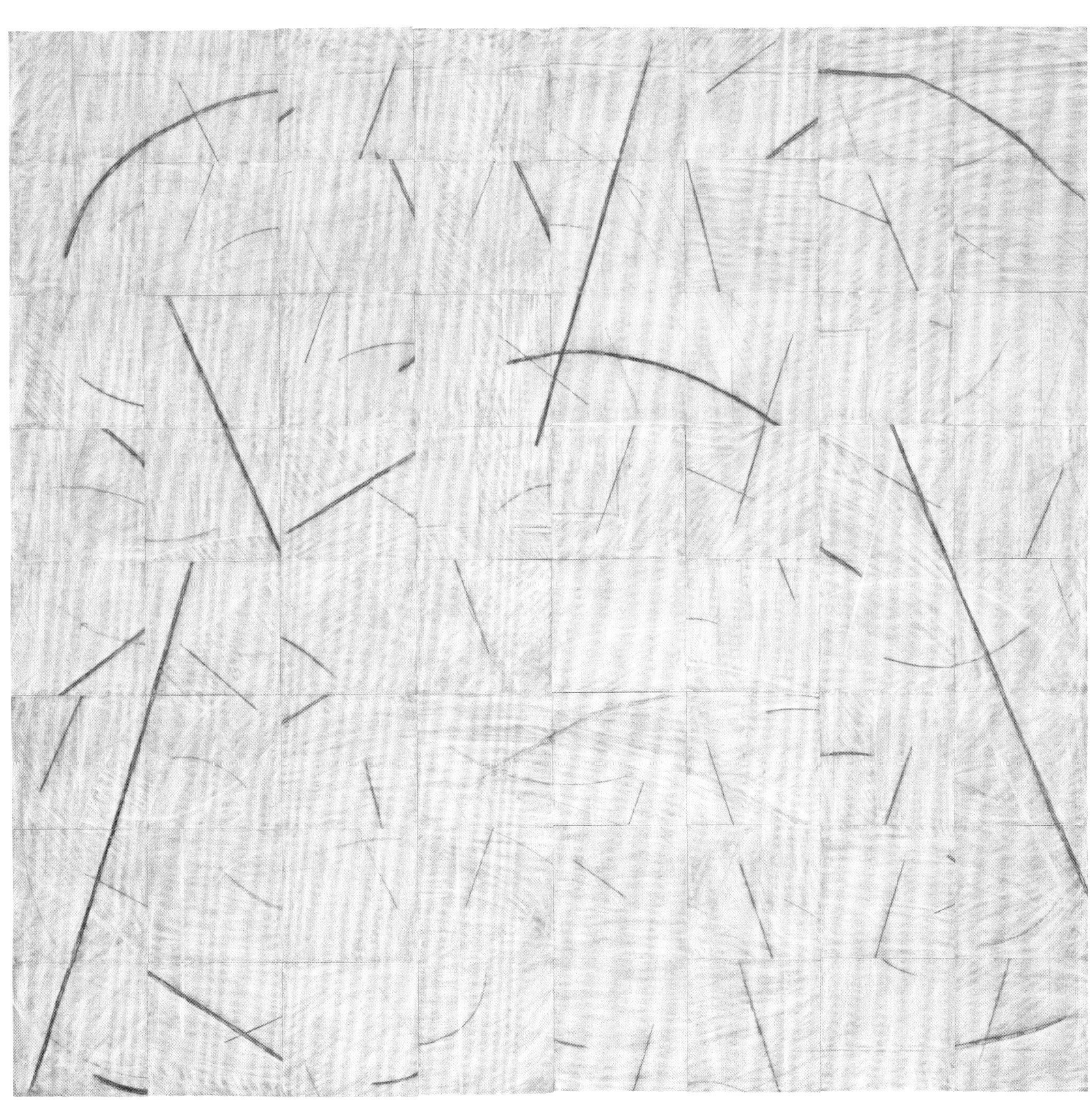

27 *Untitled F68,* 1990
mixed media on paper
16 × 16 in. (40.6 × 40.6 cm)

28 *Untitled G3,* 1990
mixed media on paper
11¾ × 11¾ in. (30 × 30 cm)

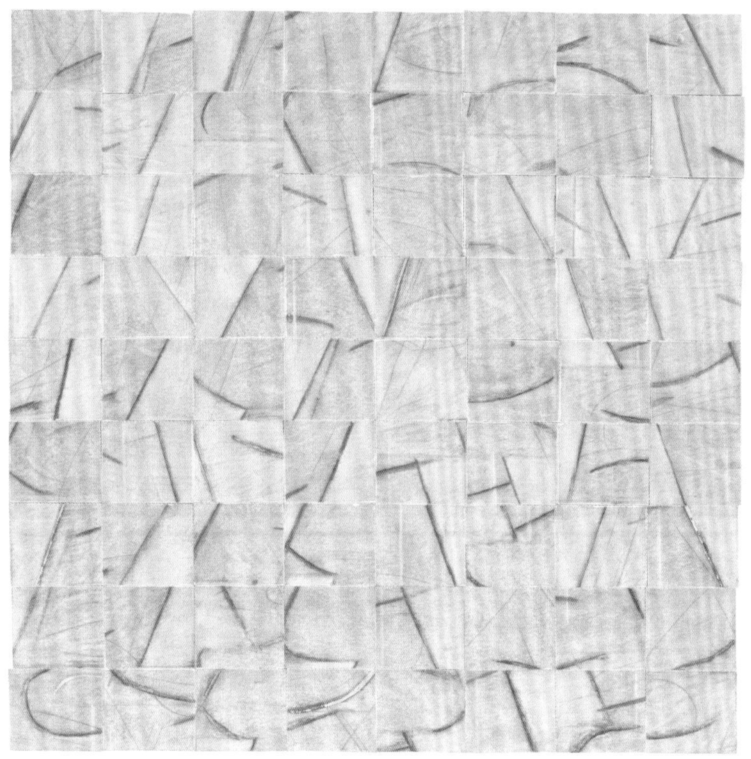

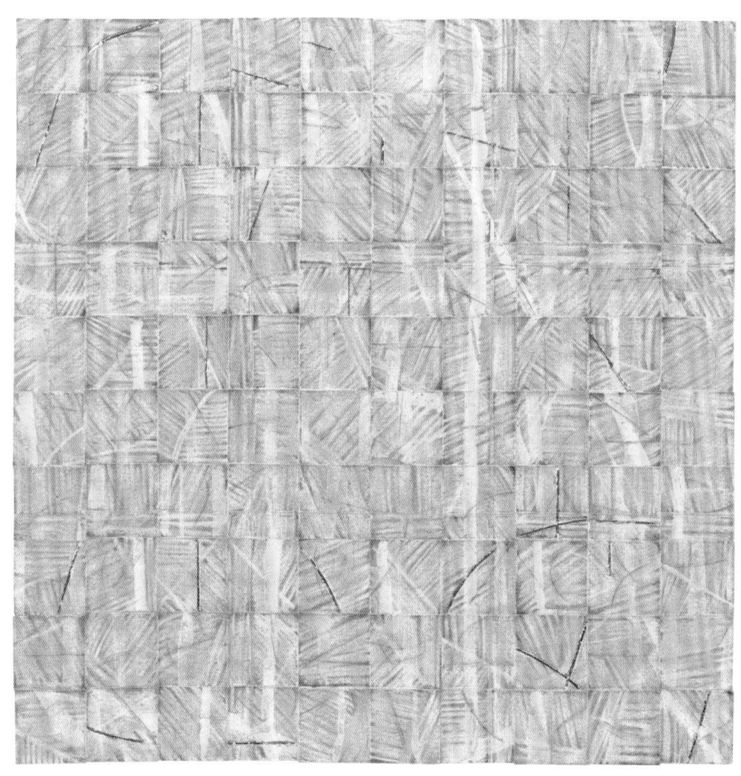

29 *Untitled F121*, 1990
mixed media on paper
14⅛ × 14⅛ in. (36 × 36 cm)

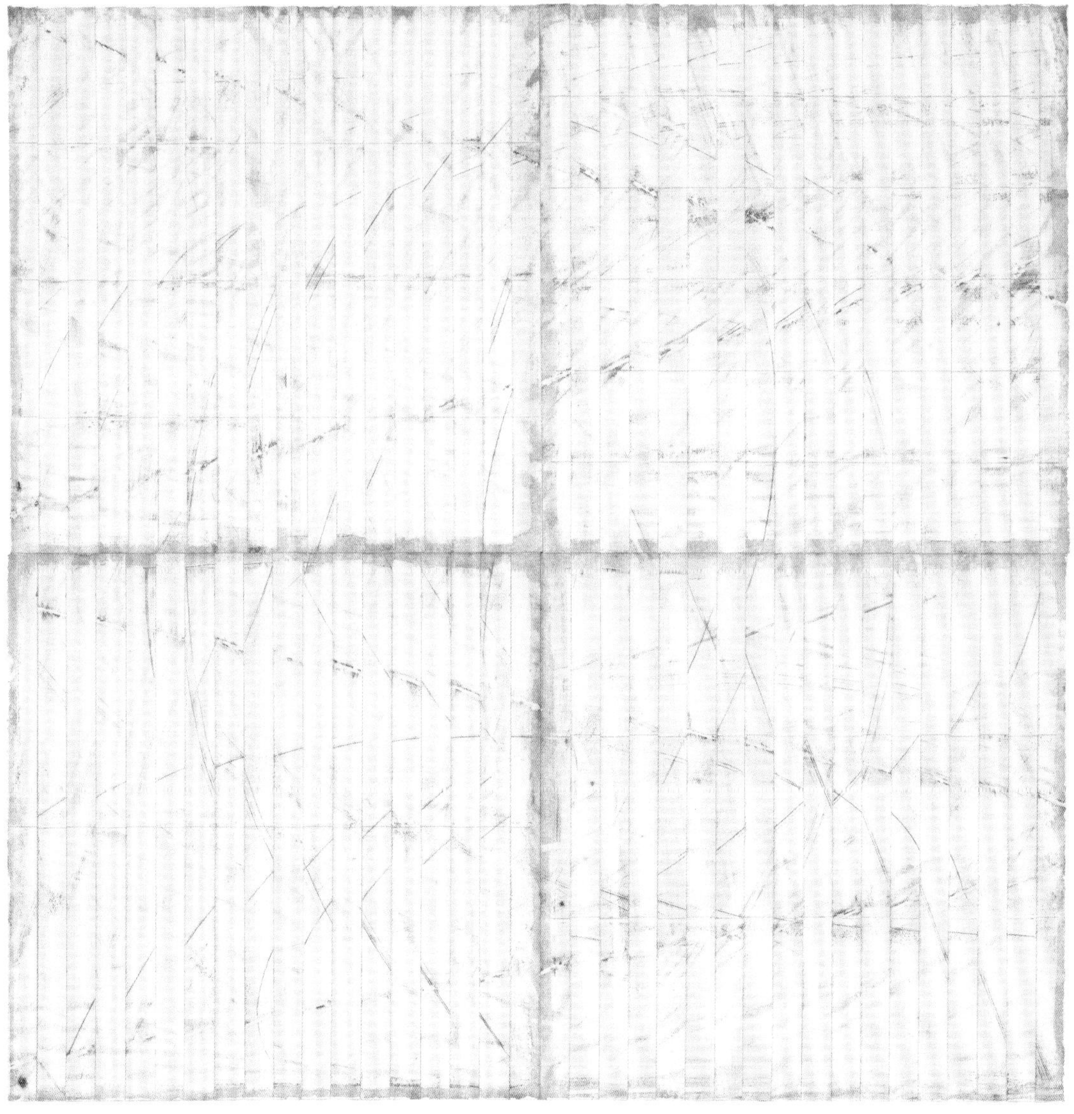

30 *Untitled F40*, 1990
mixed media on paper
29⅛ × 29⅞ in. (74 × 76 cm)

31 *Untitled F133*, 1990
mixed media on Japanese paper
32¼ × 35⅜ in. (82 × 90 cm)

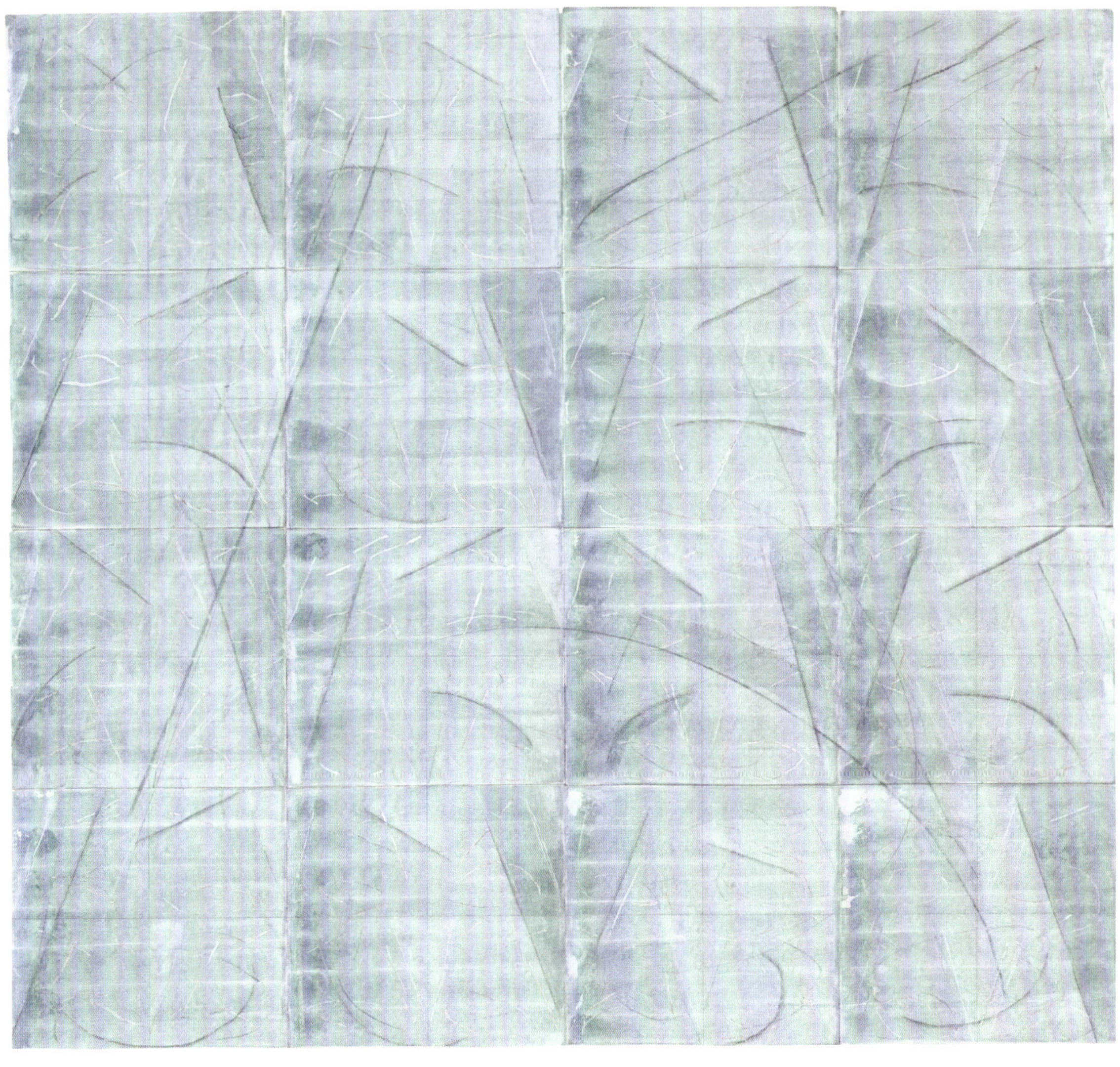

32 *Untitled D58*, 1988
mixed media on Japanese paper
18½ × 23¼ in. (47 × 59 cm)

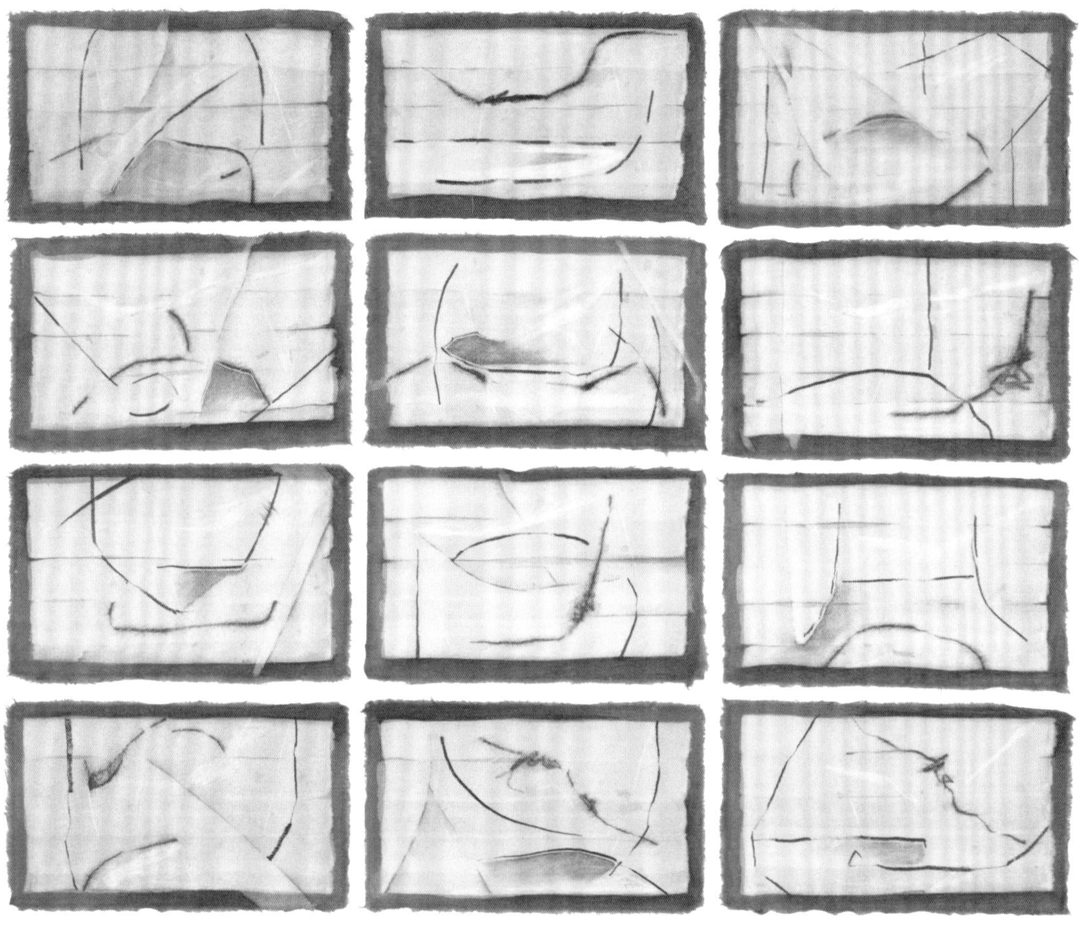

33 *Untitled D90*, 1989
mixed media on Japanese paper
25¼ × 37⅜ in. (64 × 95 cm)

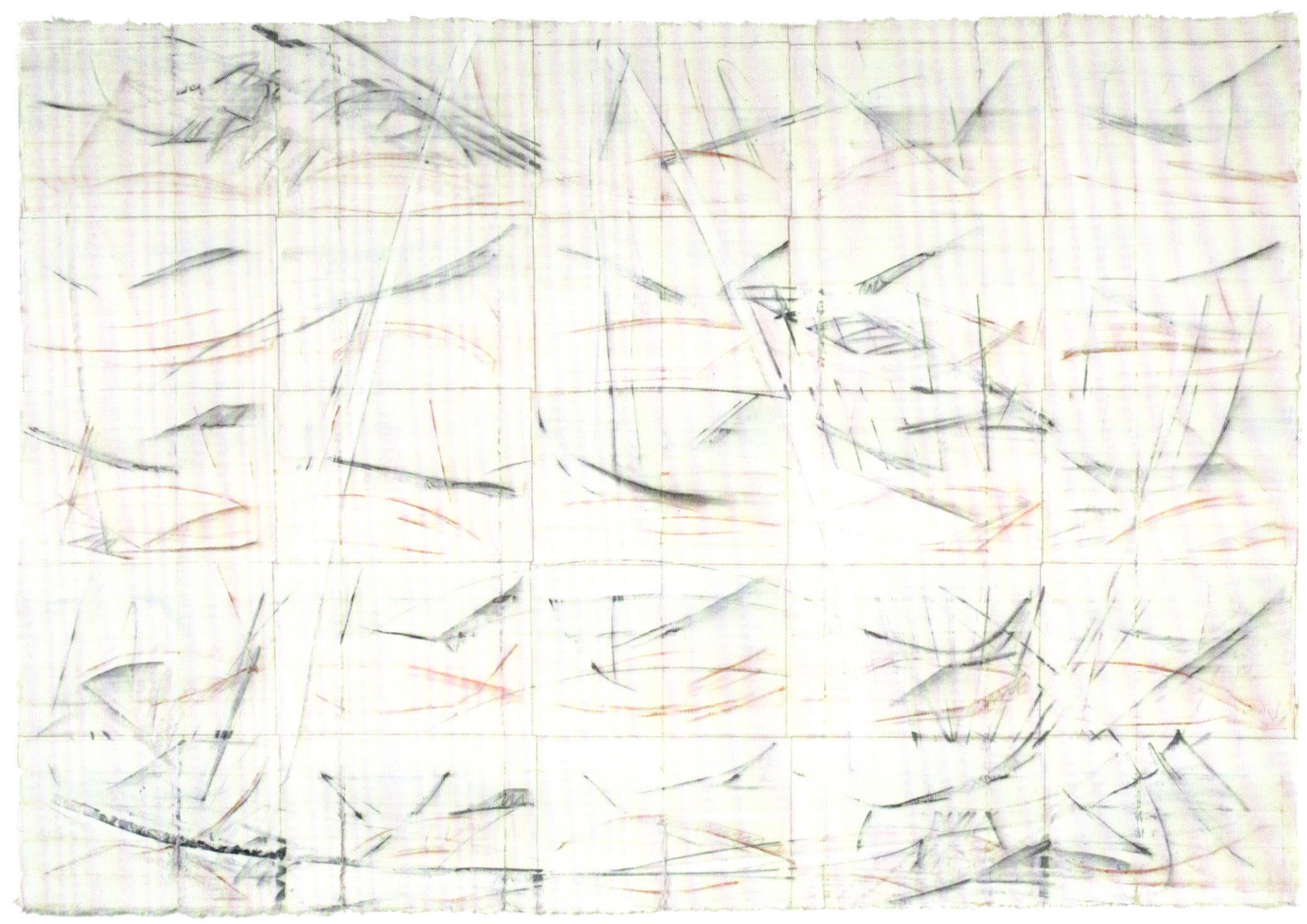

34 *Untitled D79*, 1988
mixed media on Japanese paper
18⅛ × 61⅜ in. (46 × 156 cm)

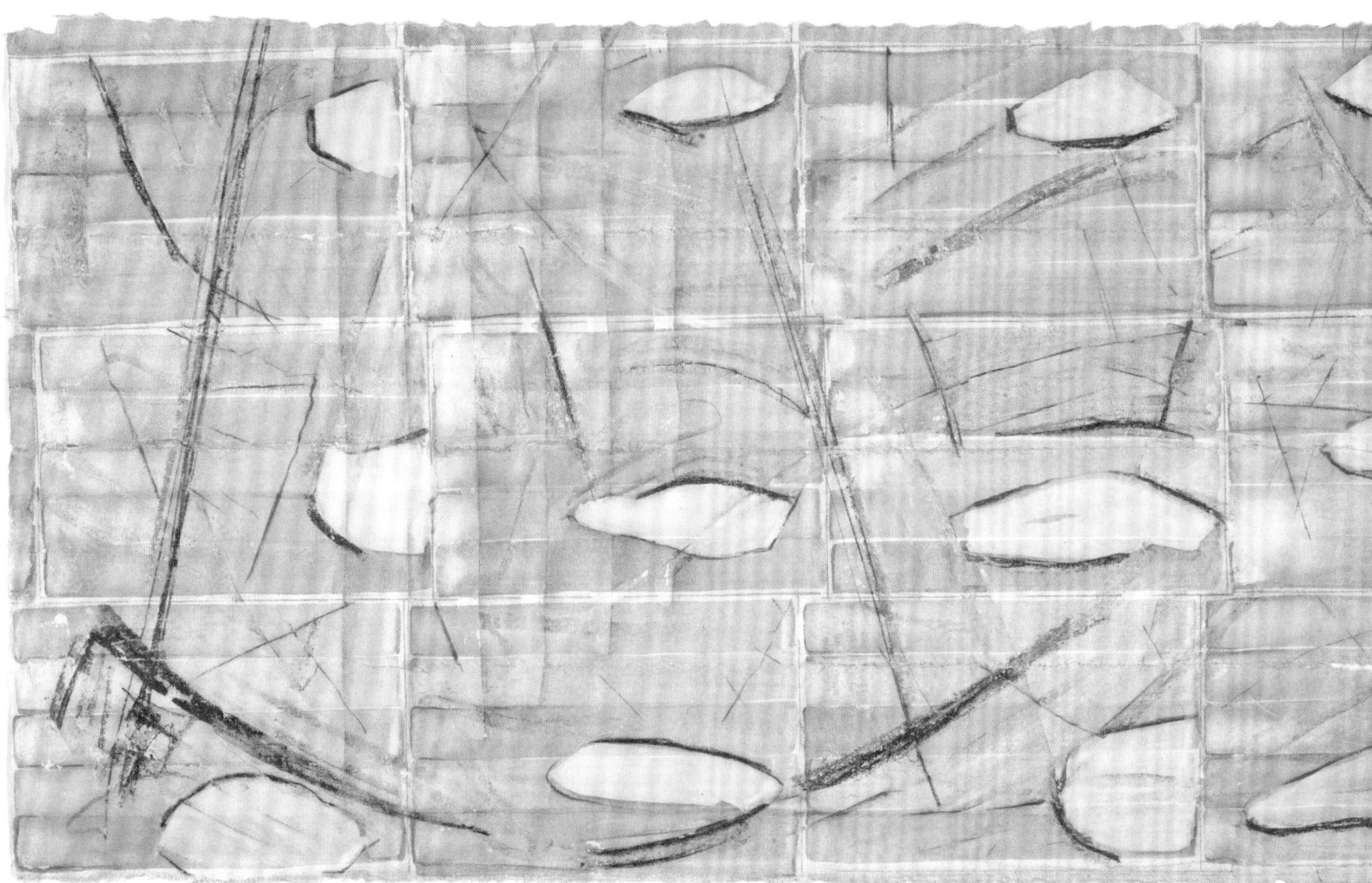

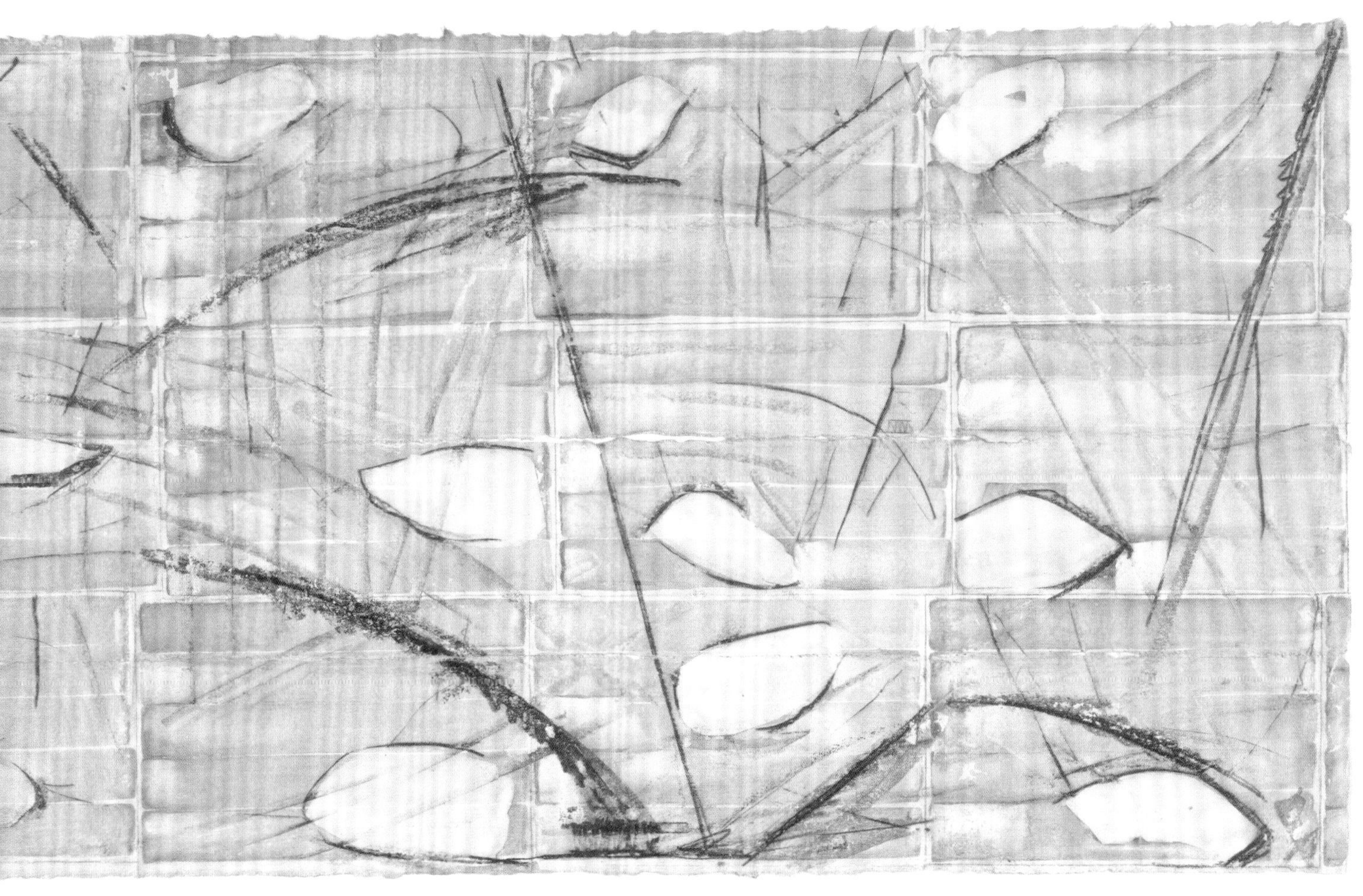

35 *Untitled D86*, 1988
mixed media on Japanese paper
11⅜ × 52¾ in. (29 × 134 cm)

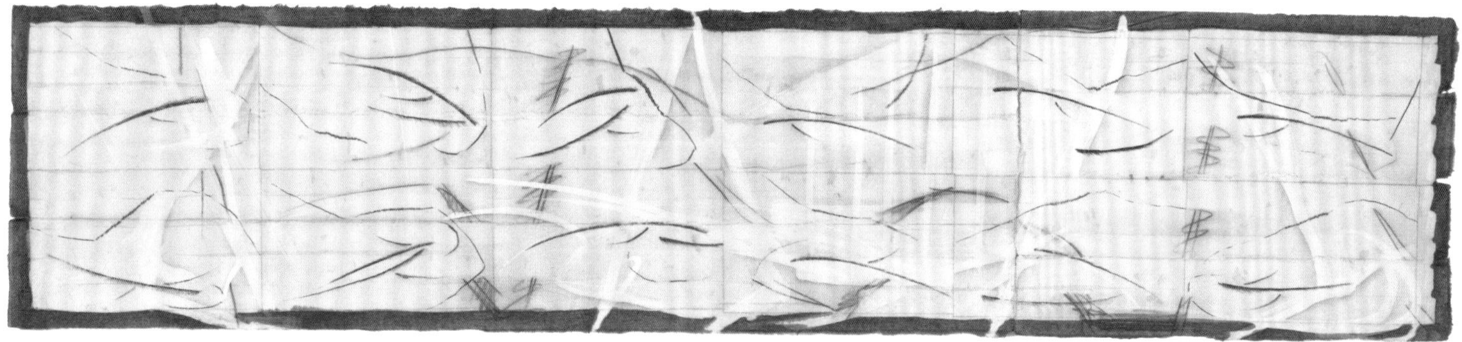

36 *Untitled D137*, 1989
mixed media on Japanese paper
63 × 28⅜ in. (160 × 72 cm)

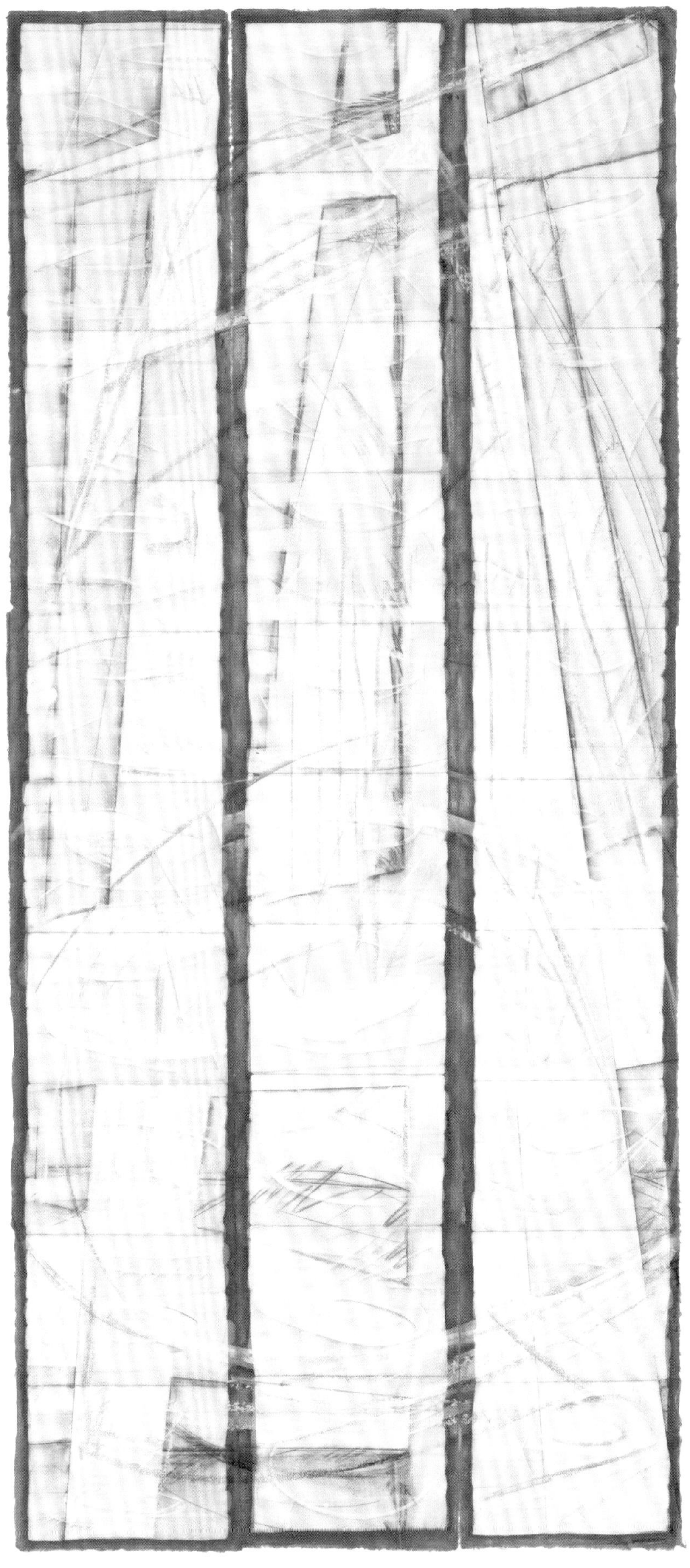

37 *Square 1*, 1991
mixed media on canvas
7⅛ × 7⅛ in. (18 × 18 cm)

Square 2, 1991
mixed media on canvas
7⅛ × 7⅛ in. (18 × 18 cm)

Square 3, 1991
mixed media on canvas
7⅛ × 7⅛ in. (18 × 18 cm)

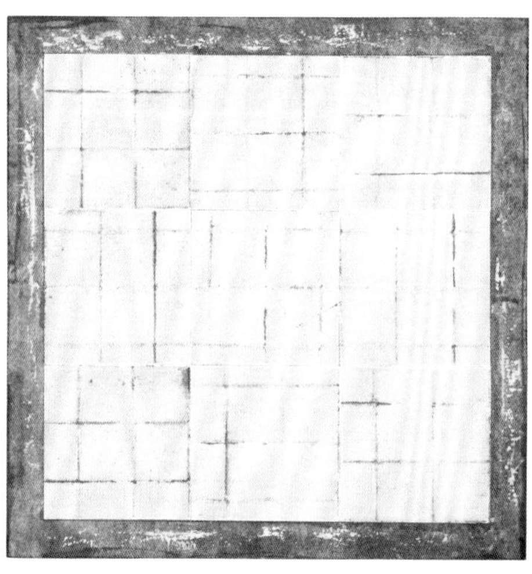

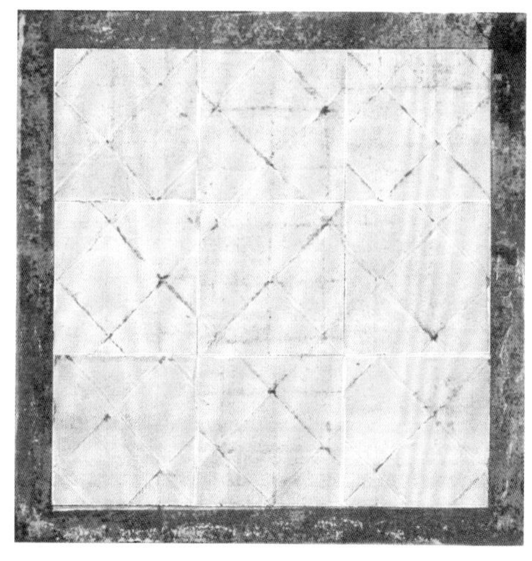

Square 4, 1991
mixed media on canvas
7⅛ × 7⅛ in. (18 × 18 cm)

Square 5, 1991
mixed media on canvas
7⅛ × 7⅛ in. (18 × 18 cm)

Square 6, 1991
mixed media on canvas
7⅛ × 7⅛ in. (18 × 18 cm)

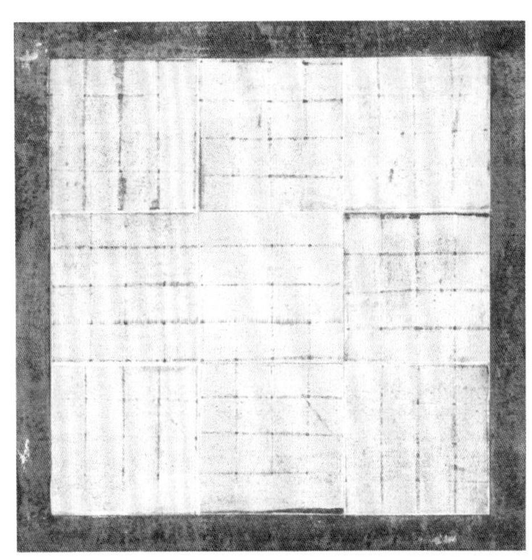

38 *Untitled F112*, 1990
mixed media on canvas
30 × 30 in. (76.2 × 76.2 cm)

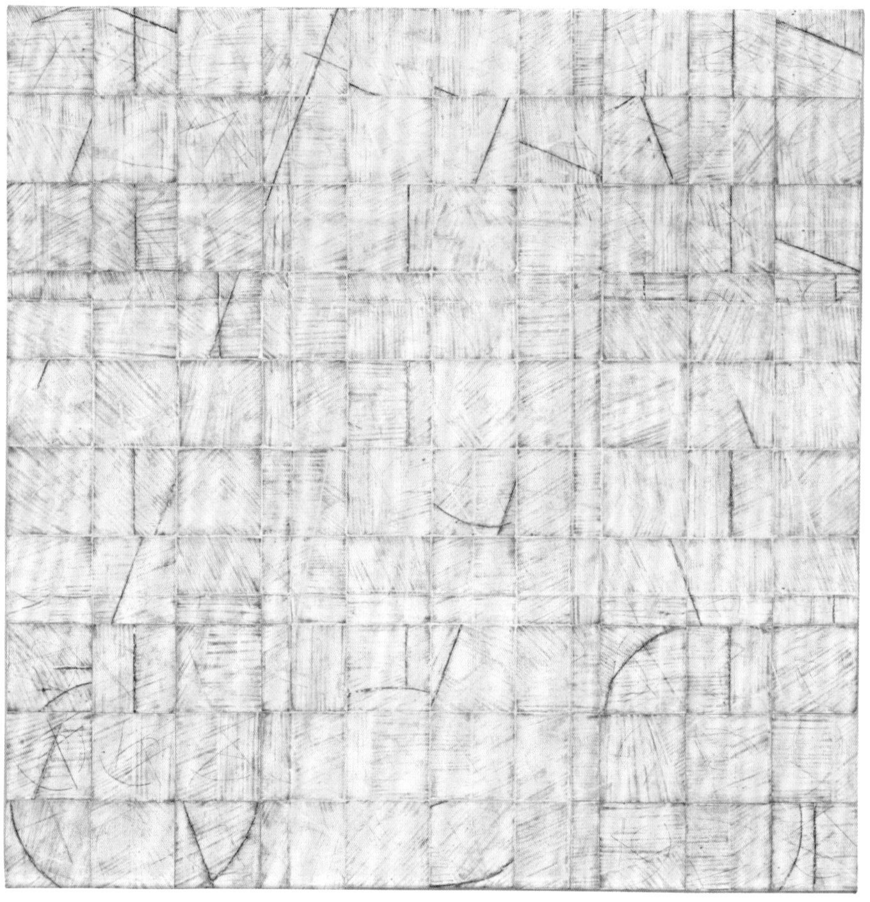

39 *Untitled G14*, 1991
mixed media on canvas
48 × 48 in. (121.9 × 121.9 cm)

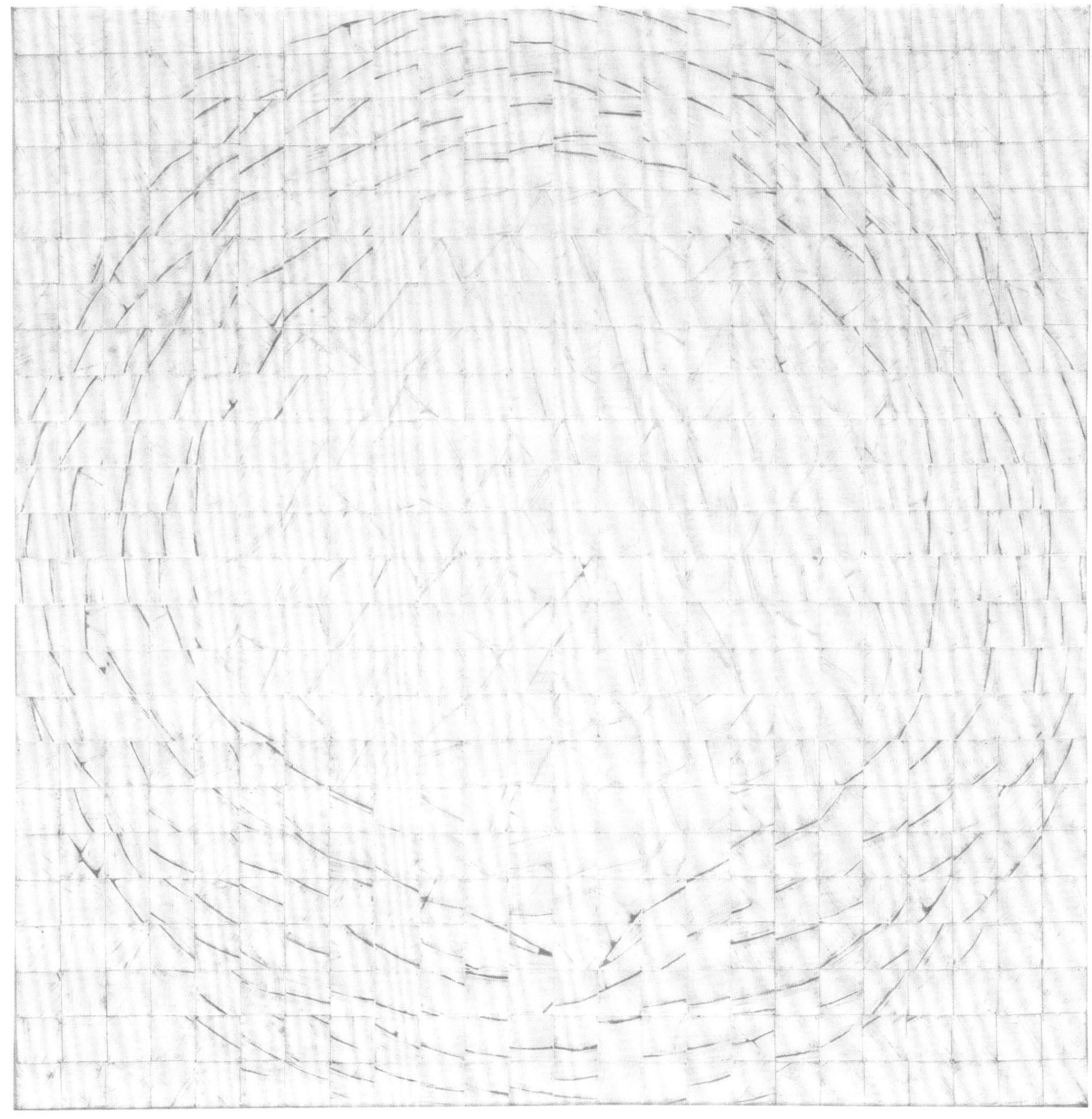

40 *Untitled G86*, 1992
mixed media on canvas
48 × 48 in. (121.9 × 121.9 cm)
Collection of Delia Smith

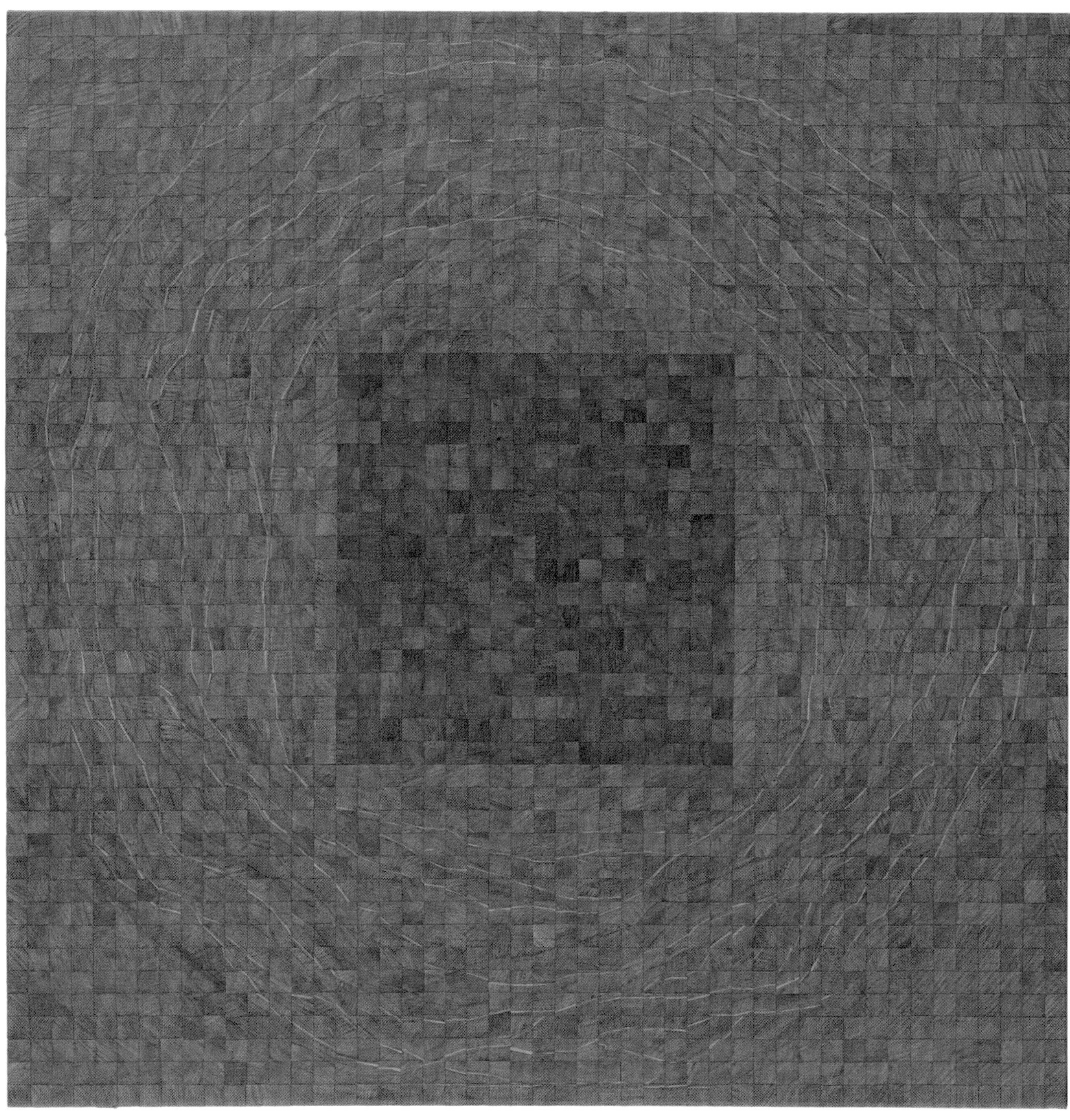

41 *Untitled H49*, 1993
mixed media on canvas
22 × 22 in. (55.9 × 55.9 cm)

42 *Untitled G27*, 1991
mixed media on canvas
48 × 48 in. (121.9 × 121.9 cm)

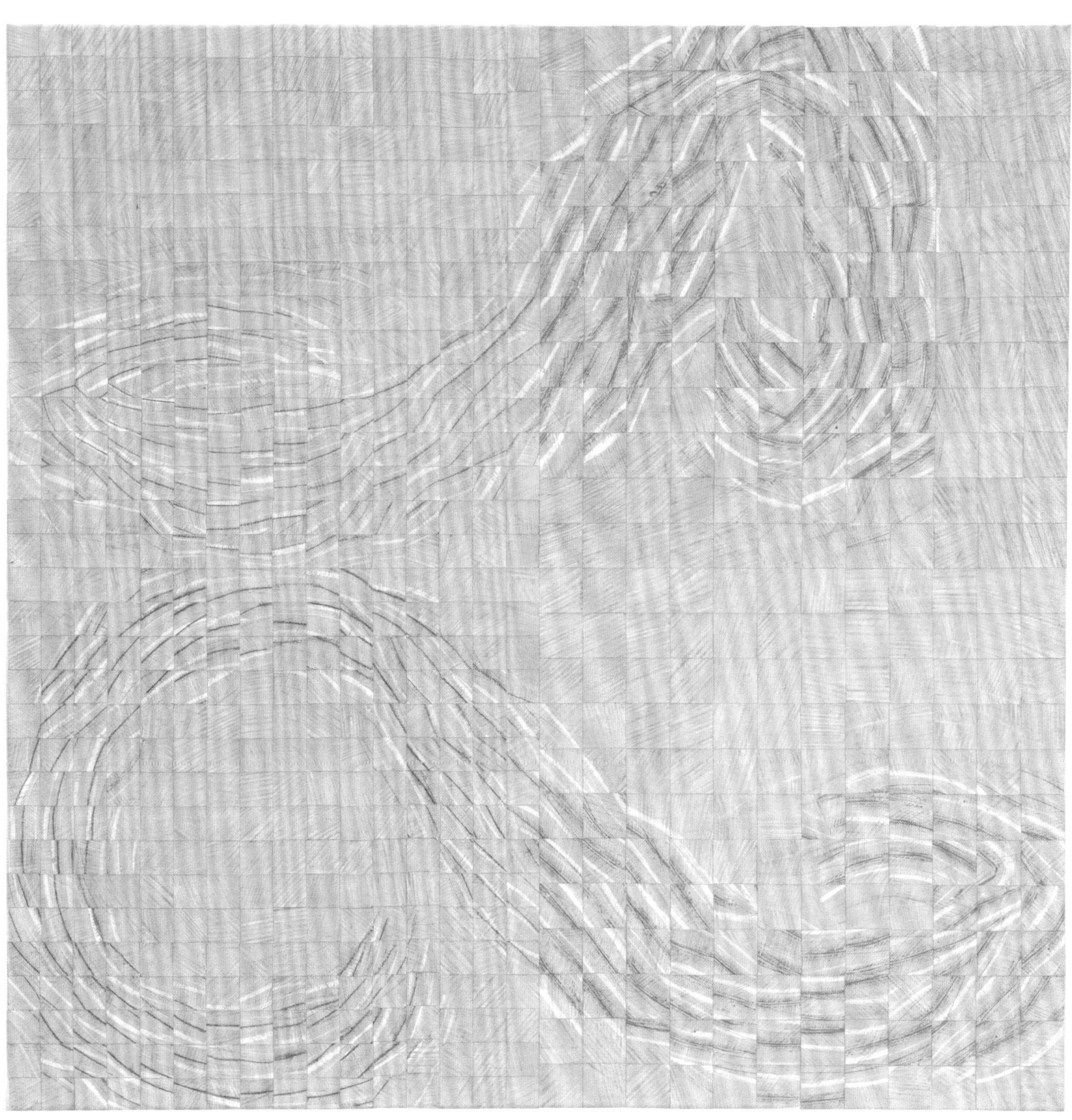

43 *Untitled H32*, 1993
mixed media on canvas
60 × 60 in. (152.4 × 152.4 cm)
Private collection

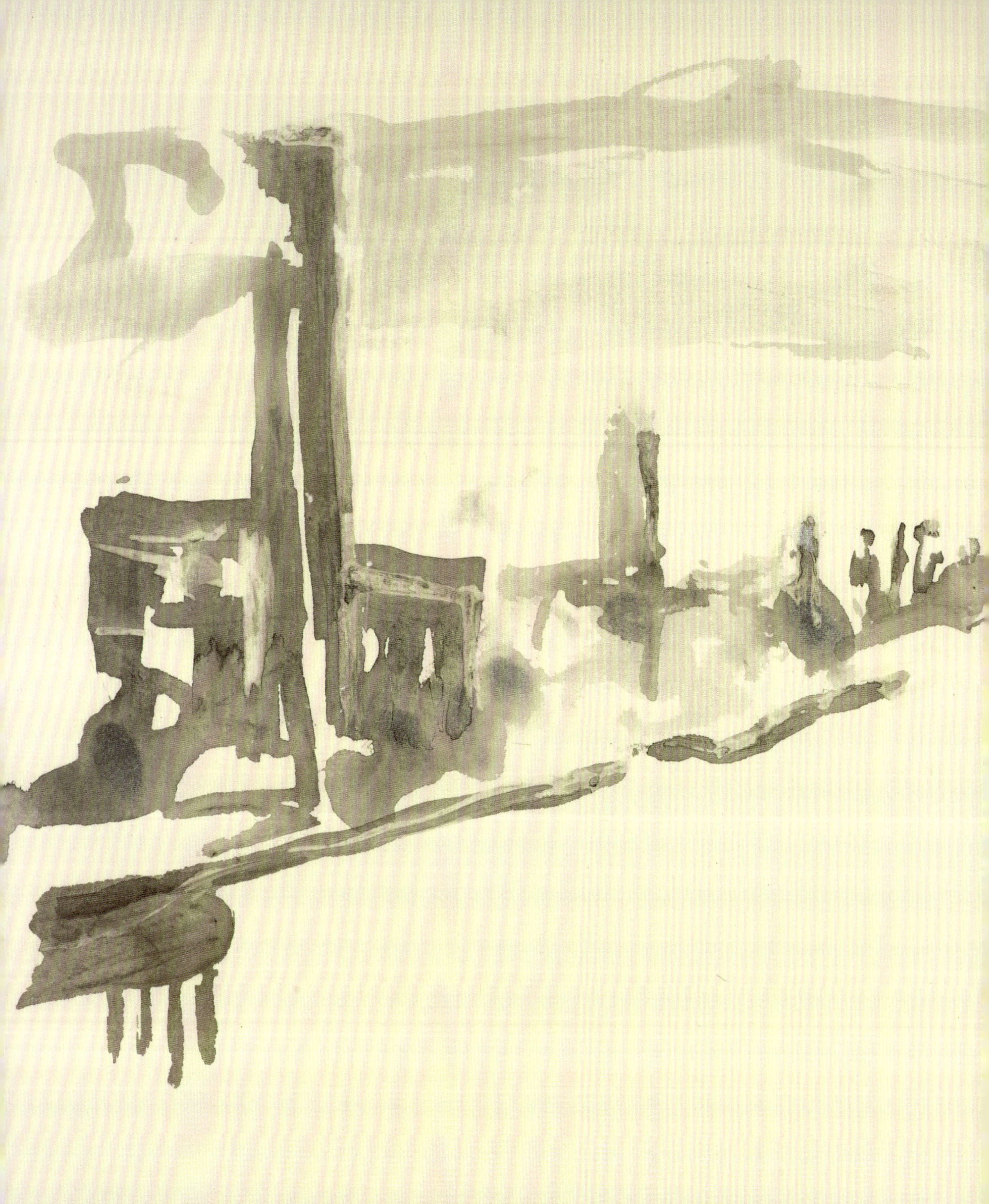

44 London sketchbook, 1990
sumi ink on paper
12½ × 9½ in. (31.8 x 24.1 cm)

top to bottom:

River Thames, Bankside Power Station (now Tate Modern)

Finsbury Park Station

Roofs, north London

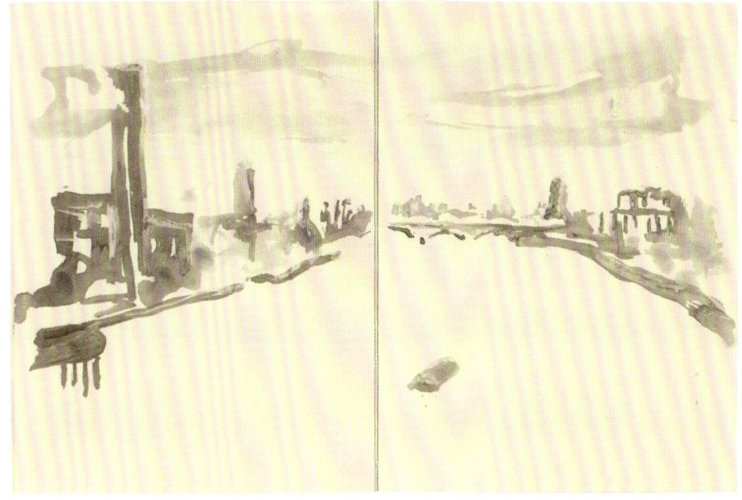

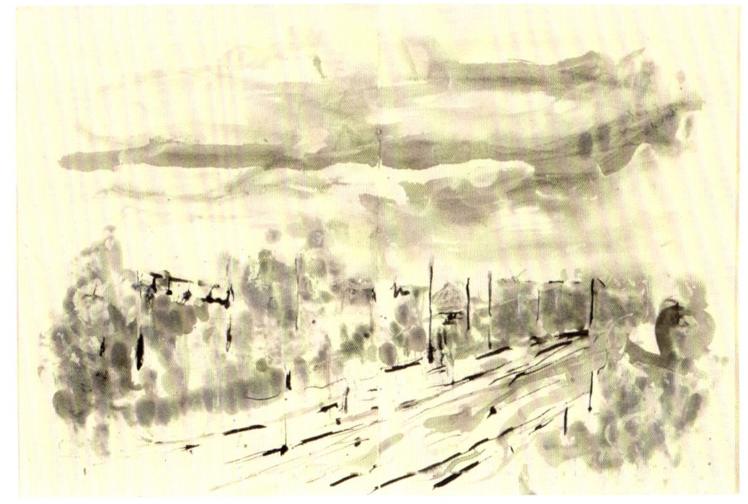

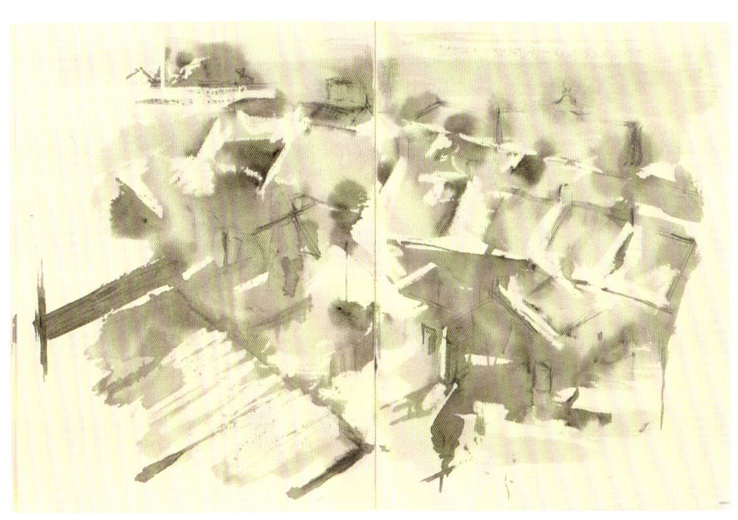

45 Lake District sketchbook, 1990
pastel on paper
12½ × 9½in. (31.8 x 24.1 cm)
Yale Center for British Art, gift of the artist in memory of her father Samuel Salter (1922–2005)

this page
top to bottom:

Crummock Water

Honister Pass

Ullswater

opposite page
top to bottom:

Ullswater

High Row

High Row

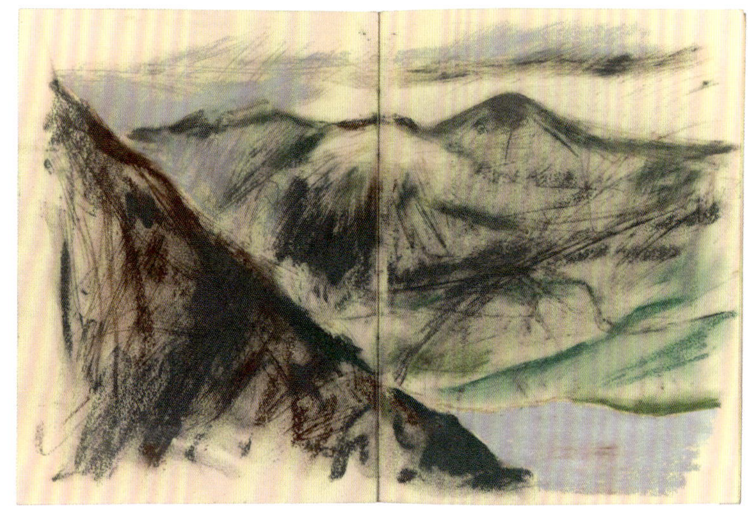

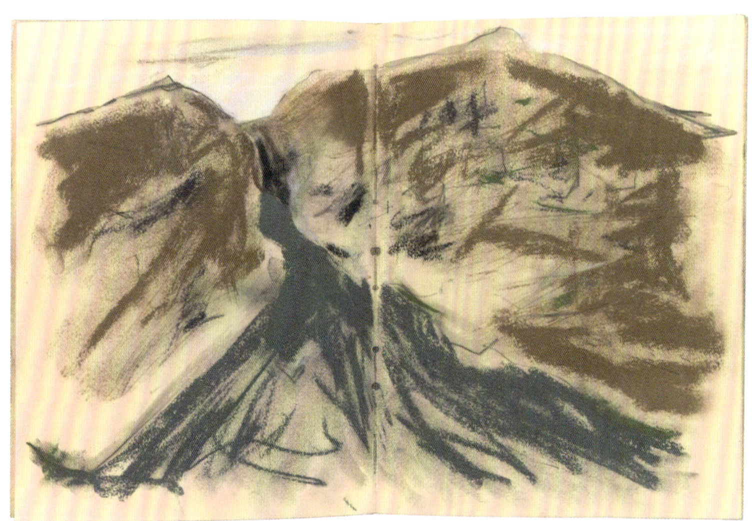

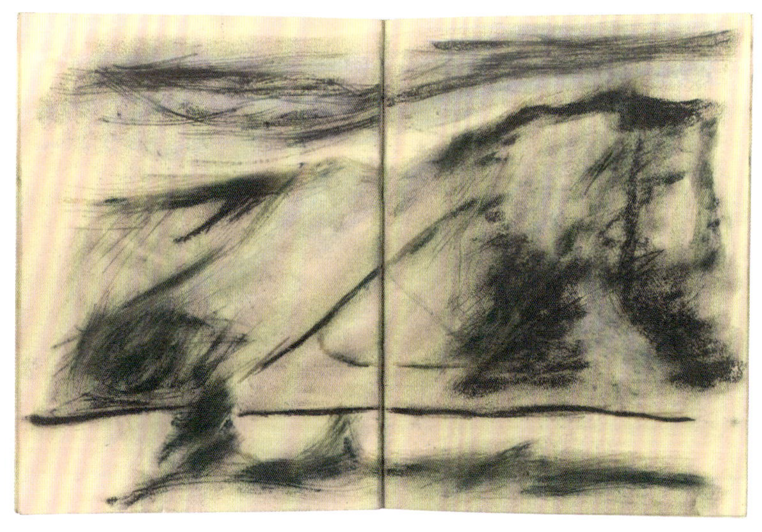

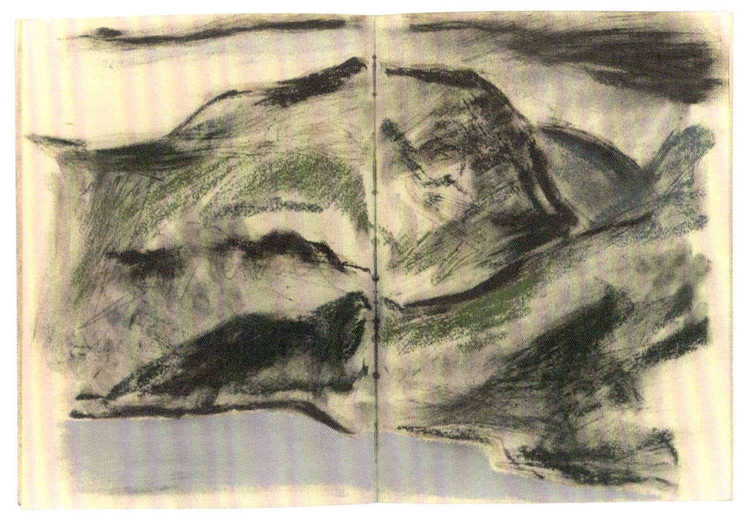

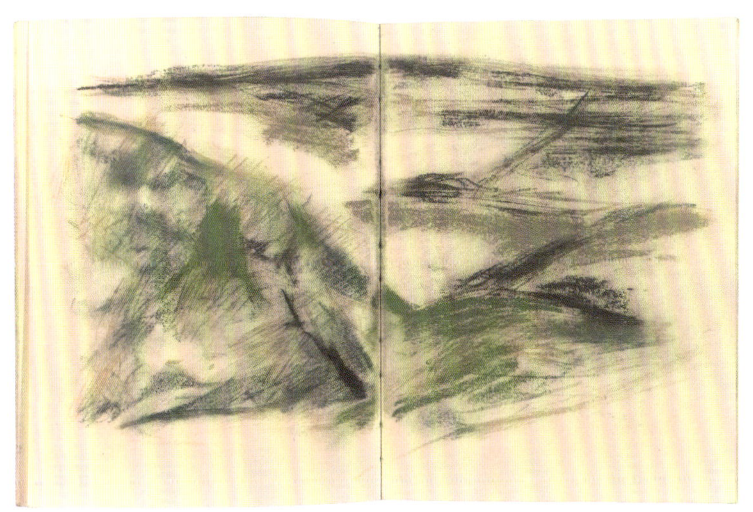

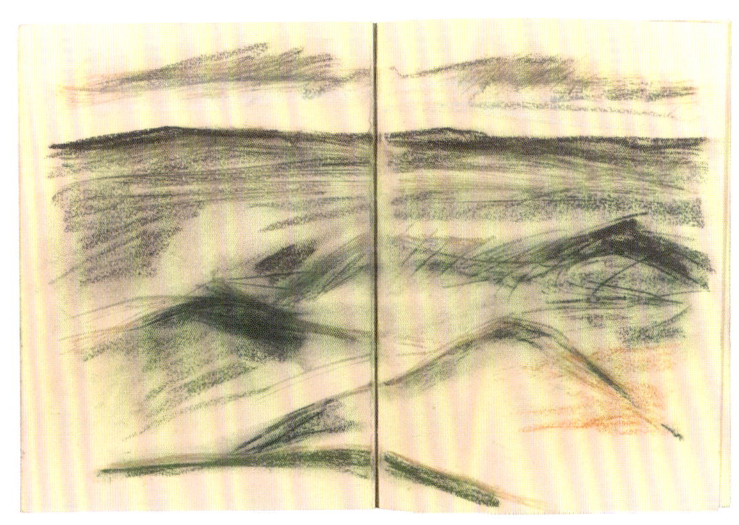

46 *Untitled H43*, 1993
mixed media on paper
48⅞ × 46½ in. (124 × 118 cm)
detail overleaf

47 *Untitled J1*, 1994
mixed media on canvas
72 × 108 in. (182.9 × 274.3 cm)

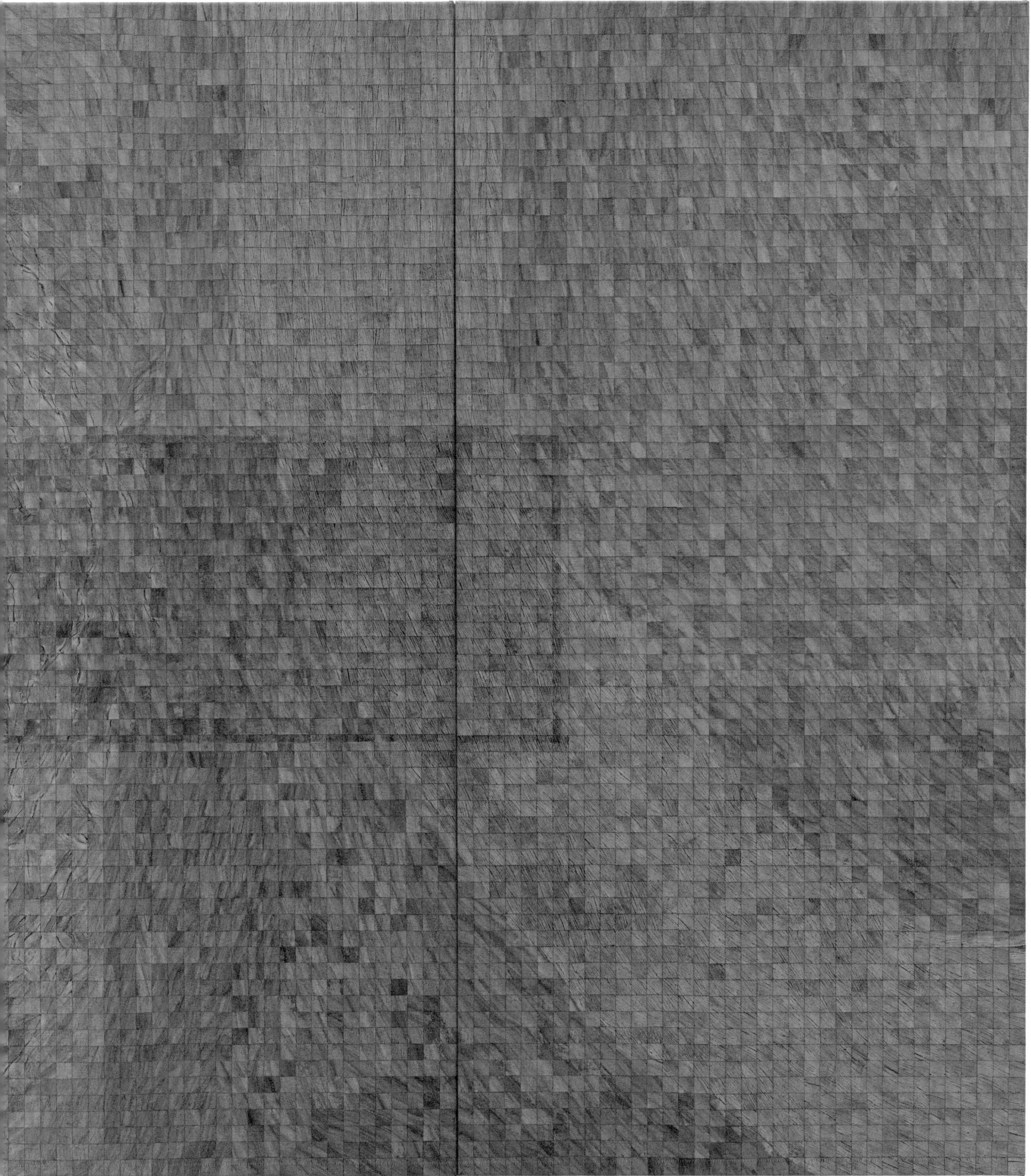

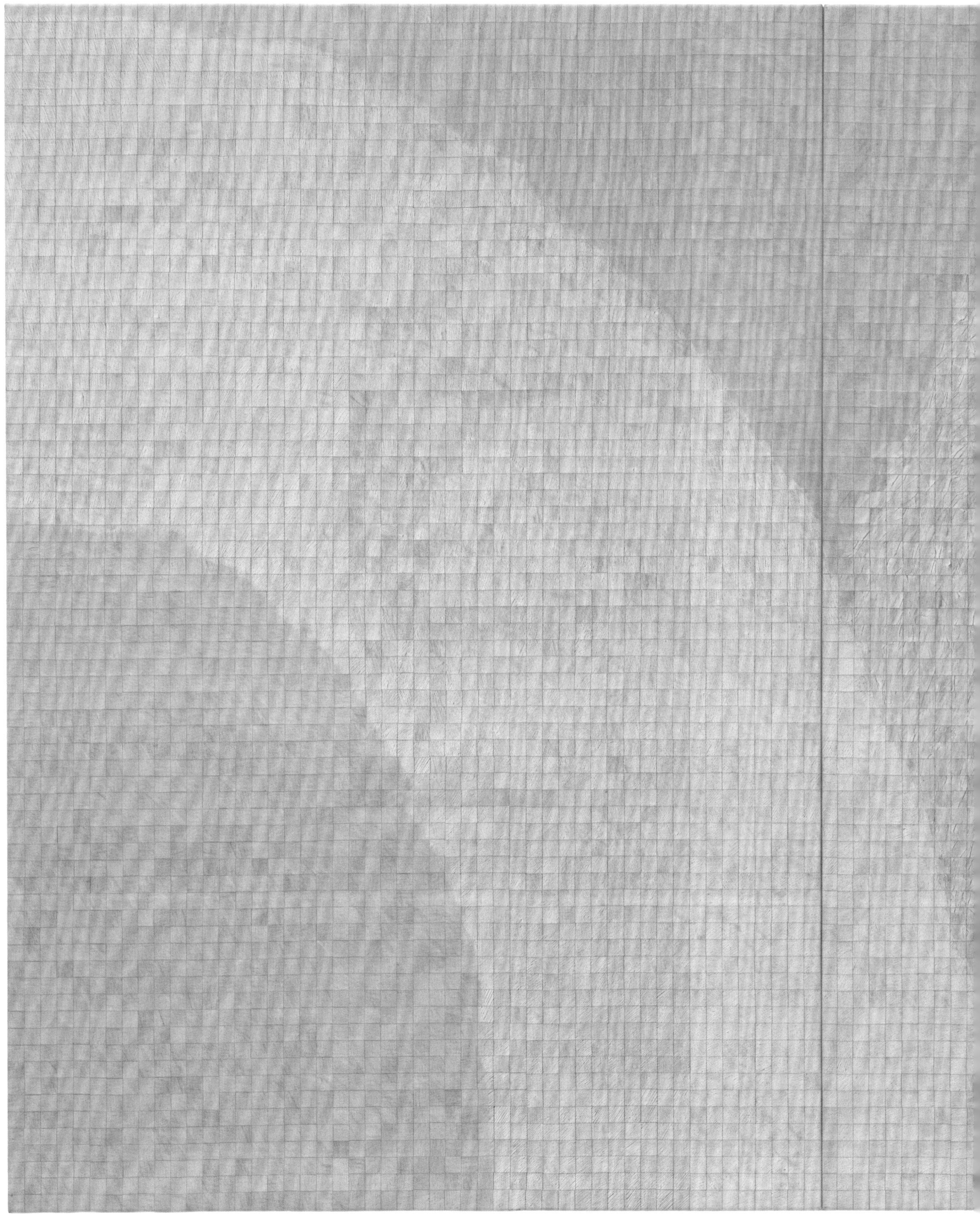

48 *Untitled J23,* 1995
mixed media on canvas
72 × 86 in. (182.9 × 218.4 cm)

49 Lake District sketchbooks
Wastwater Screes
1994, 2007
2008, 2009
watercolor on paper
12½ × 9½ in. (31.8 × 24.1 cm)

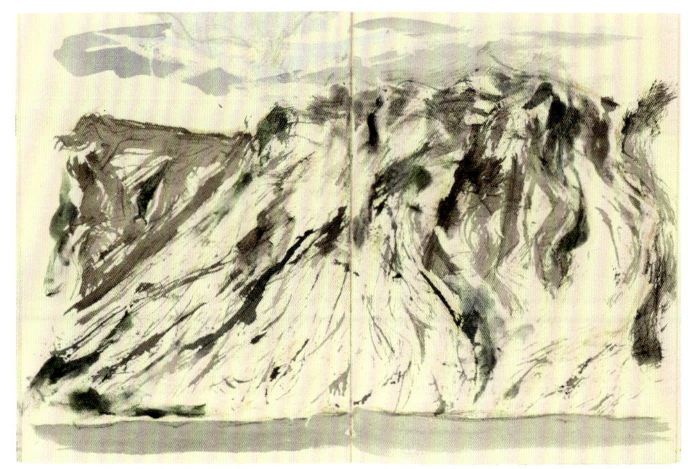

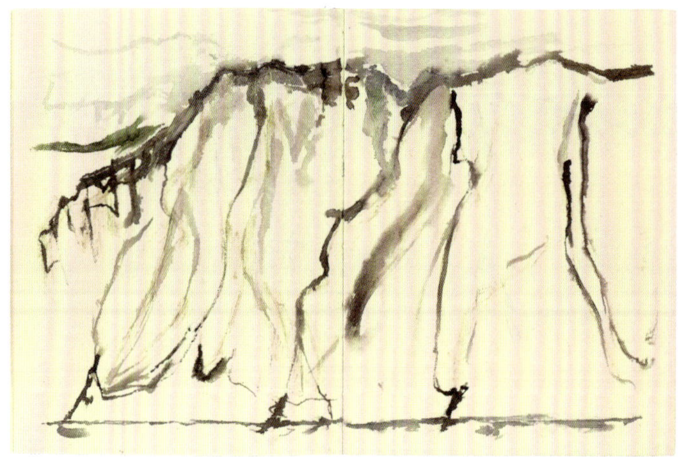

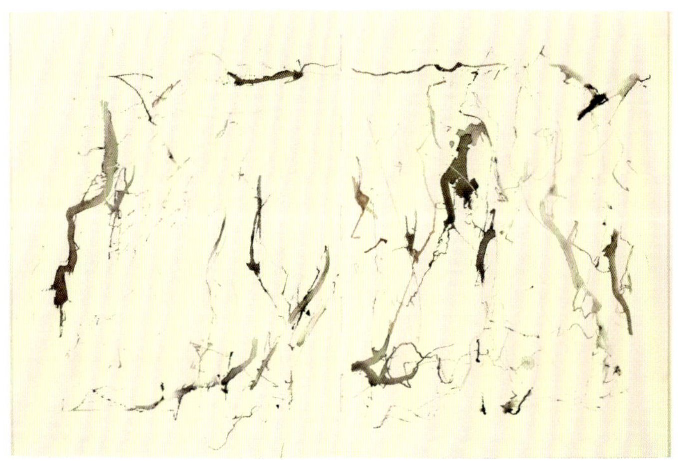

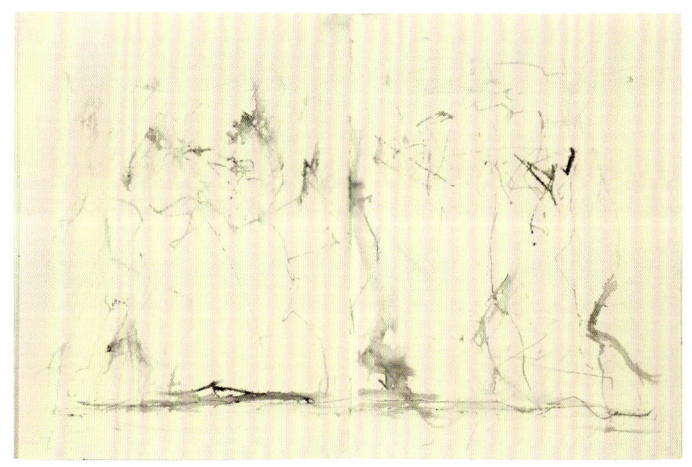

50 Lake District sketchbooks
Wastwater, Looking East
1994, 2008
2008, 2009
watercolor on paper
12½ × 9½ in. (31.8 × 24.1 cm)

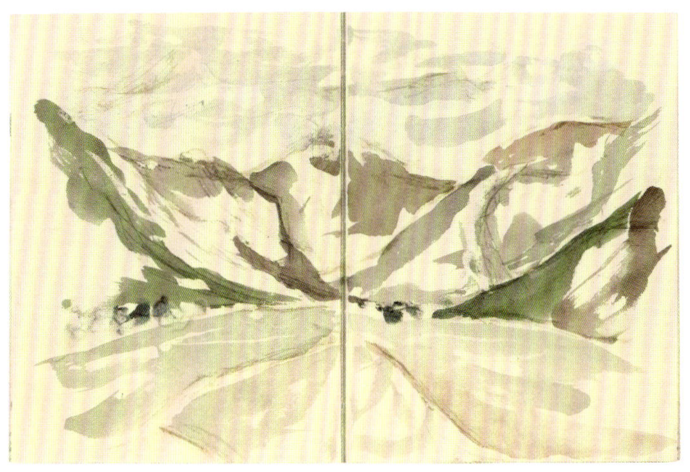

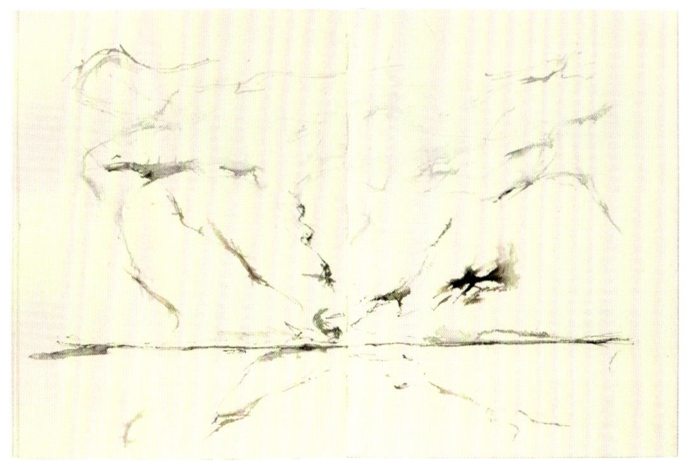

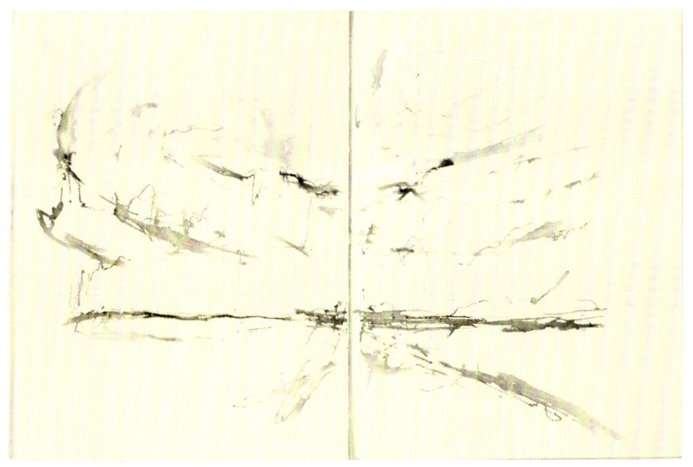

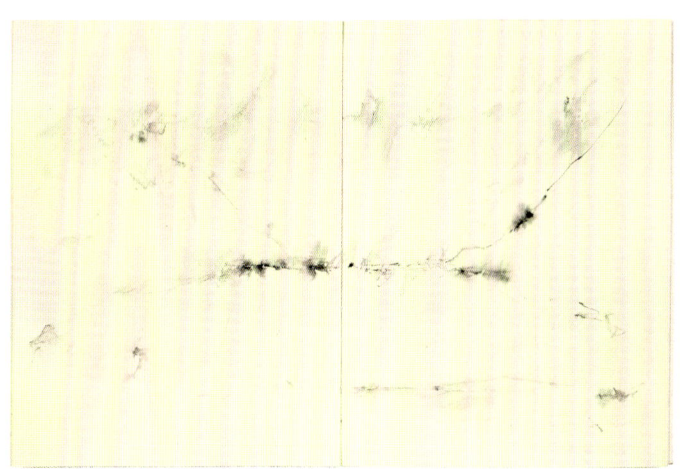

51 *Lingmell/Kirkfell*, 2009
watercolor on paper
5 × 19¼ in. (12.7 × 48.9 cm)

52 *Lingmell/Kirkfell/Screes*, 2009
watercolor on paper
5½ × 43¾ in. (14 × 111.1 cm)

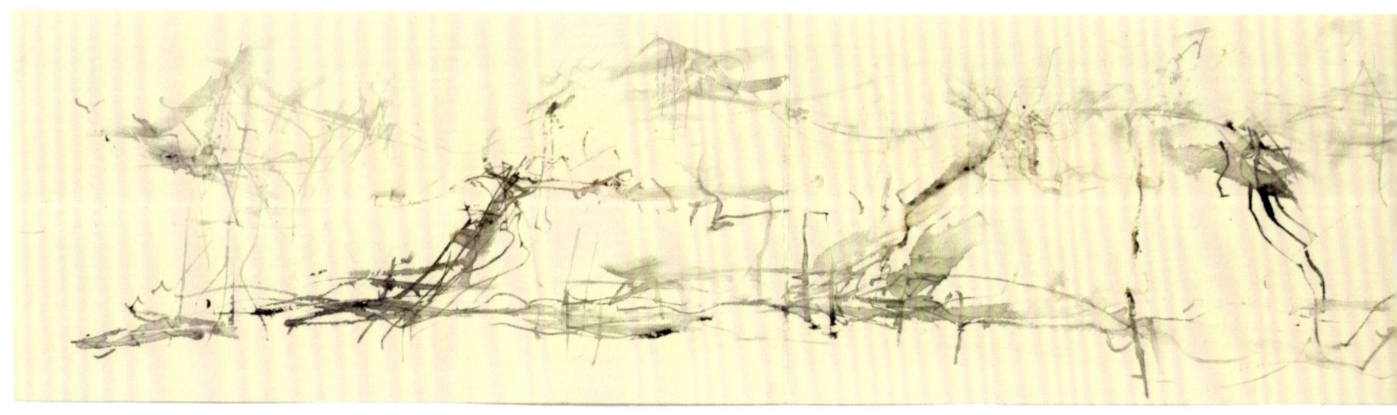

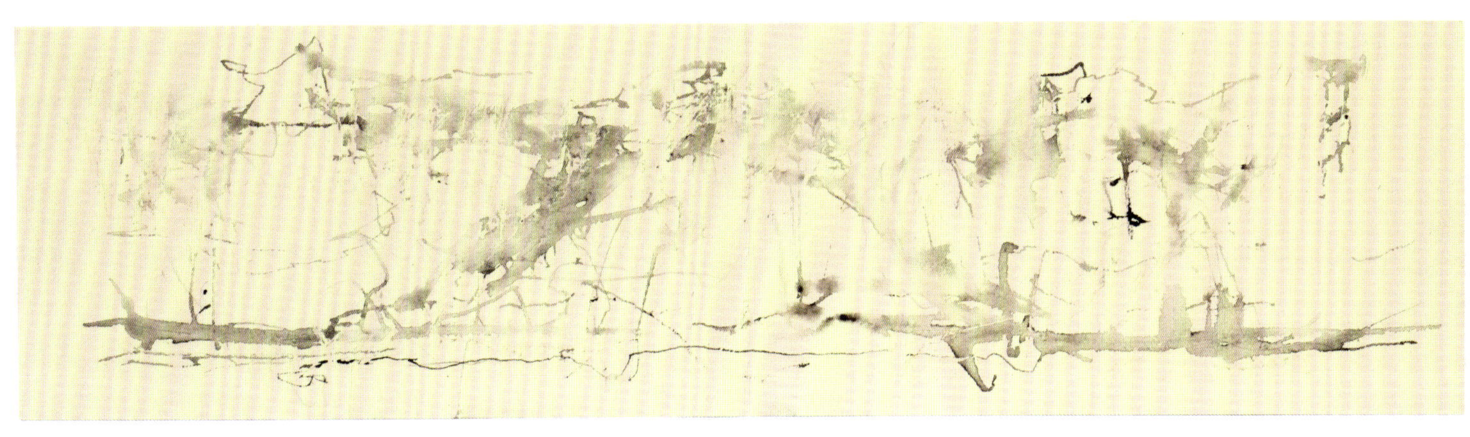

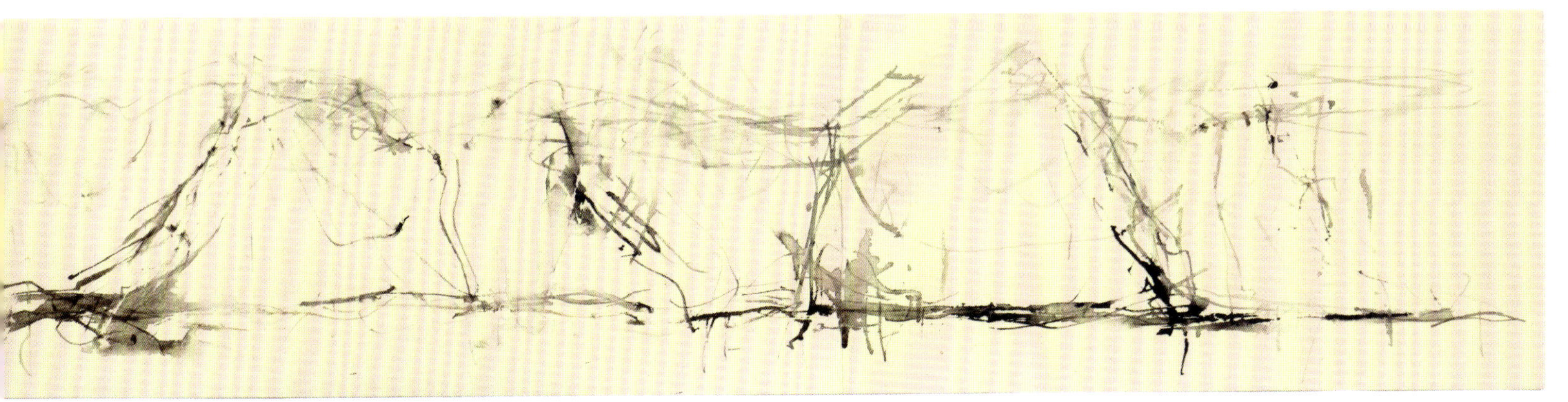

53 *Untitled J47*, 1995
mixed media on paper
29½ × 28¾ in. (75 × 73 cm)

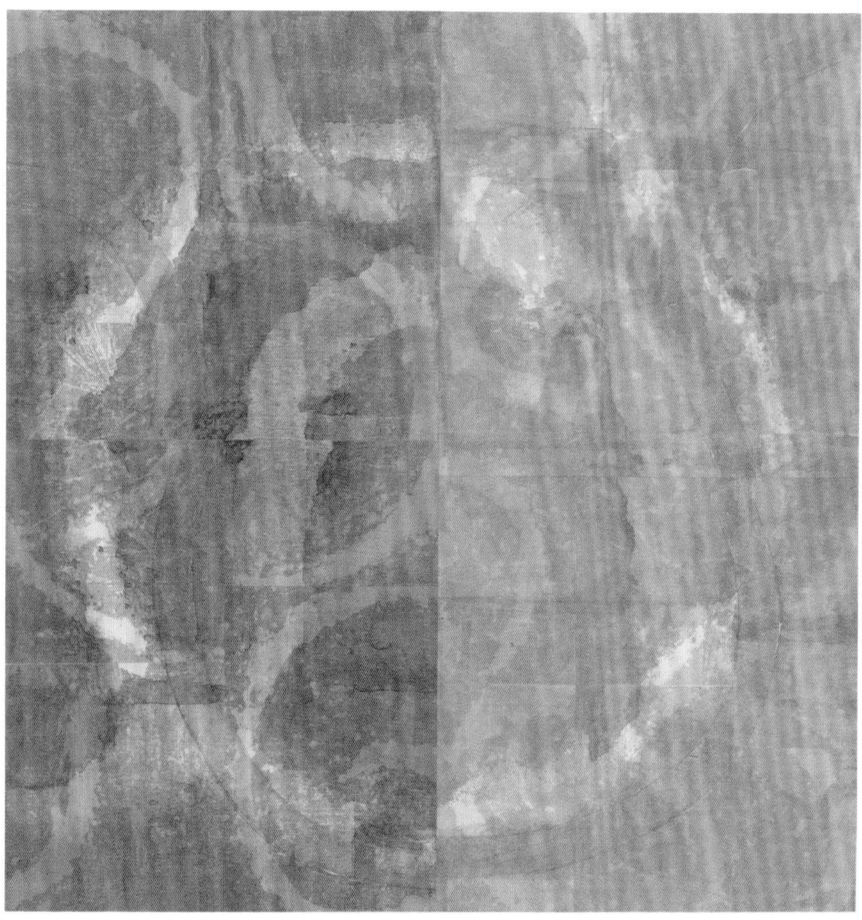

54 *Untitled J48*, 1995
mixed media on paper
29⅞ × 44⅛ in. (76 × 112 cm)

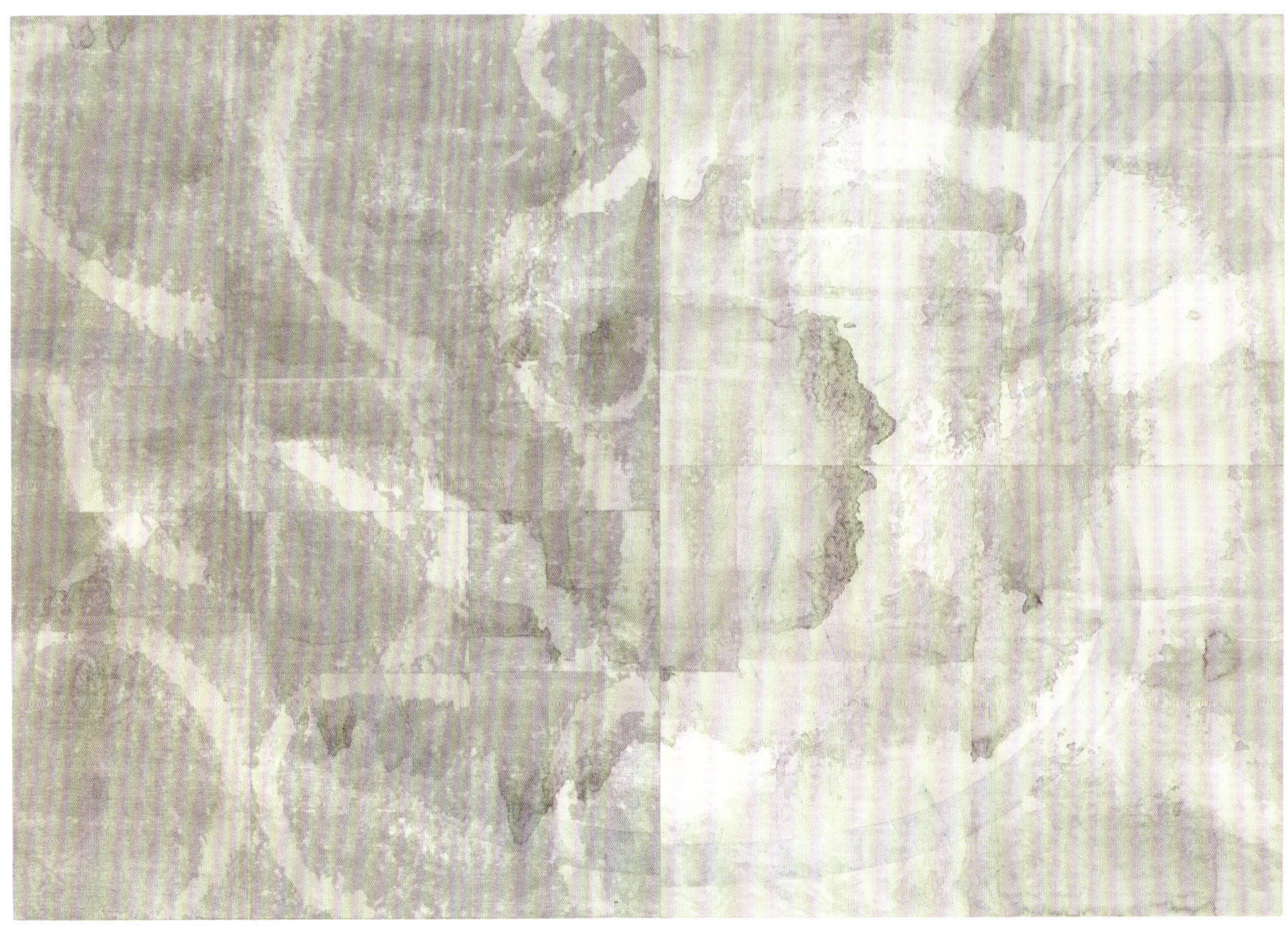

55 *Untitled K20*, 1996
mixed media on canvas
29 × 39 in. (73.7 × 99.1 cm)

56 *Arc I*, 1995
woodblock on Japanese paper
22½ × 22½ in. (57 × 57 cm)

57 *Arc II*, 1995
woodblock on Japanese paper
22½ × 22½ in. (57 × 57 cm)

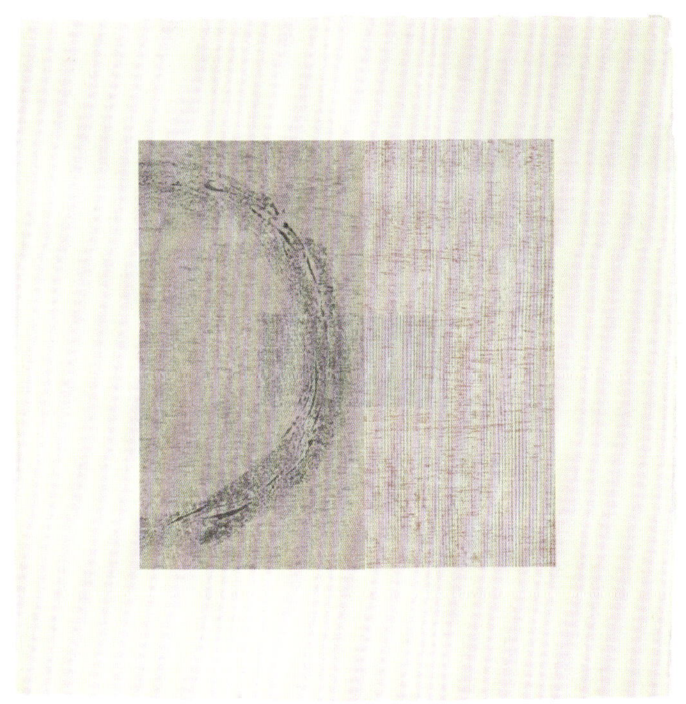

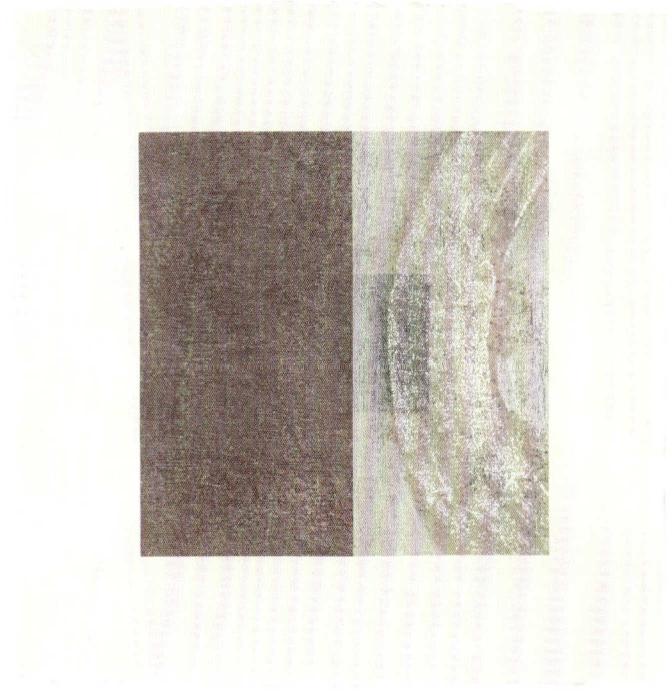

58 *Untitled J72*, 1994
mixed media on canvas
72 × 73 in. (182.8 × 185.4 cm)
Diane Davies

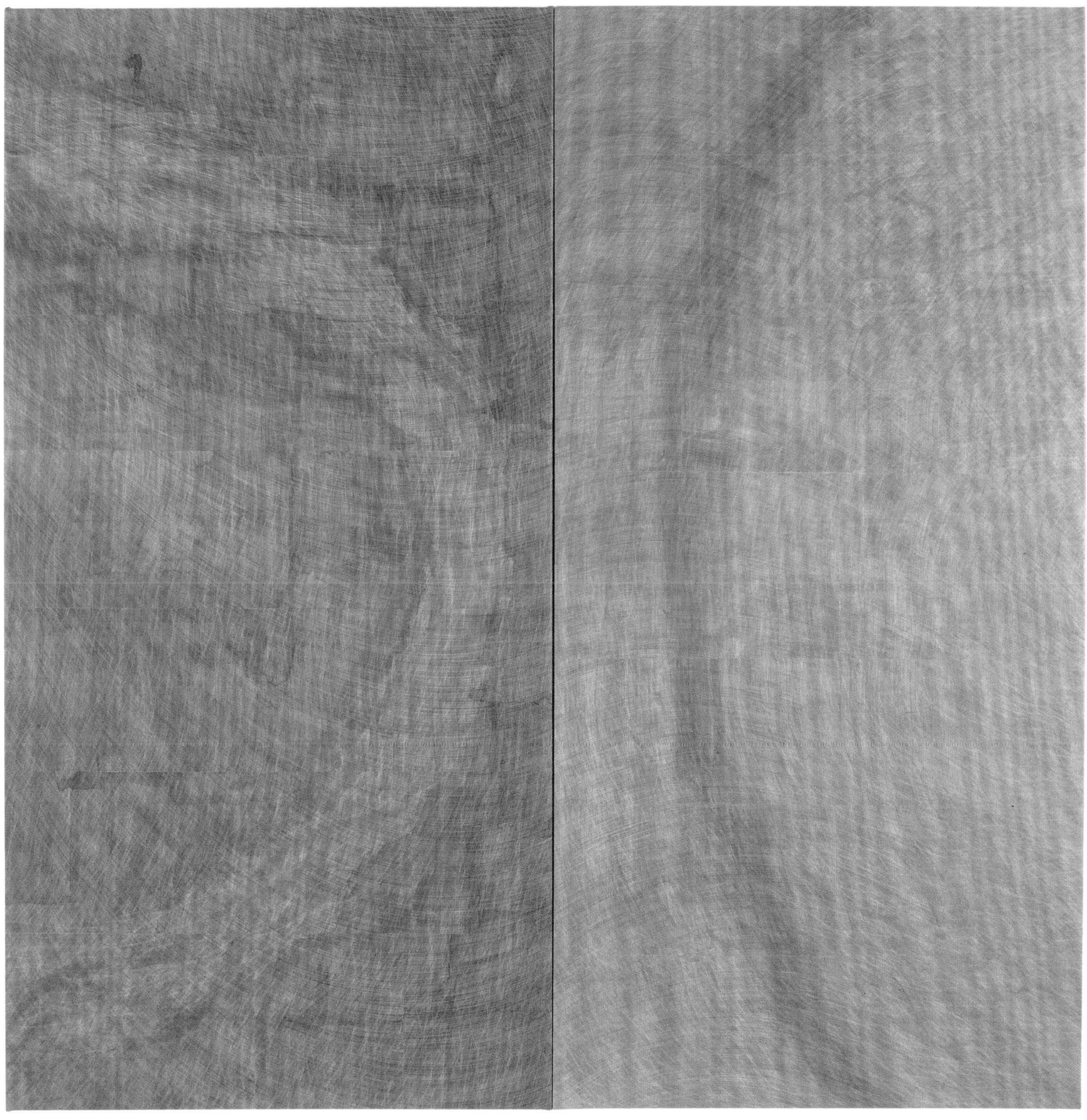

59 *Untitled K37*, 1996
mixed media on canvas
72 × 96 in. (182.9 × 243.8 cm)

60 *Untitled K78*, 1996
mixed media on paper
44⅛ × 60⅝ in. (112 × 154 cm)

61 *Untitled M20*, 1997
mixed media on canvas
48 × 54 in. (121.9 × 137.2 cm)

62 *March Monoprint I*, 1998
woodblock on Japanese paper
15 × 16⅛ in. (38 × 41 cm)

63 *March Monoprint III*, 1998
woodblock on Japanese paper
15 × 16⅛ in. (38 × 41 cm)

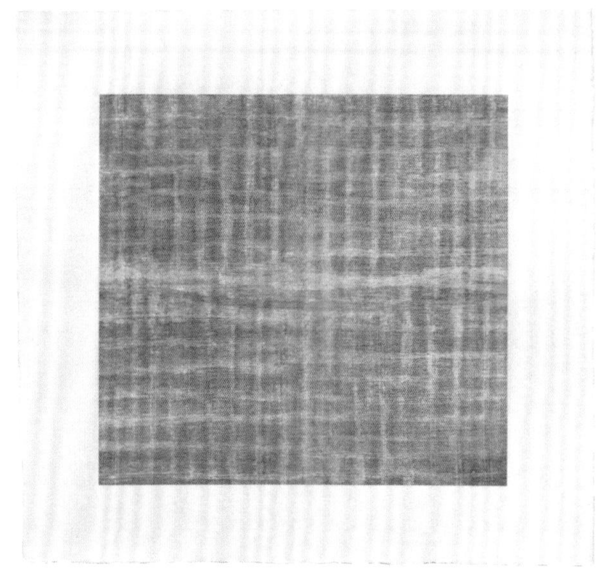

64 *March Monoprint VII*, 1998
woodblock on Japanese paper
15 × 16⅛ in. (38 × 41 cm)

65 *March Monoprint IX*, 1998
woodblock on Japanese paper
15 × 16⅛ in. (38 × 41 cm)

66 *Untitled M108*, 1998
mixed media on canvas
32 × 34 in. (81.3 × 86.4 cm)

67 *Untitled M43*, 1997
mixed media on canvas
32 × 36 in. (81.3 × 91.4 cm)
Lea Babcock Scherer and
Jeffrey A. Scherer

68 *Untitled M72*, 1997
mixed media on canvas
68 × 72 in. (172.7 × 182.9 cm)
Private collection

69 *Untitled M134*, 1998
mixed media on canvas
66 × 66 in. (167.6 × 167.6 cm)

70 *Sana'a*, 1998
watercolor on paper
5¾ × 8¼ in. (14.6 × 21 cm)

71 *Sana'a*, 1998
watercolor on paper
5½ × 6⅛ in. (14 × 15.6 cm)

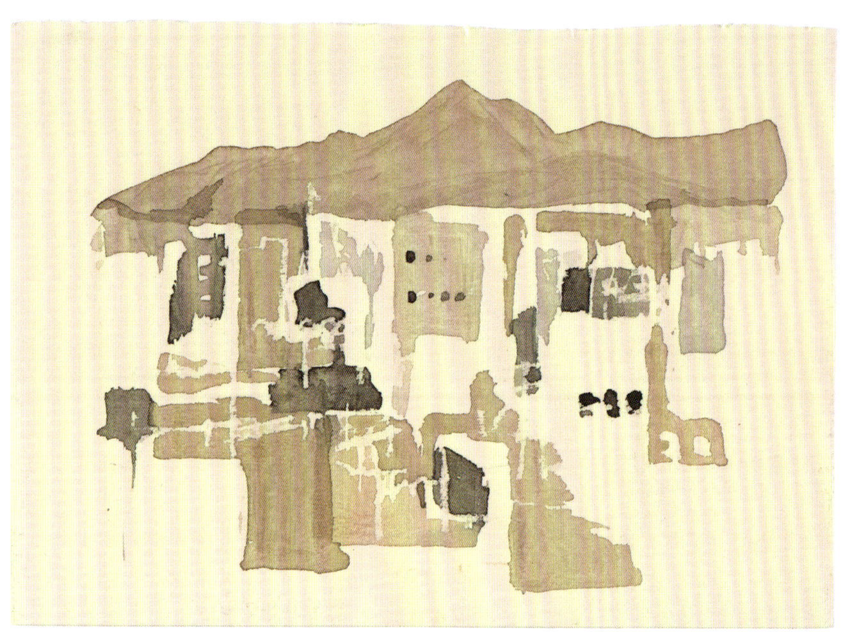

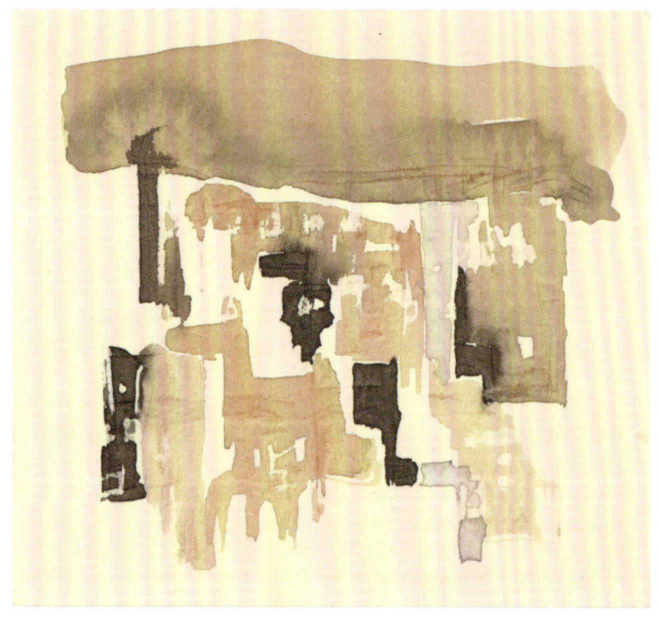

72 *Shibam*, 1998
watercolor on paper
5¼ × 7 in. (13.3 × 17.8 cm)

73 *Shibam*, 1998
watercolor on paper
5¼ × 7 in. (13.3 × 17.8 cm)

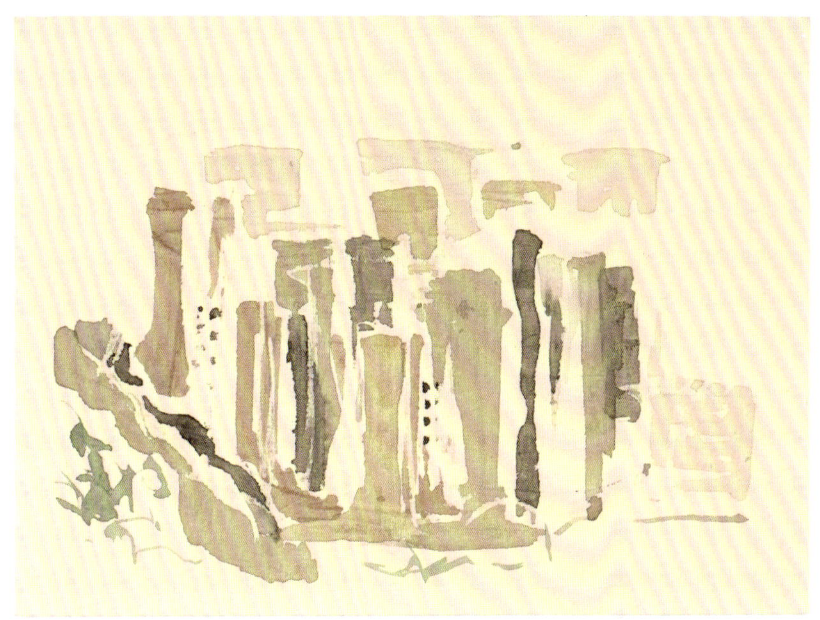

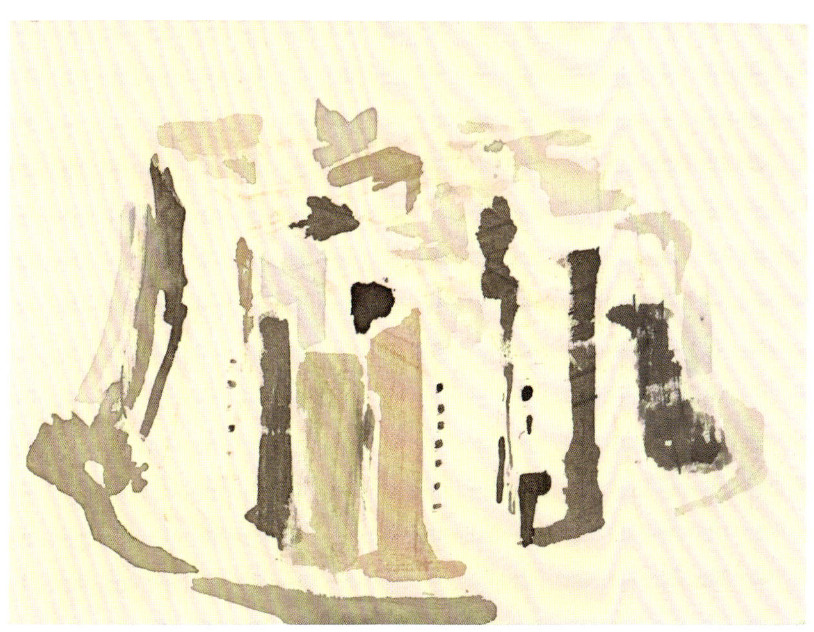

74 *Untitled R72*, 1999
mixed media on canvas
72 × 61 in. (182.9 × 154.9 cm)

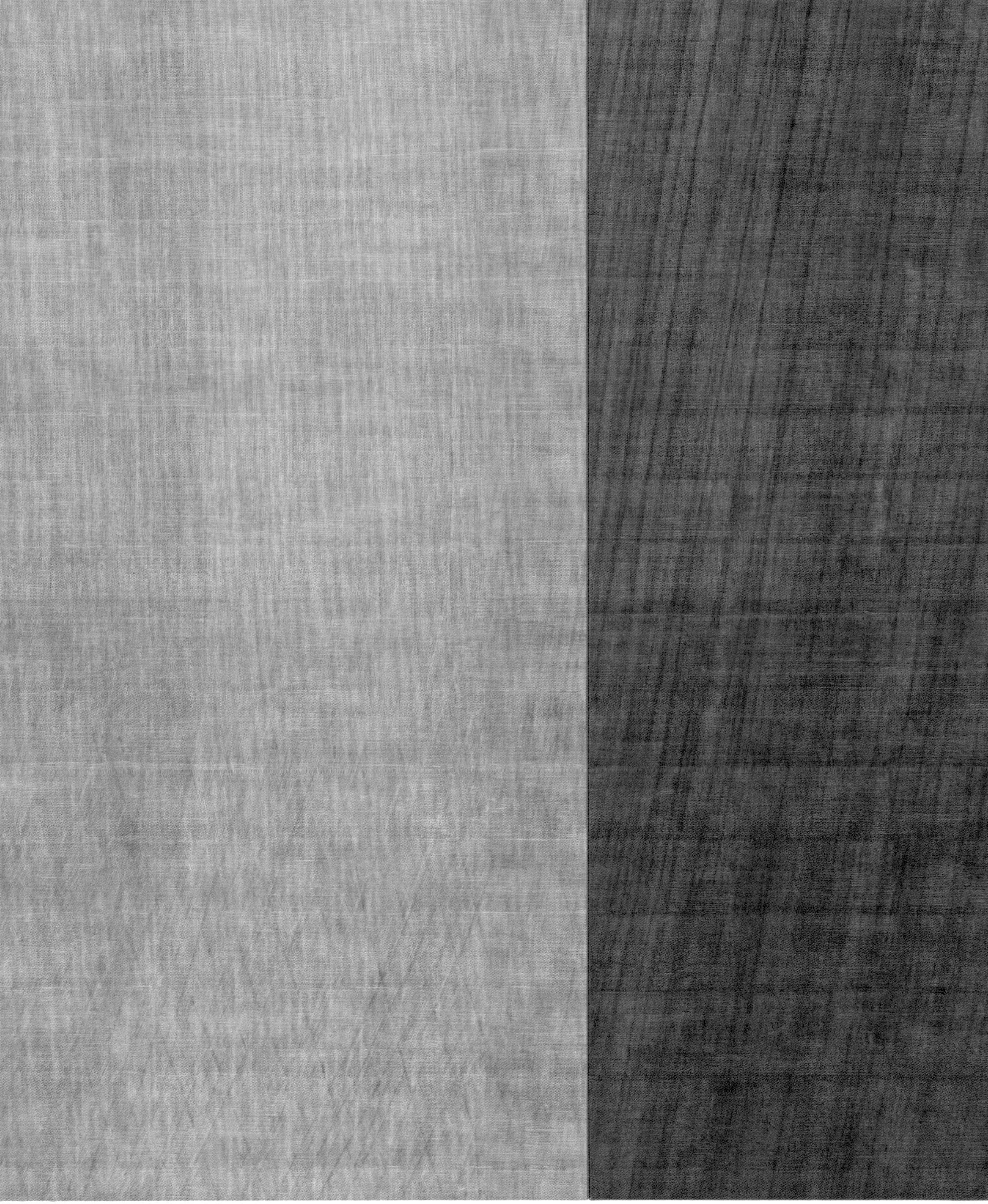

75 *Untitled R76*, 1999
mixed media on canvas
21 × 18 in. (53.3 × 45.7 cm)
Private collection

76 *Untitled AA29*, 2000
mixed media on canvas
22½ × 19 in. (57.2 × 48.3 cm)
Gill and Steve Evans

77 *Untitled AA3*, 2000
mixed media on canvas
96 × 34 in. (243.8 × 86.4 cm)

78 *Untitled AA52*, 2000
mixed media on canvas
96 × 34 in. (243.8 × 86.4 cm)

79 *Untitled BB32*, 2001
mixed media on canvas
28 × 7 in. (71.1 × 17.8 cm)
Private collection, Cambridge, UK

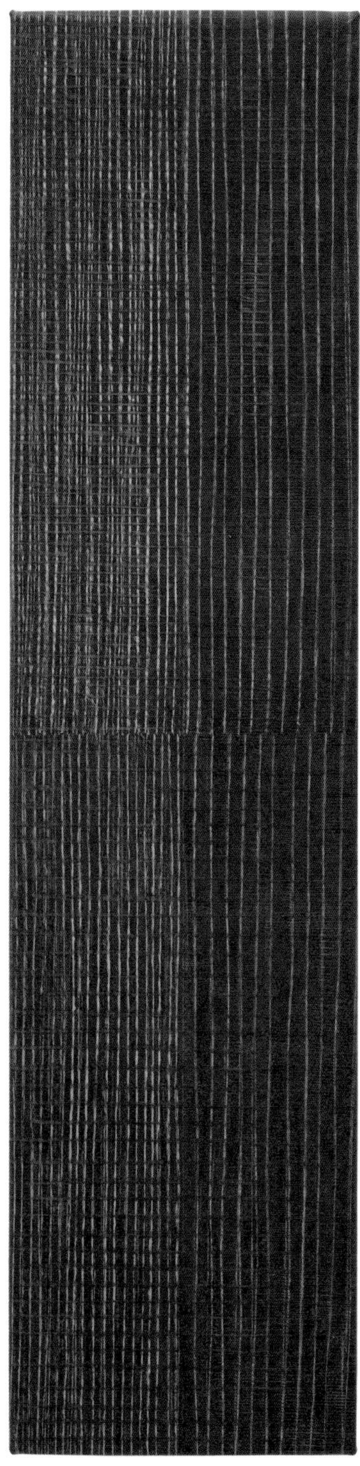

80 *Untitled BB45*, 2001
mixed media on canvas
28 × 7 in. (71.1 × 17.8 cm)
Private collection, Cambridge, UK

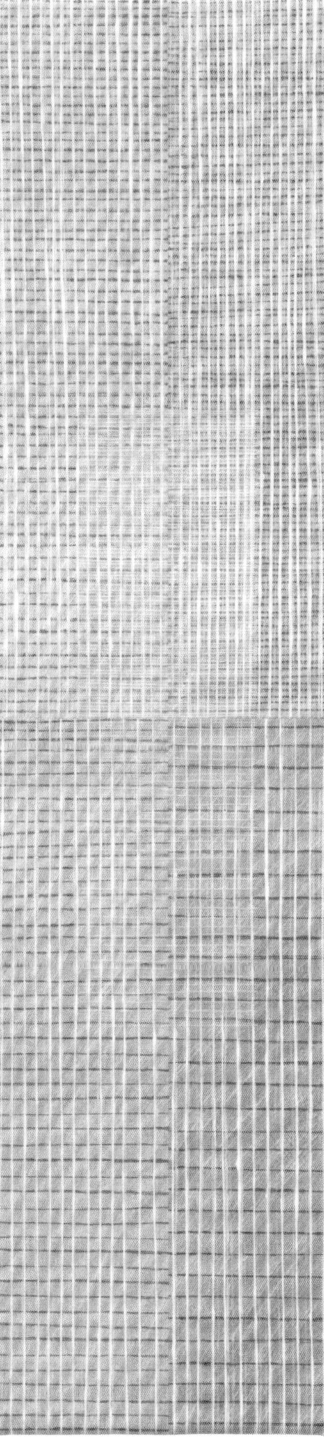

81 *Untitled BB47*, 2001
mixed media on canvas
50 × 72 in. (127 × 182.9 cm)
Collection of Holly Lennihan
and Robert Cox, Washington, DC

82 *Untitled BB60*, 2001
mixed media on canvas
10 × 9 in. (25.4 × 22.9 cm)
Private collection

83 *Untitled DD25*, 2003
mixed media on canvas
10 × 9 in. (25.4 × 22.9 cm)

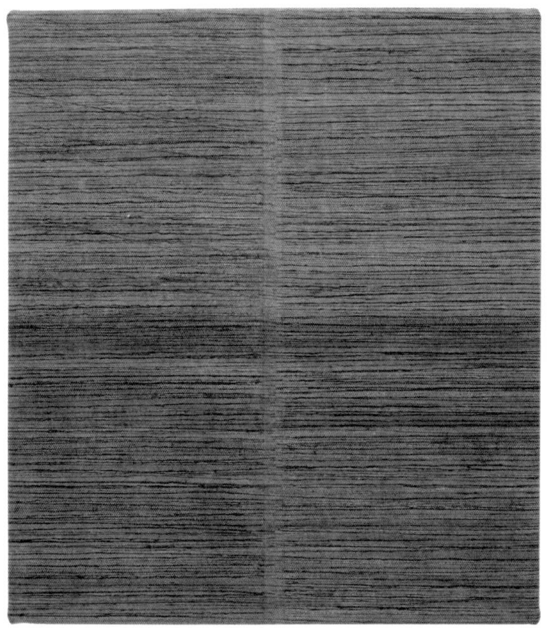

84 *Untitled DD14*, 2003
mixed media on canvas
10 × 9 in. (25 × 23 cm)

85 *Untitled JJ2*, 2006
mixed media on canvas
10 × 9 in. (25.4 × 22.9 cm)

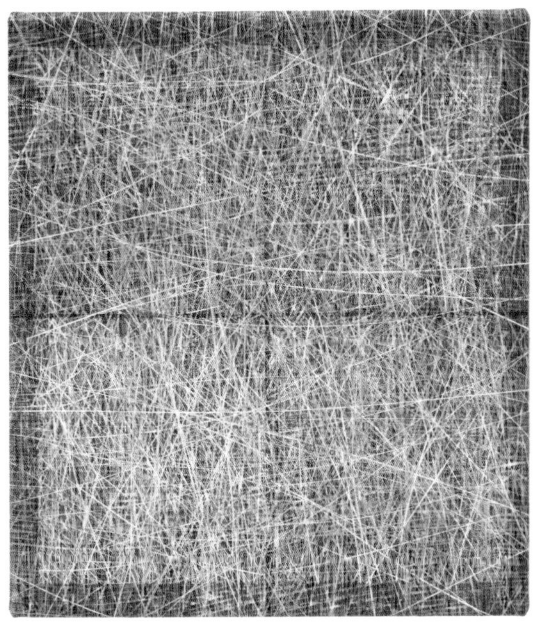

86 *Untitled CC40*, 2002
mixed media on paper
20⅞ × 31⅛ in. (53 × 79 cm)

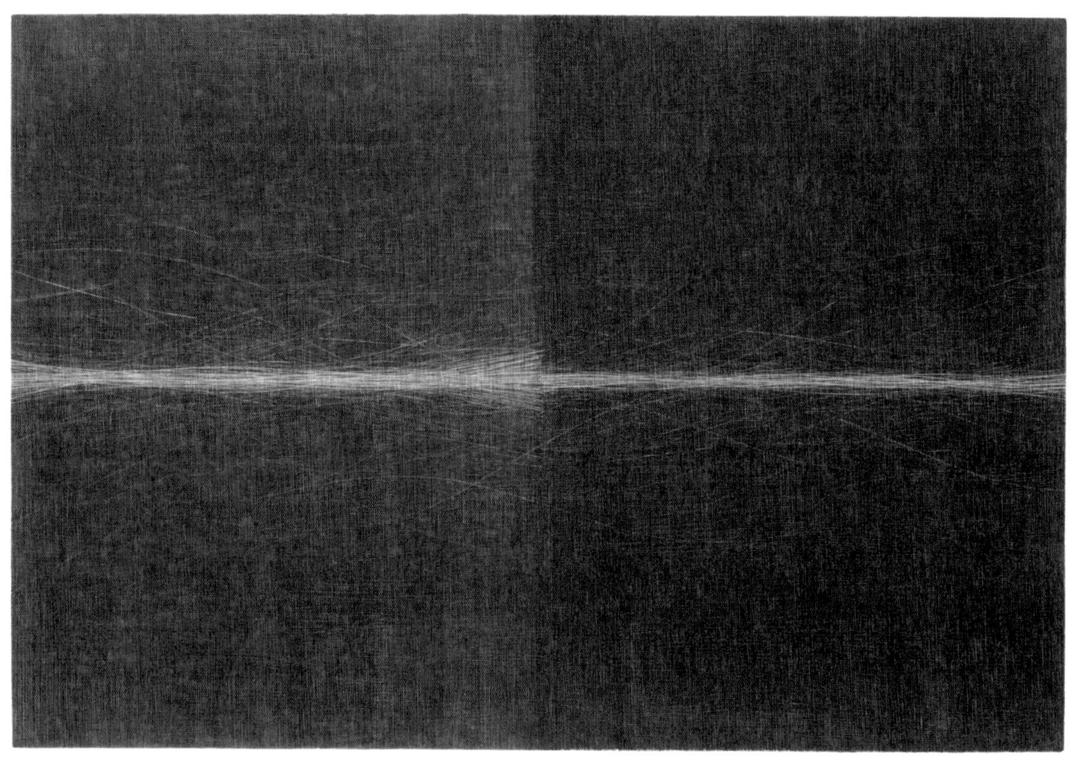

87 *Untitled CC43*, 2002
mixed media on canvas
14 × 56 in. (35.6 × 142.2 cm)

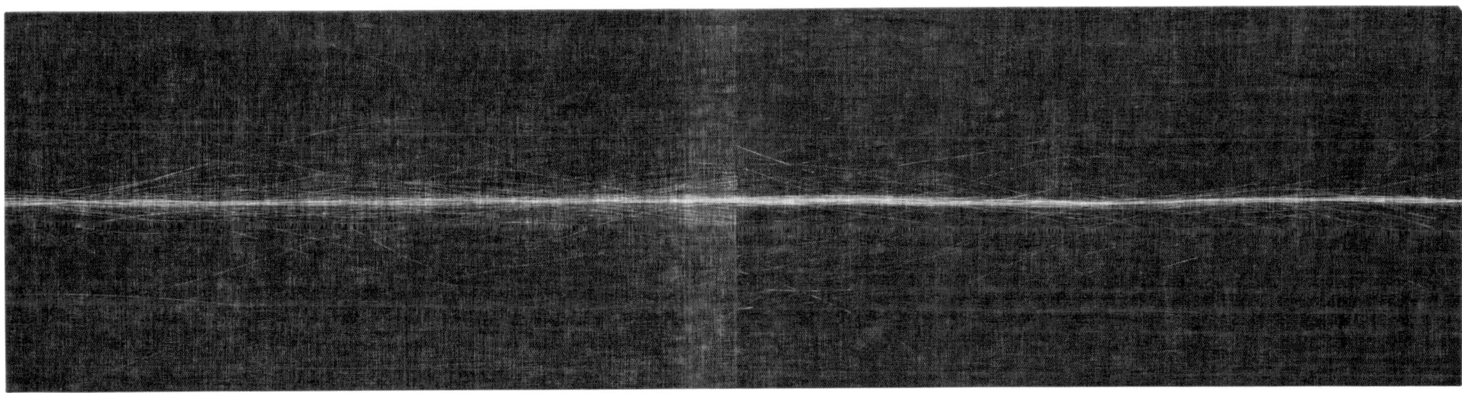

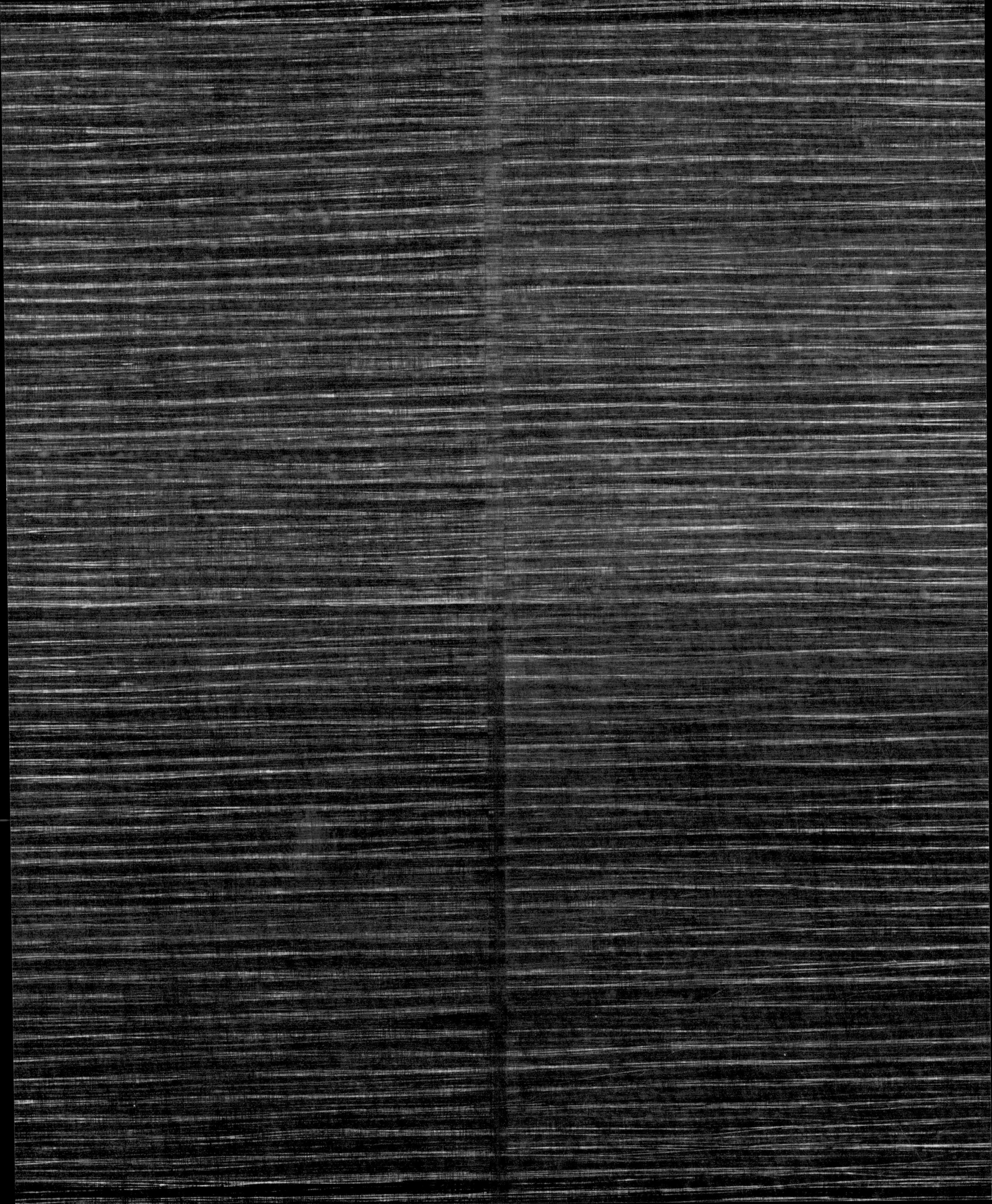

88 *Untitled CC45*, 2002
mixed media on canvas
36 × 35½ in. (91.4 × 90.2 cm)

89 *Untitled DD1*, 2003
mixed media on paper
39 × 59⅞ in. (99 × 152 cm)

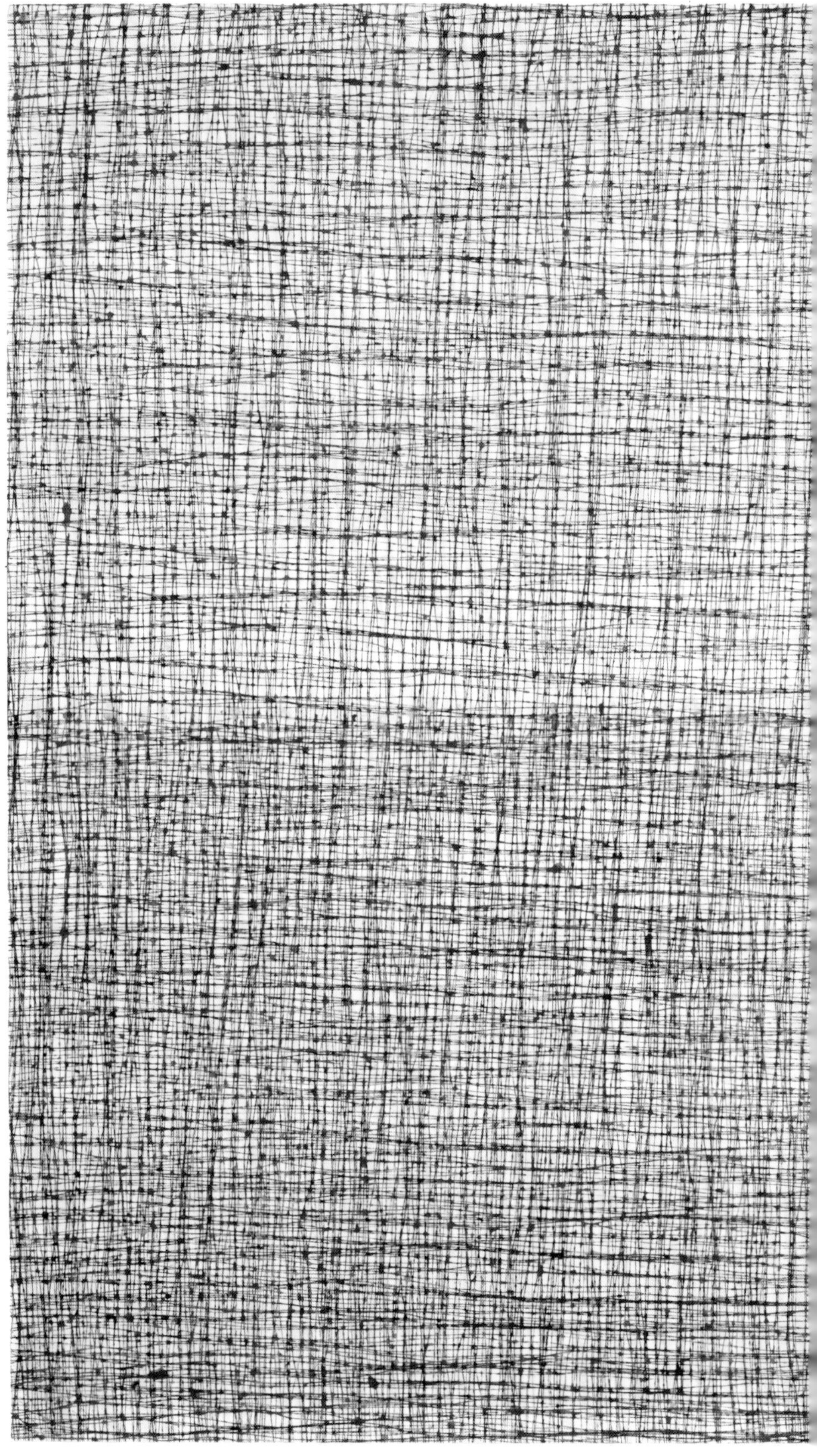

90 *Untitled DD8*, 2003
mixed media on canvas
24 × 22 in. (61 × 55.9 cm)

91 *Untitled DD21*, 2003
mixed media on canvas
24 × 20 in. (61 × 50.8 cm)

92 *Untitled DD20*, 2003
mixed media on canvas
43 × 67 in. (109.2 × 170.2 cm)
Collection of Vicky and Terry Klein

93 *Bethany 3*, 2003
mixed media on paper
19¼ × 25¼ in. (49 × 64 cm)

94 *Bethany 6*, 2003
mixed media on paper
19¼ × 25¼ in. (49 × 64 cm)

95 *Bethany 8*, 2003
mixed media on paper
19¼ × 25¼ in. (49 × 64 cm)

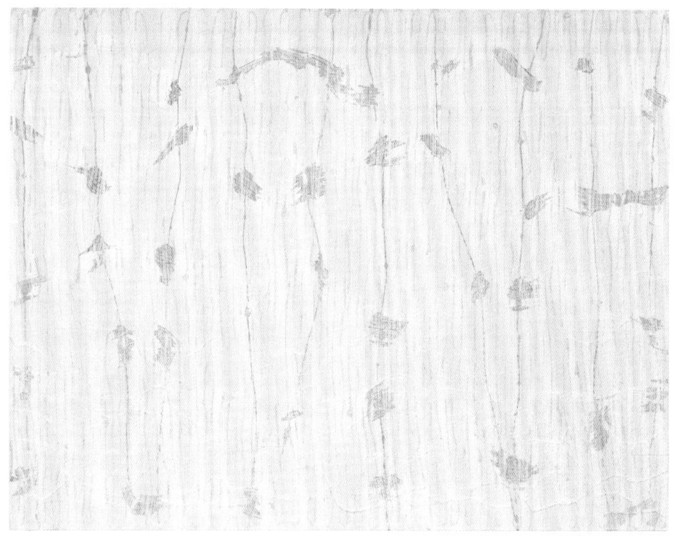

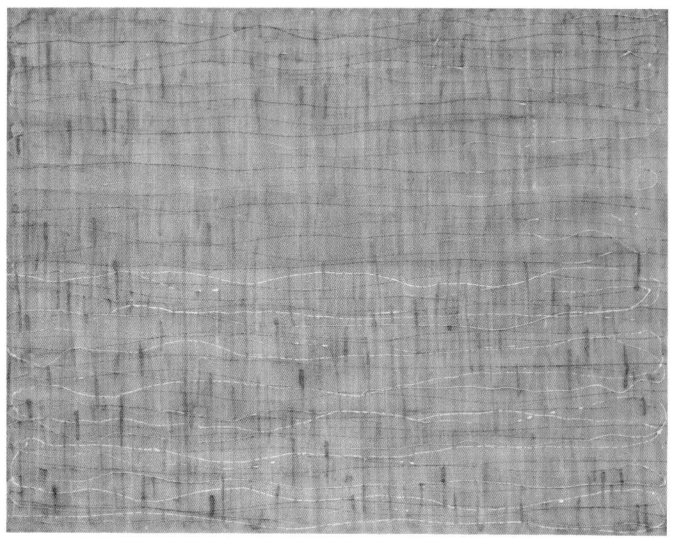

96 *Bethany 10*, 2003
mixed media on paper
19¼ × 25¼ in. (49 × 64 cm)

97 *Bethany 11*, 2003
mixed media on paper
19¼ × 25¼ in. (49 × 64 cm)

98 *Bethany 21*, 2003
mixed media on paper
19¼ × 25¼ in. (49 × 64 cm)

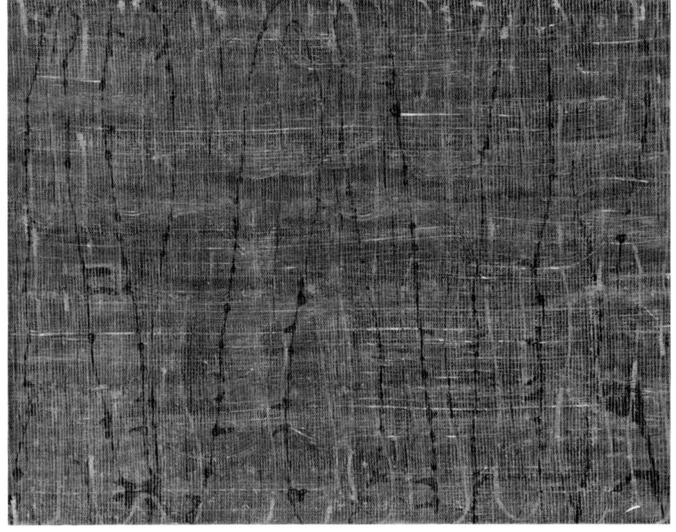

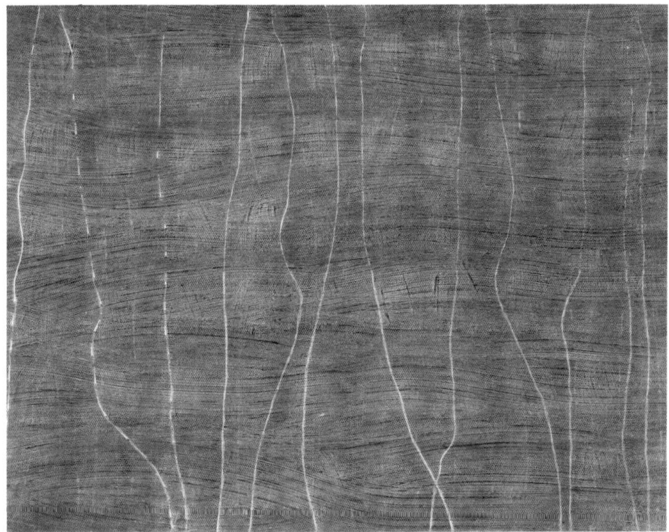

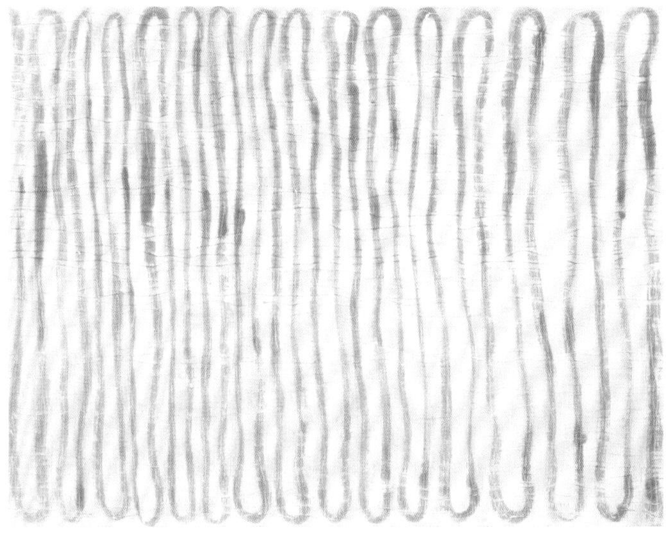

99 *Bethany Squares*, 2003
mixed media on paper mounted
on aluminum
each square: 9 × 9 in. (23 × 23 cm)
Yale Center for British Art, Friends of
British Art Fund

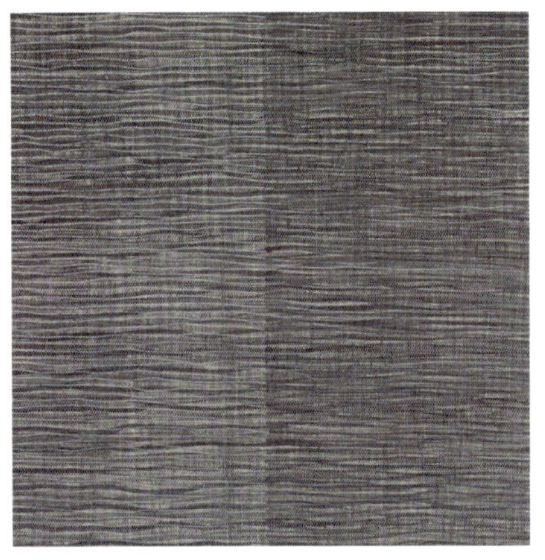
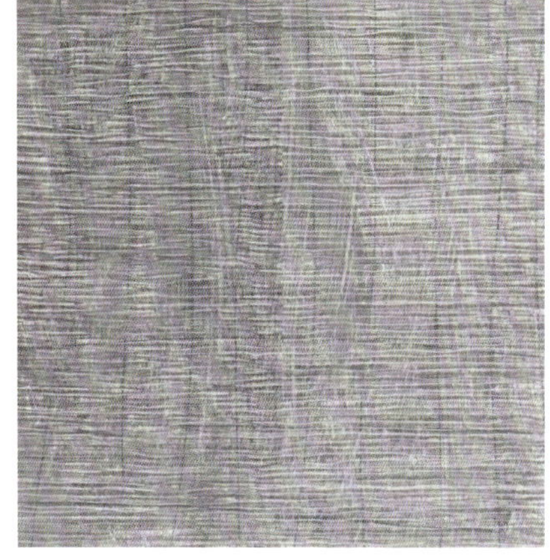
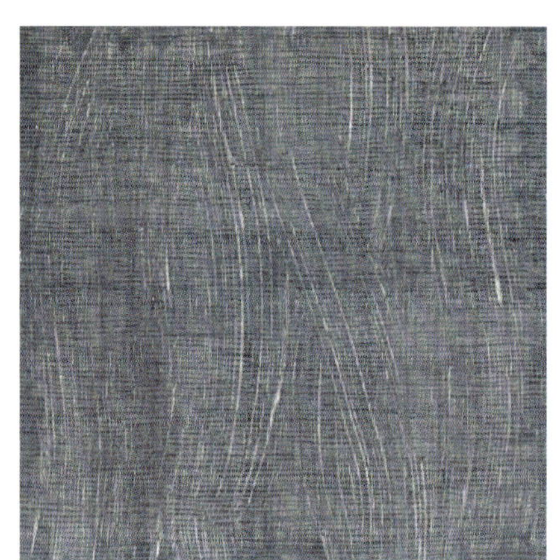

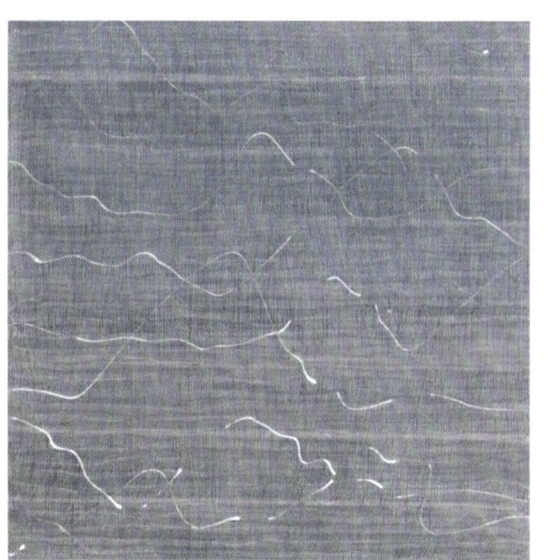

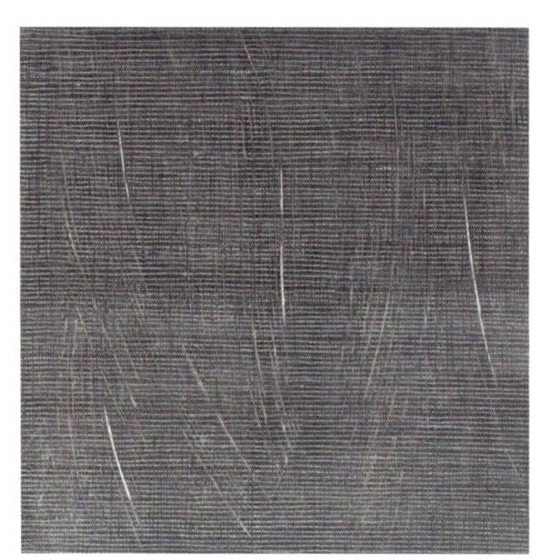

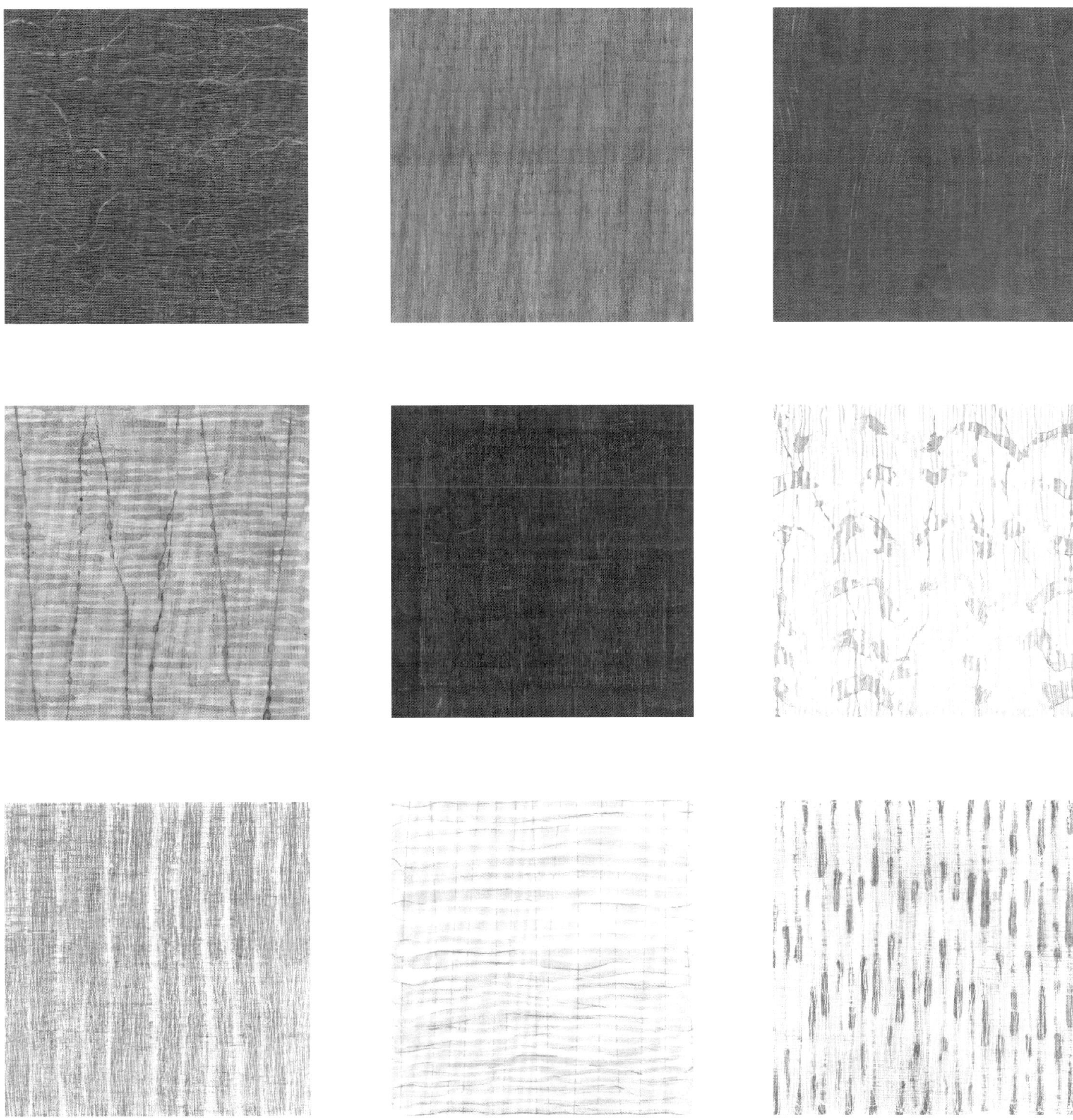

100 *Untitled EE12*, 2004
mixed media on canvas
36 × 36 in. (91.4 × 91.4 cm)

101 *Untitled EE13*, 2004
mixed media on canvas
36 × 36 in. (91.4 × 91.4 cm)

102 *Untitled HH4*, 2005
mixed media on linen
30 × 30 in. (76.2 × 76.2 cm)
Private collection

103 *Untitled HH5*, 2005
mixed media on linen
38 × 44 in. (96.5 × 111.8 cm)
Private collection

104 *Untitled HH30*, 2005
mixed media on linen
72 × 96 in. (182.9 × 243.8 cm)

105 *Eminence I*, 2005
woodblock on tissue on
Japanese paper
10⅝ × 15 in. (27 × 38 cm)

106 *Eminence II*, 2005
woodblock on tissue on
Japanese paper
10⅝ × 15 in. (27 × 38 cm)

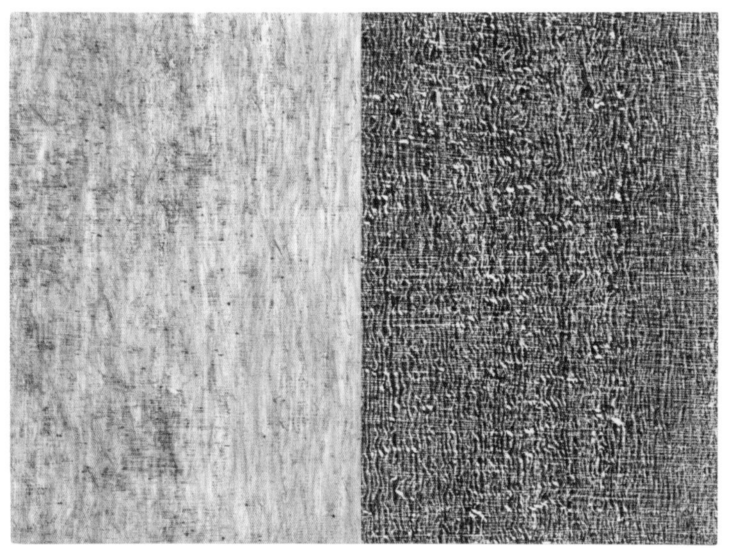

107 *Eminence III*, 2005
woodblock on tissue on
Japanese paper
10⅝ × 15 in. (27 × 38 cm)

108 *Eminence IV*, 2005
woodblock on tissue on
Japanese paper
10⅝ × 15 in. (27 × 38 cm)

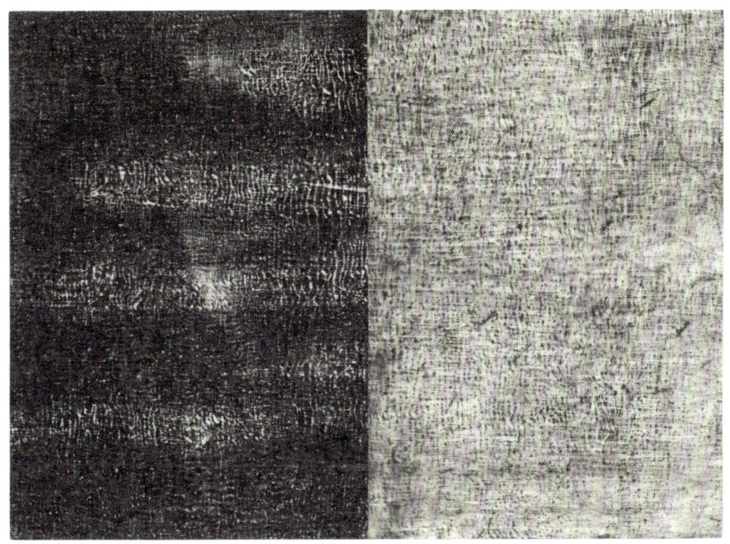

109 *Untitled JJ11*, 2006
mixed media on paper
37⅜ × 57⅛ in. (95 × 145 cm)

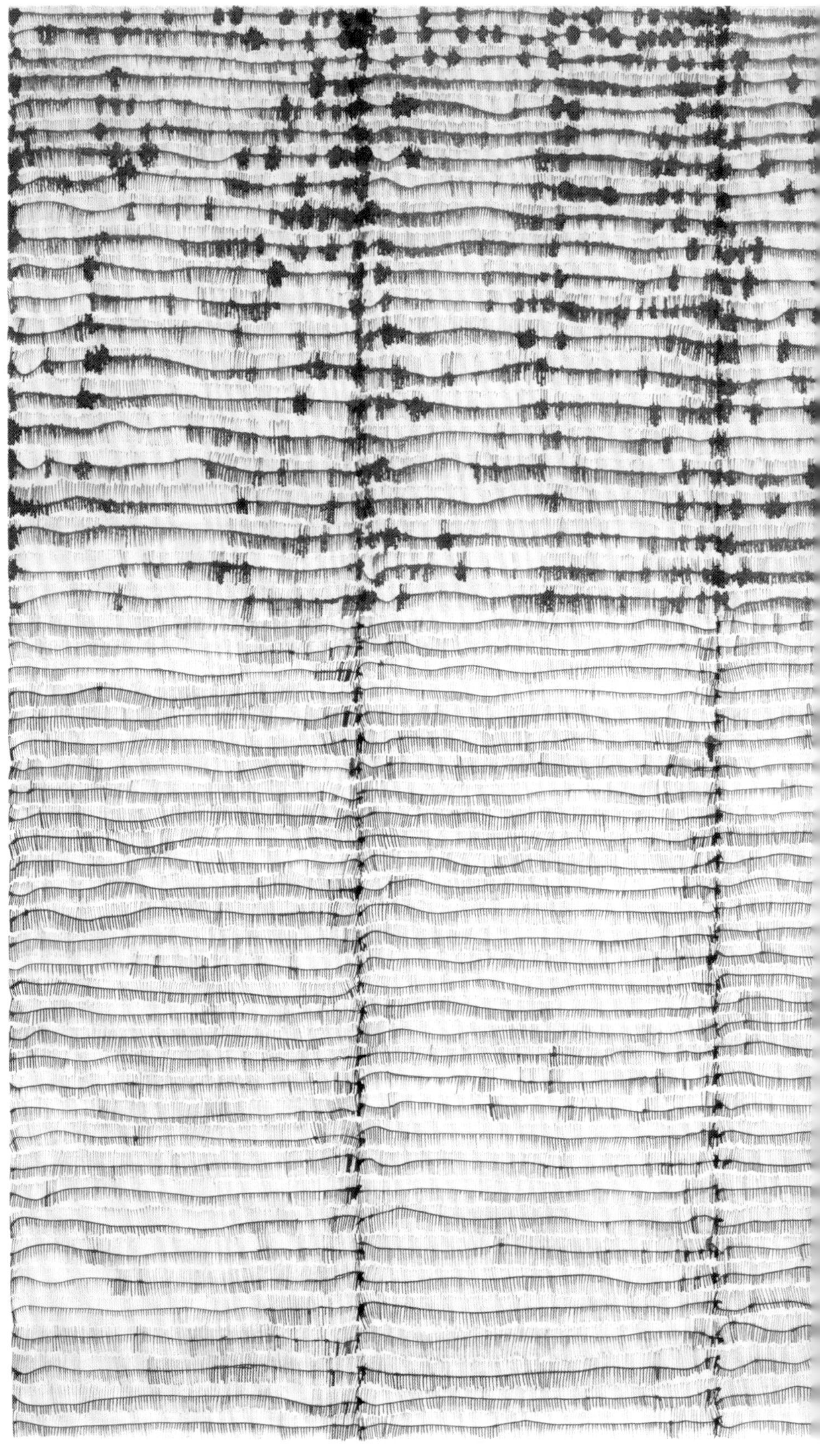

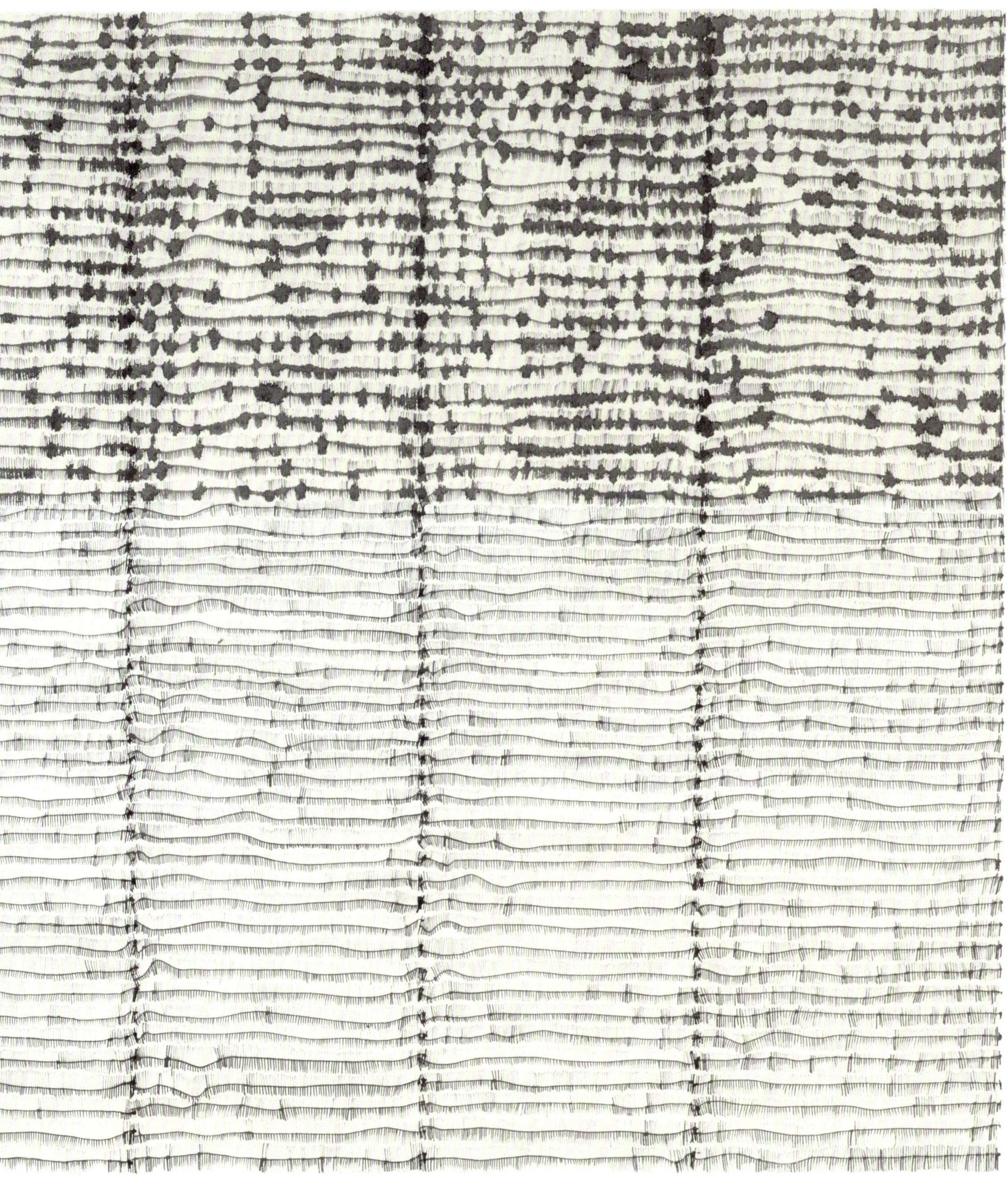

above, left to right:

110 *Untitled 2006-27*, 2006
mixed media on paper
10⅝ × 12⅝ in. (27 × 32 cm)
Andrew Lambirth Collection

111 *Untitled JJ47*, 2006
mixed media on paper
11 × 11¾ in. (28 × 30 cm)

112 *Untitled 2006-28*, 2006
mixed media on paper
14¼ × 15 in. (36 × 38 cm)

below, left to right:

113 *Untitled 2006-35*, 2006
mixed media on paper
11¼ × 11¾ in. (28.5 × 30 cm)

114 *Untitled JJ46*, 2006
mixed media on paper
11 × 11¾ in. (28 × 30 cm)

115 *Untitled 2006-29*, 2006
mixed media on paper
14¼ × 15 in. (36 × 38 cm)

116 *Untitled JJ21*, 2006
mixed media on paper
14¼ × 15 in. (36 × 38 cm)

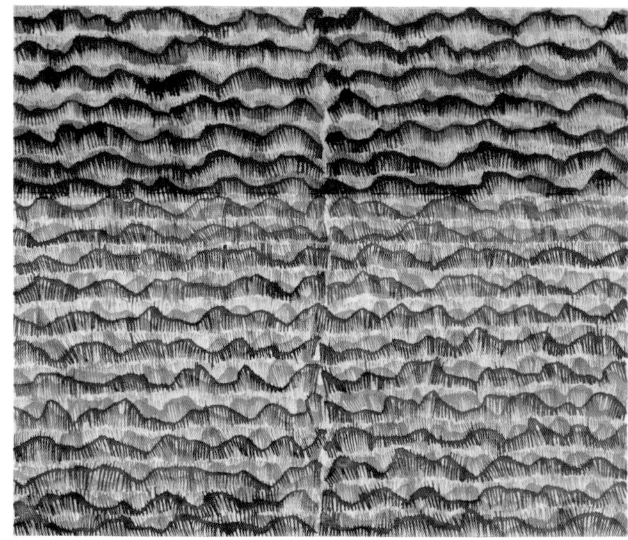

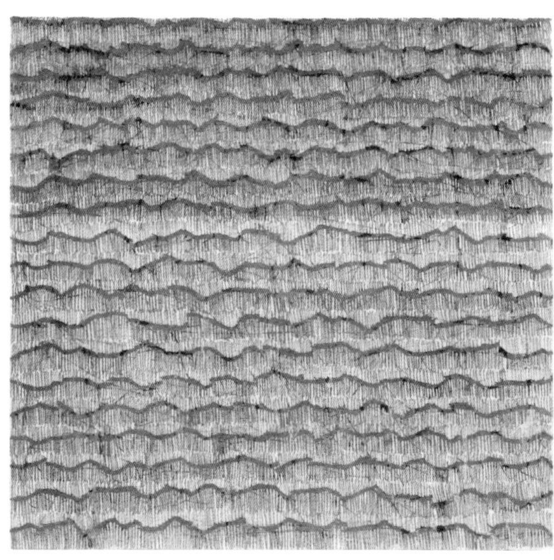

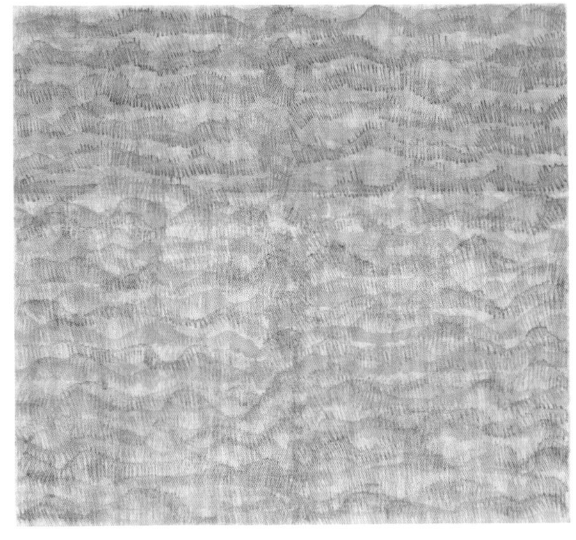

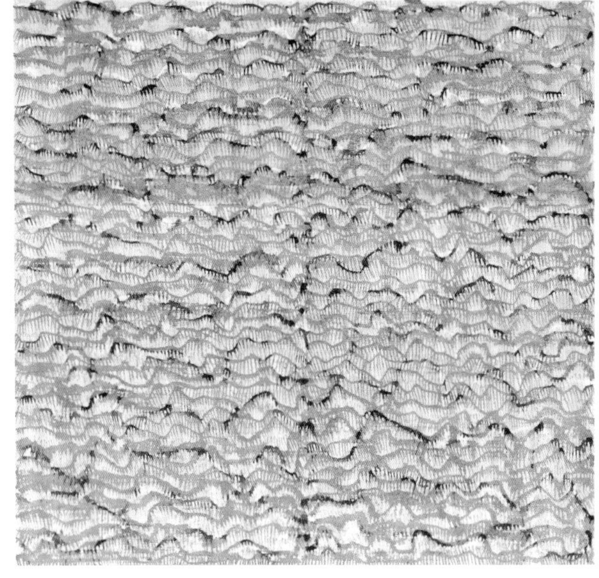

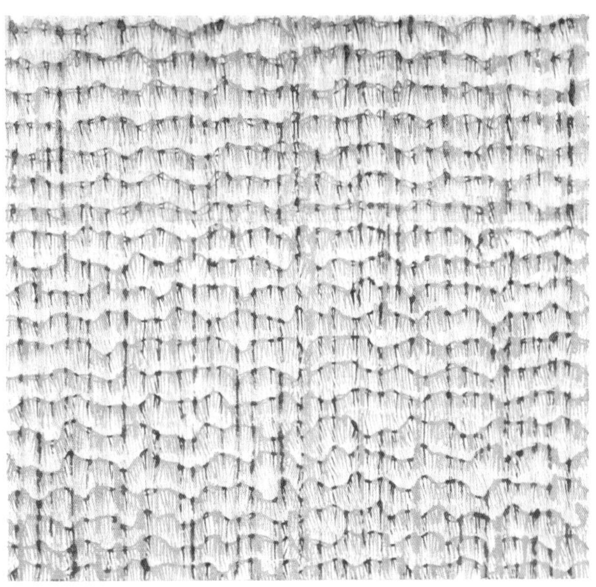

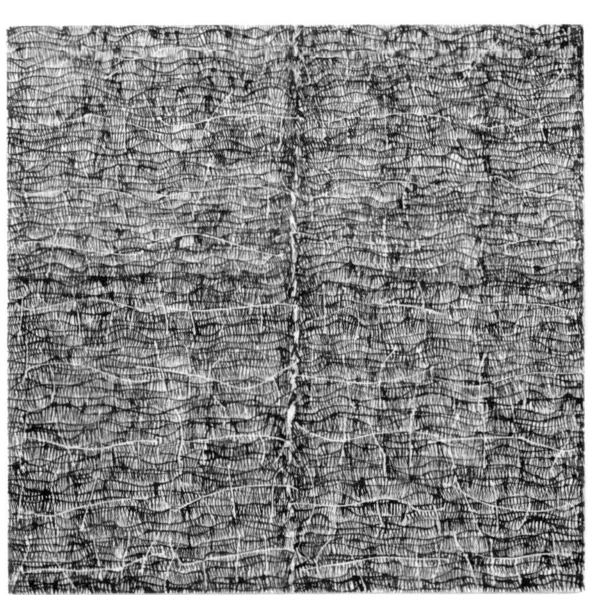

117 *Untitled KK12*, 2007
mixed media on linen
10 × 10 in. (25.4 × 25.4 cm)
Collection of Richard and
Janet Caldwell

118 *Untitled KK5*, 2007
mixed media on linen
23⅝ × 23⅝ in. (60 × 60 cm)
Howard Foote

119 *Untitled HH37*, 2005
mixed media on linen
10 × 9 in. (25.4 × 22.9 cm)
Bruce and Aphrodite Garrison

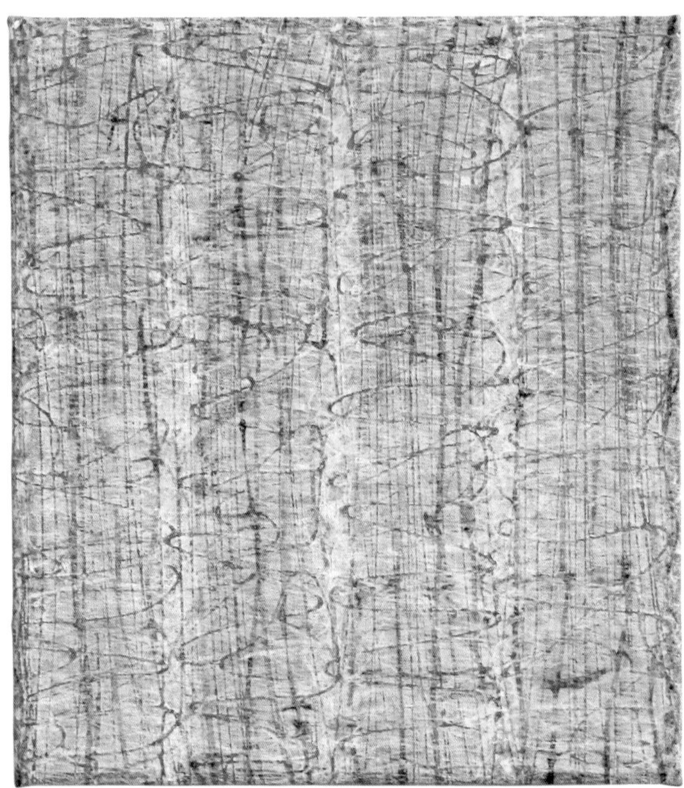

120 *Untitled KK14*, 2007
mixed media on linen
21⅝ × 23⅝ in. (55 × 60 cm)
Private collection, New York

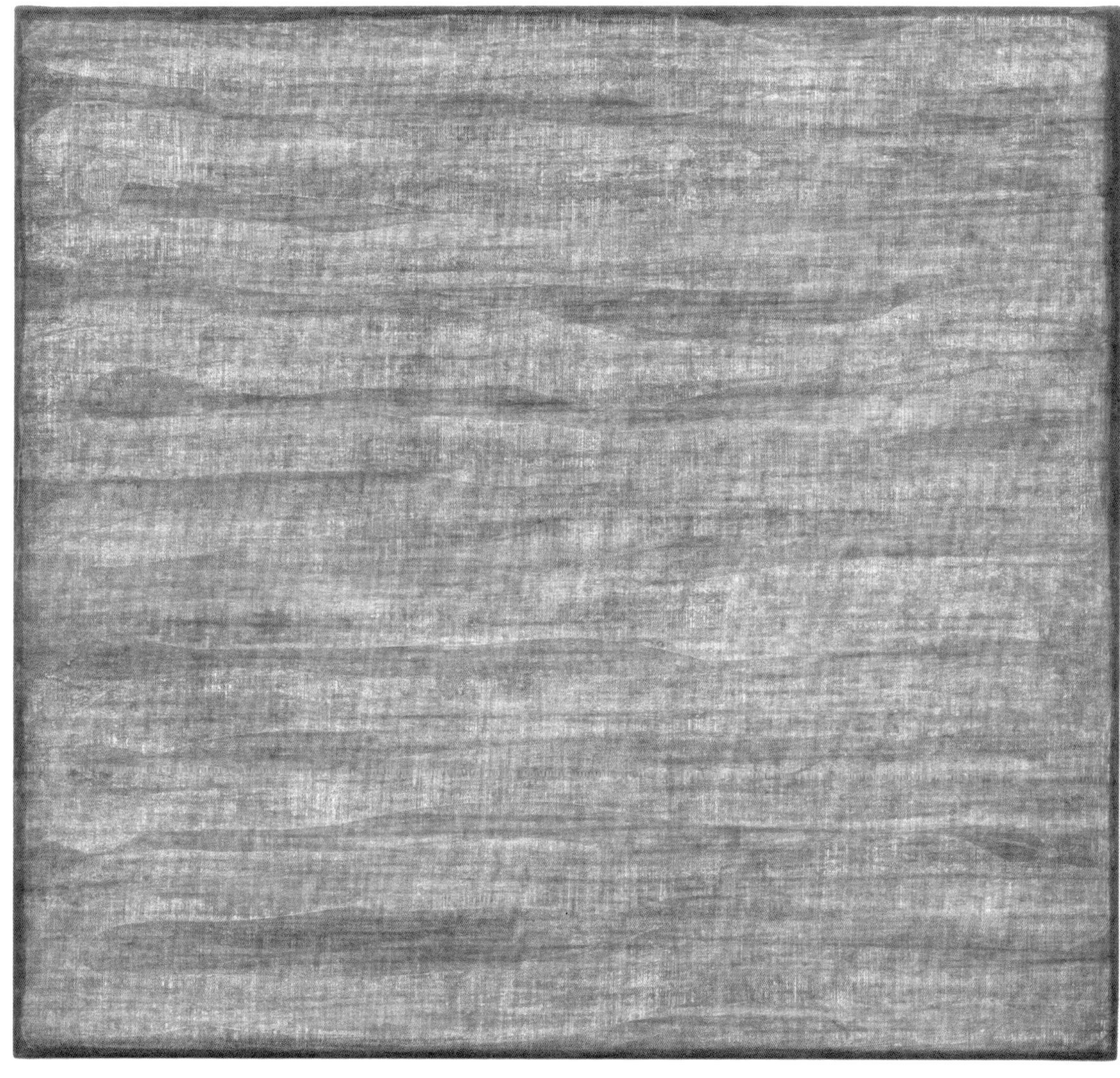

121 *Untitled 2007-18*, 2007
mixed media on paper
14⅛ × 18⅞ in. (36 × 48 cm)

122 *Untitled 2007-19*, 2007
mixed media on paper
14⅛ × 18⅞ in. (36 × 48 cm)

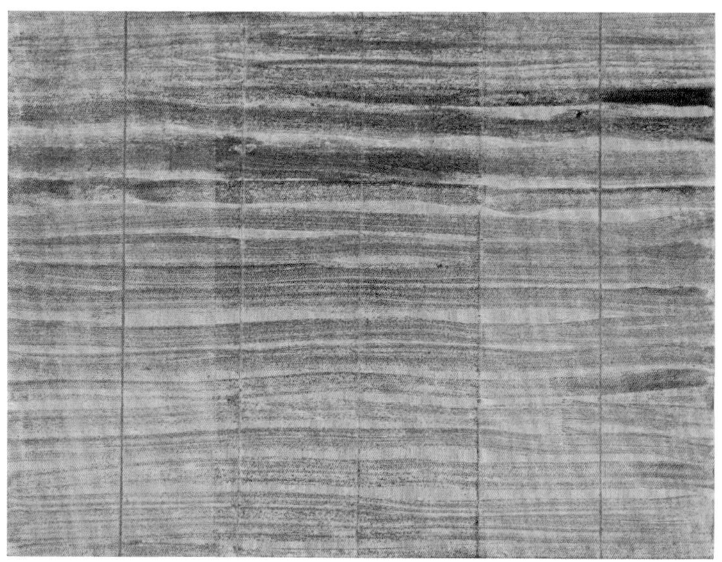

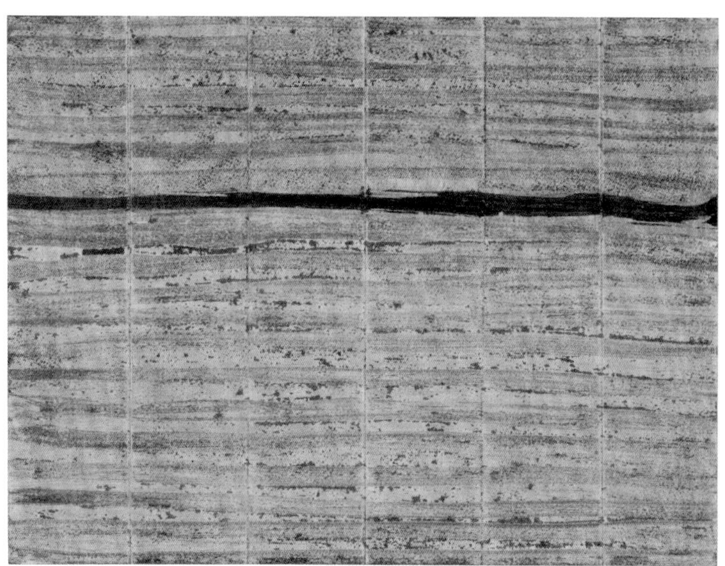

123 *Untitled KK25*, 2007
mixed media on linen
47¼ × 66⅞ in. (120 × 170 cm)
Collection of Fanchon and
Howard Hallam

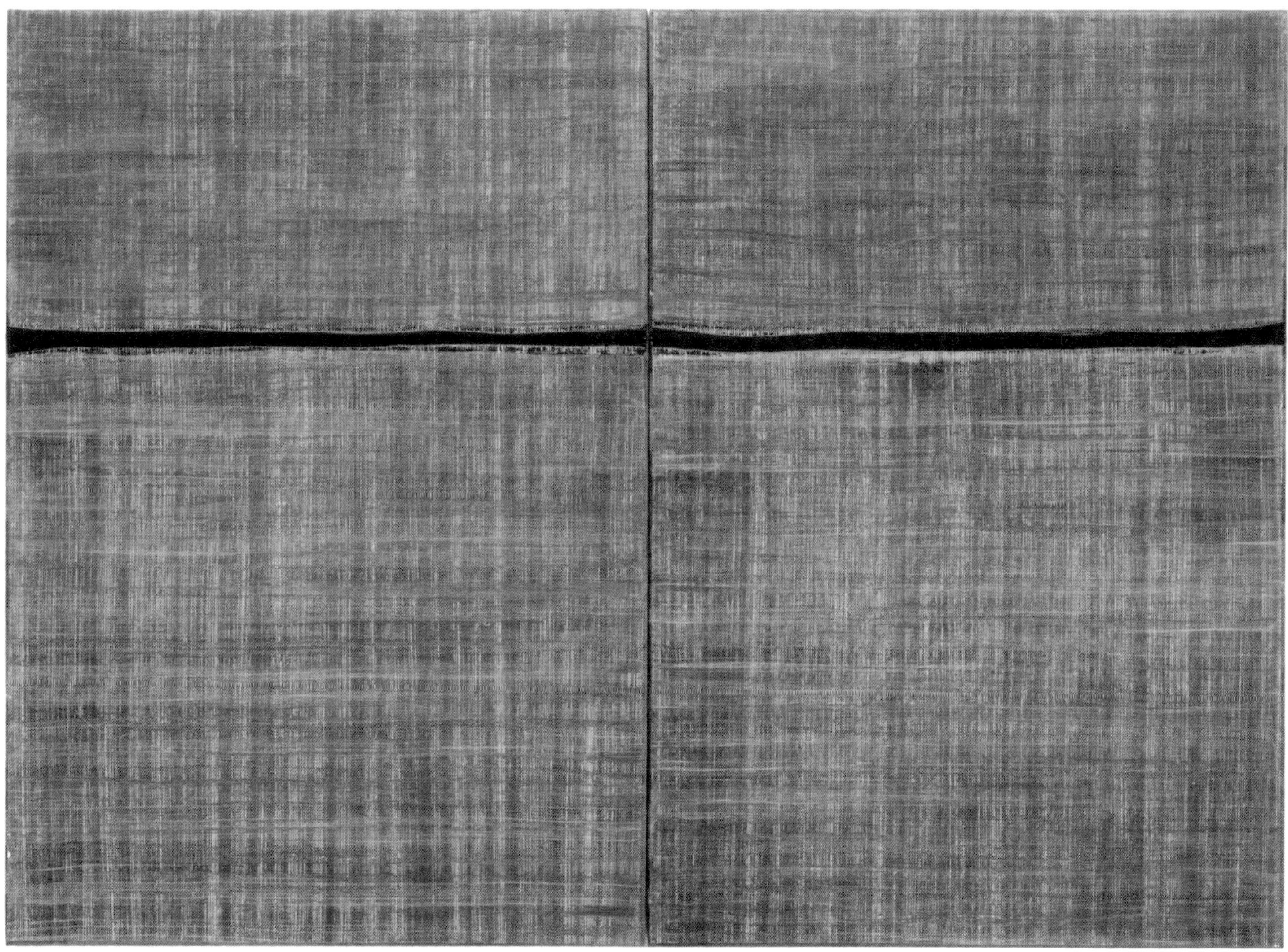

124 *Untitled, KK33*, 2007
mixed media on paper
14⅛ × 18⅞ in. (36 × 48 cm)

125 *Untitled KK40*, 2007
mixed media on paper
14⅛ × 18⅞ in. (36 × 48 cm)

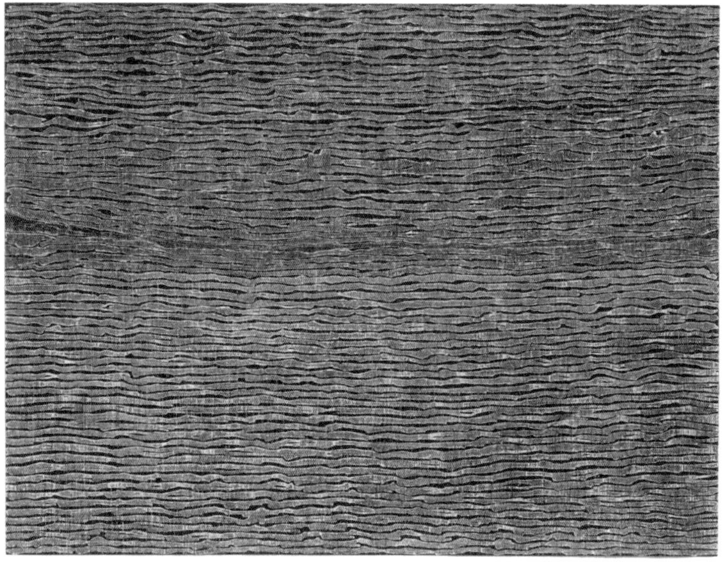

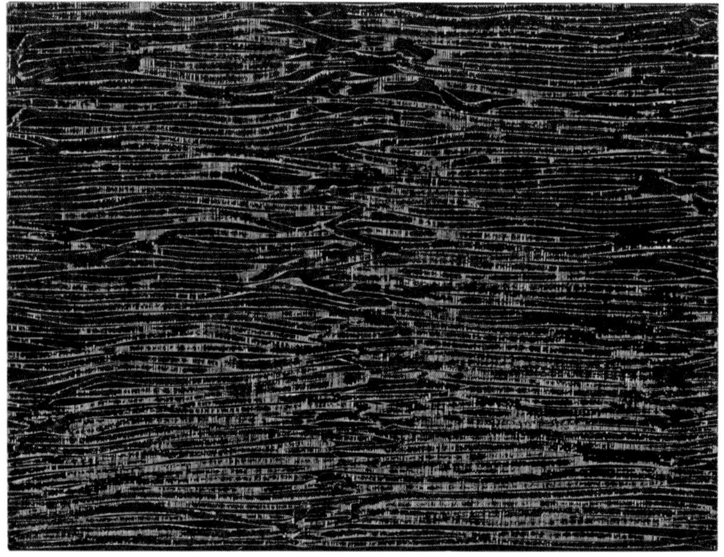

126 *Untitled KK37*, 2007
mixed media on linen
33½ × 47¼ in. (85 × 120 cm)

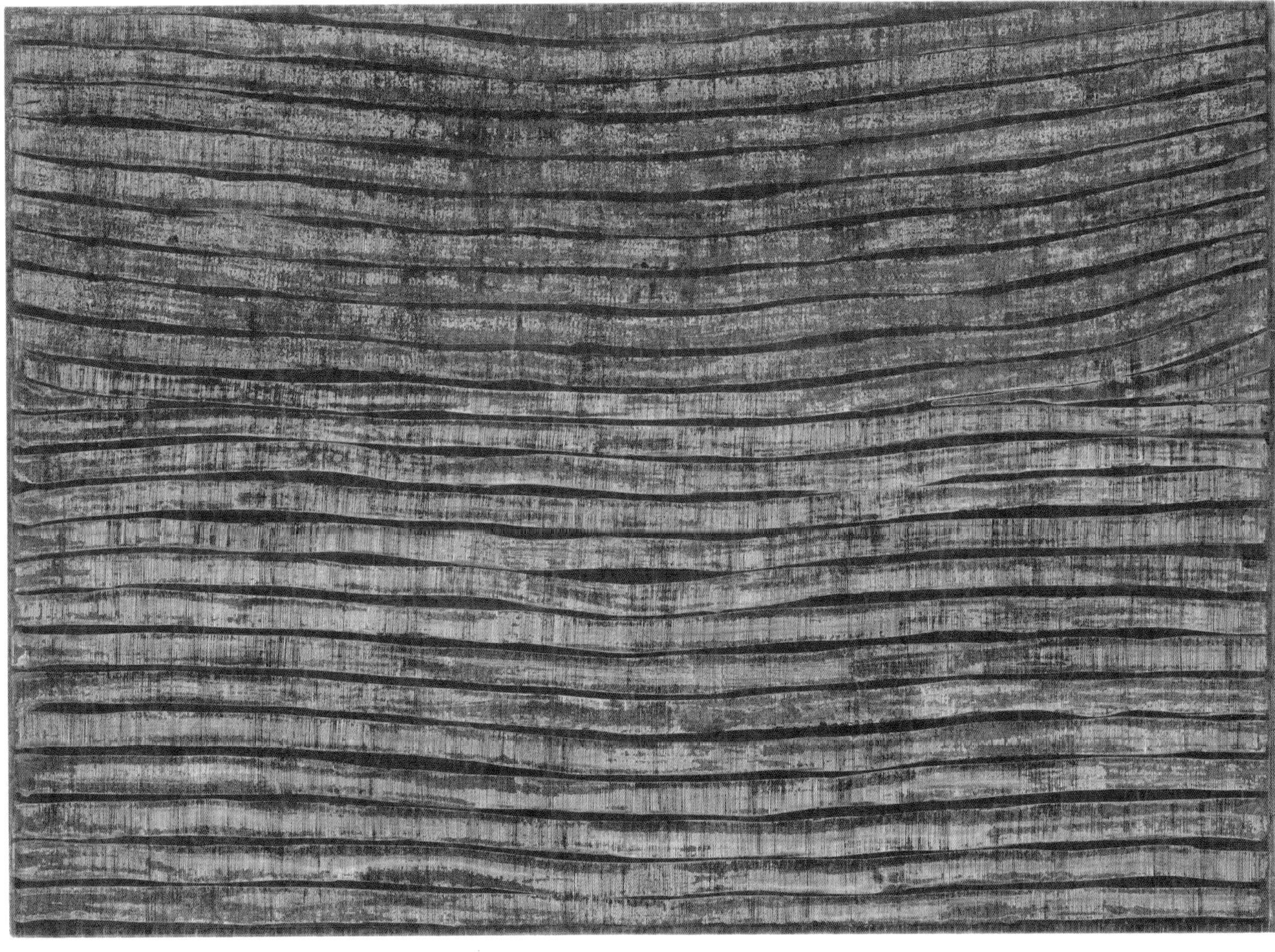

127 *Sky I*, 2008
woodblock on Japanese paper
11 × 16⅞ in. (28 × 43 cm)
commissioned by
ANA InterContinental Tokyo

128 *Sky II*, 2008
woodblock on Japanese paper
11 × 16⅞ in. (28 × 43 cm)
commissioned by
ANA InterContinental Tokyo

A suite of traditional Japanese woodblock prints by Satō Woodblock Workshop, Kyoto, translated from drawings by Rebecca Salter

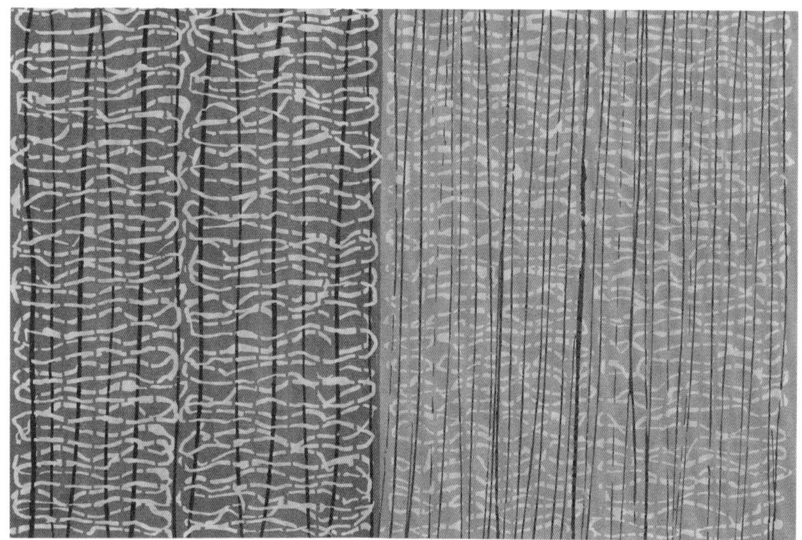

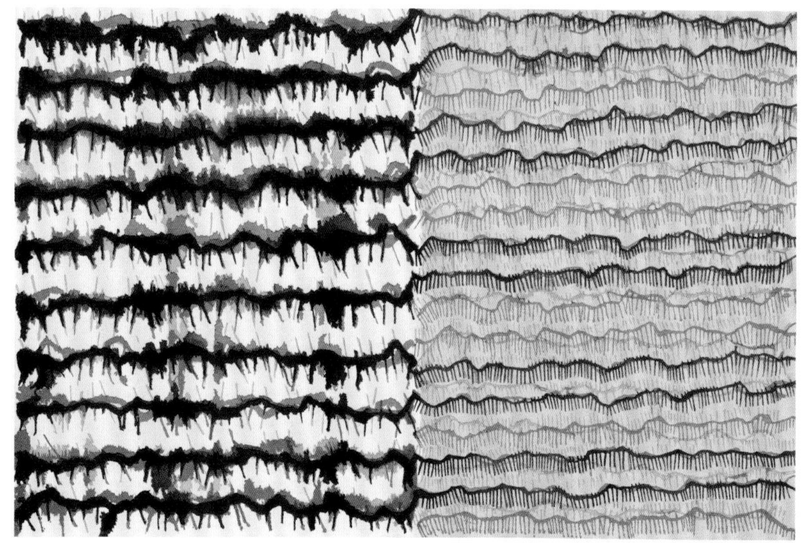

129 *Sky III*, 2008
woodblock on Japanese paper
11 × 16⅞ in. (28 × 43 cm)
commissioned by
ANA InterContinental Tokyo

130 *Sky IV*, 2008
woodblock on Japanese paper
11 × 16⅞ in. (28 × 43 cm)
commissioned by
ANA InterContinental Tokyo

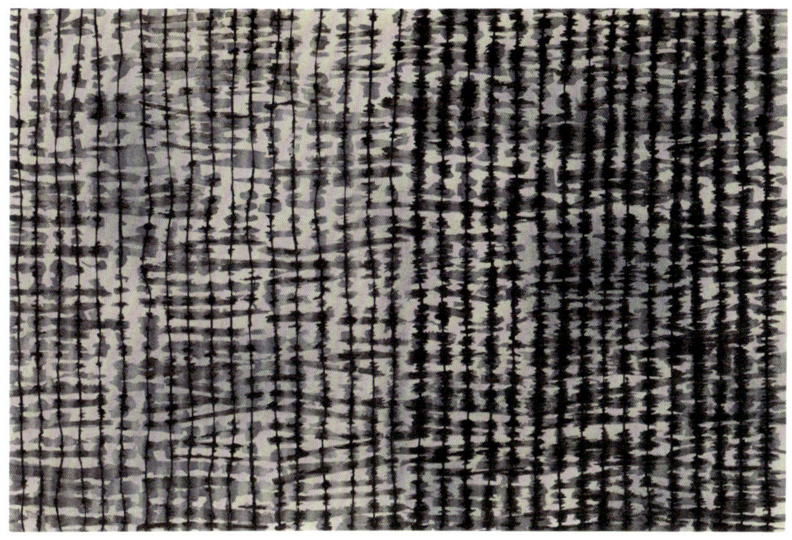

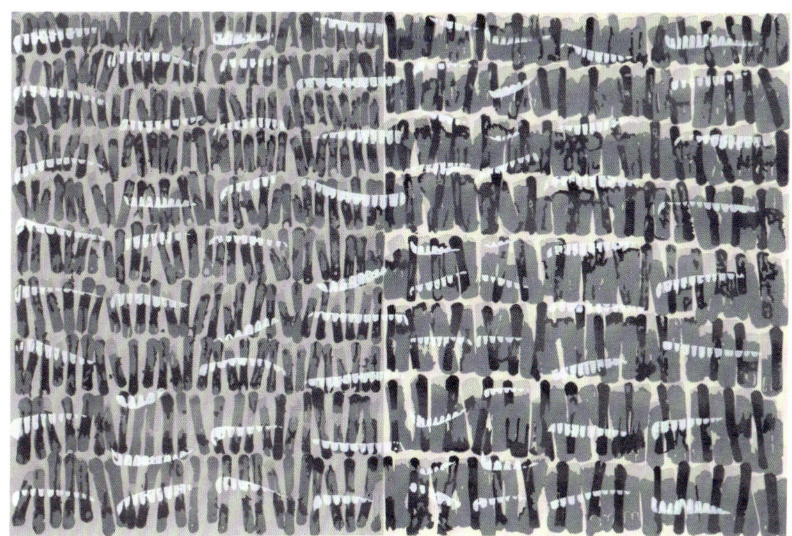

131 *Untitled MM2*, 2008
mixed media on paper
39¾ × 59⅞ in. (101 × 152 cm)

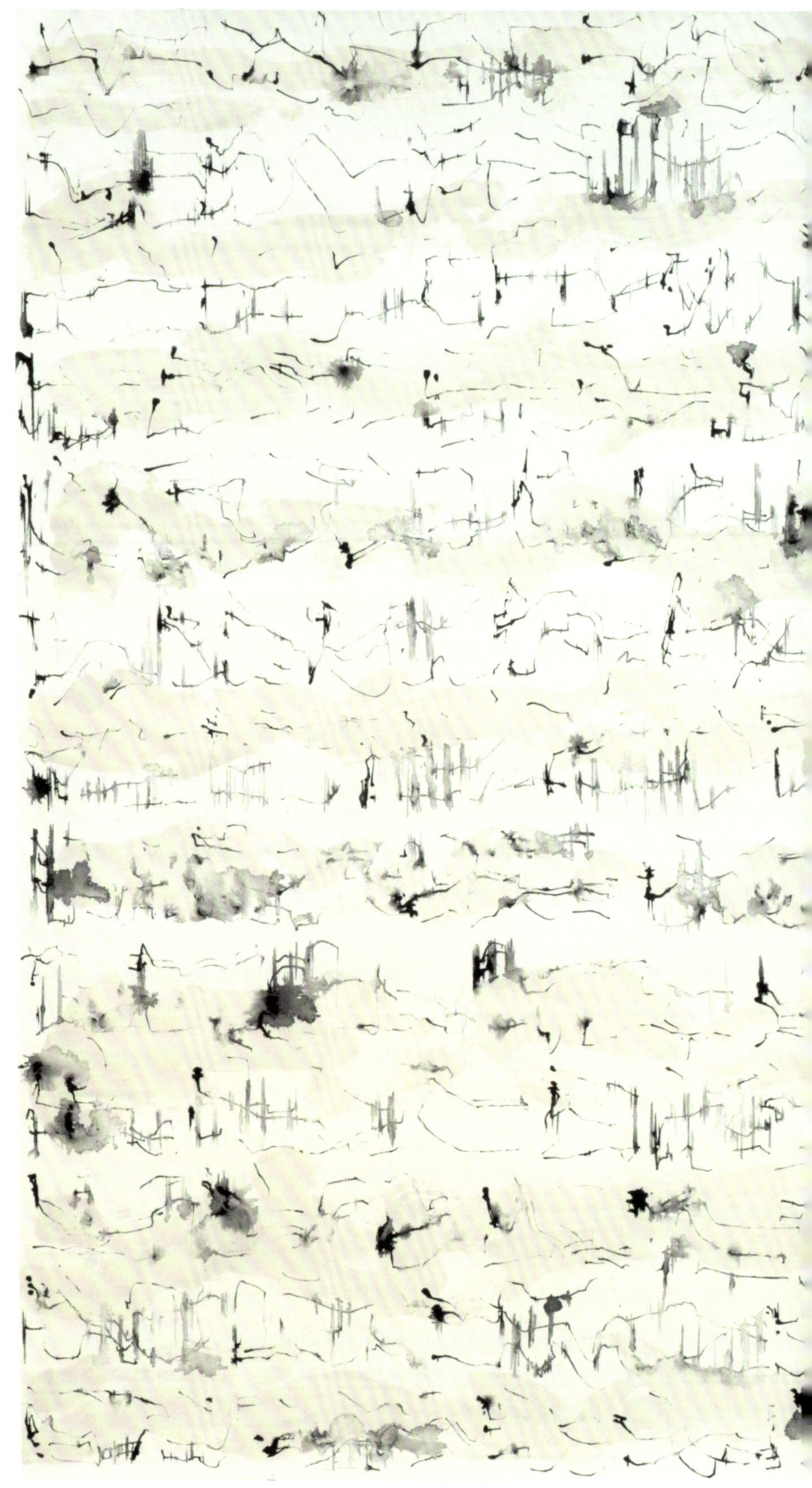

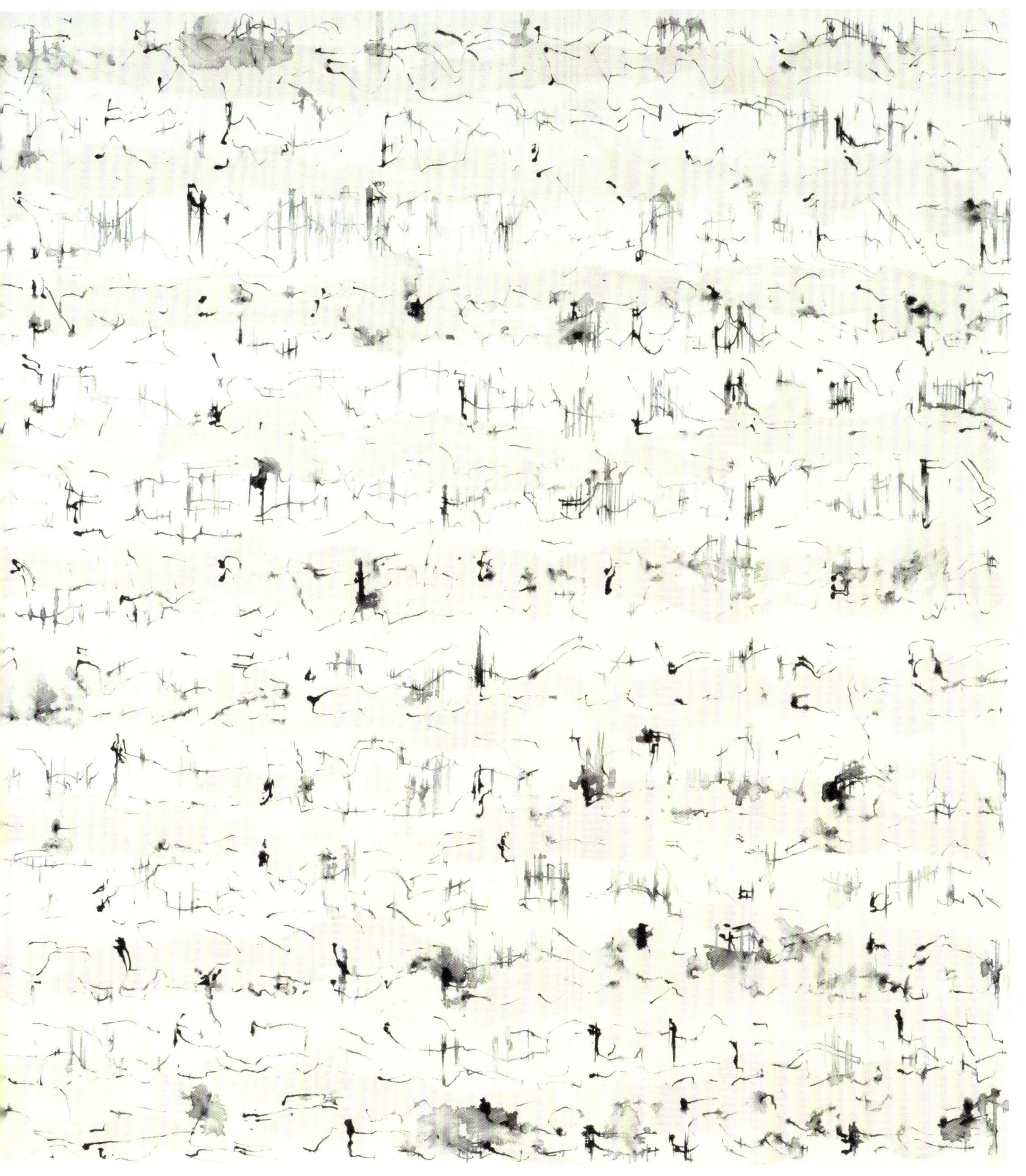

132 *Untitled MM8*, 2008
mixed media on linen
51⅛ × 55⅛ in. (130 × 140 cm)
detail overleaf

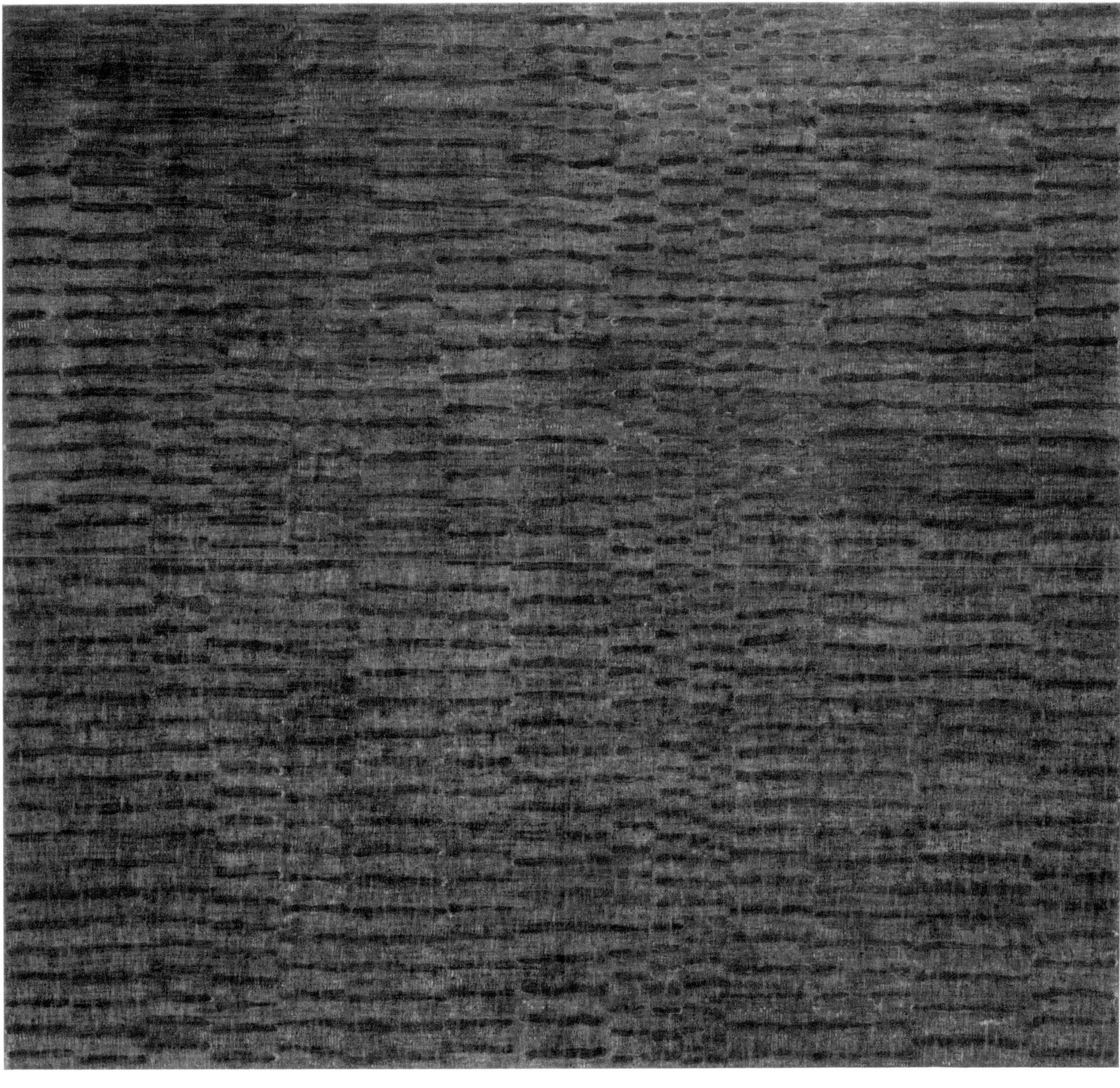

133 *Untitled MM33*, 2008
mixed media on linen
41⅜ × 43¼ in. (105 × 110 cm)
Private collection, London

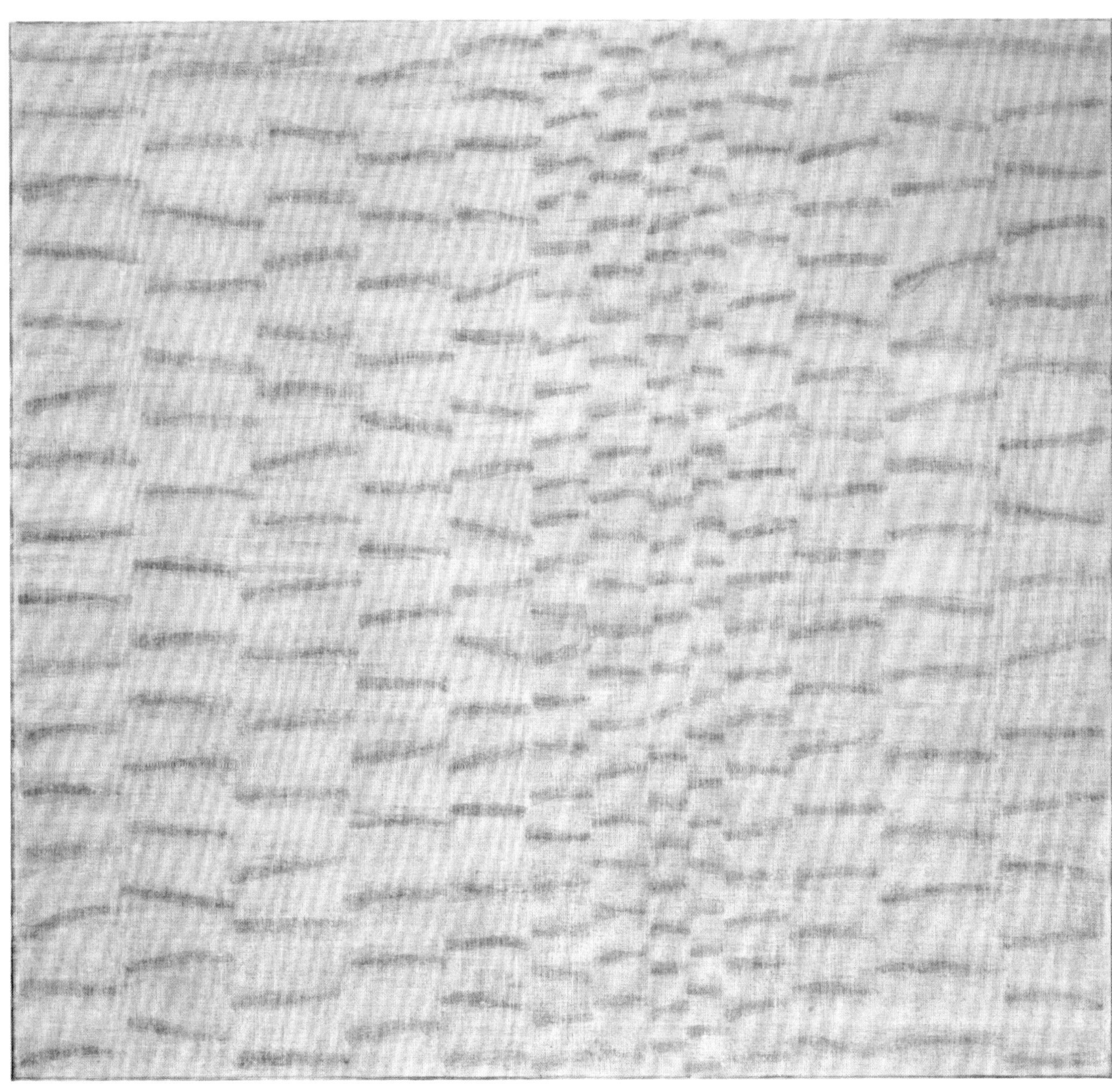

134 *Untitled MM37*, 2008
mixed media on linen
63 × 66⅞ in. (160 × 170 cm)
Private collection

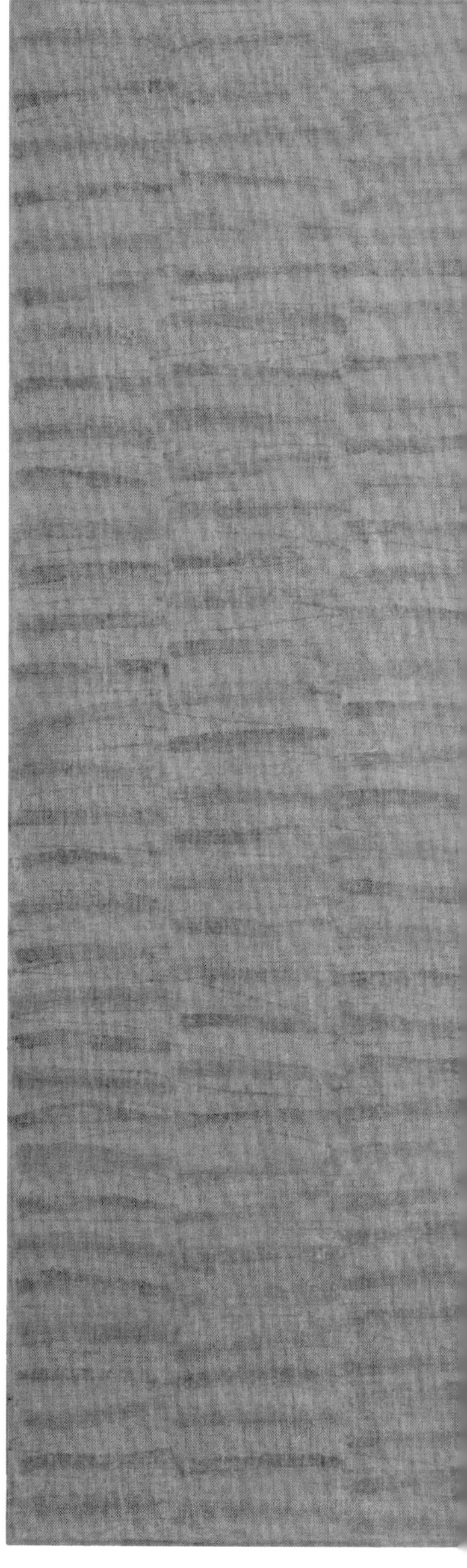

135 *Untitled RR2*, 2009
mixed media on paper
39¾ × 59⅞ in. (101 × 152 cm)

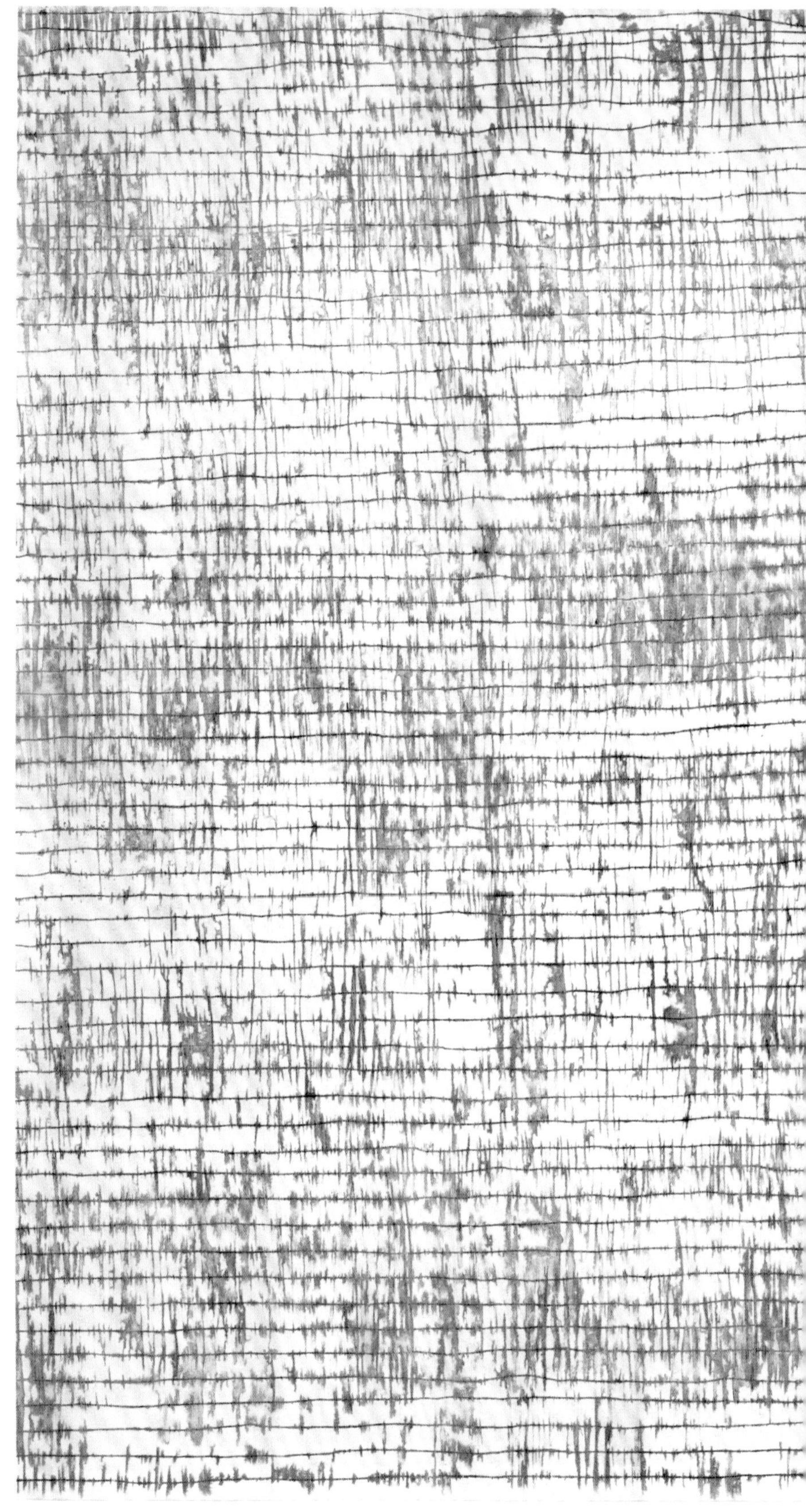

136 *Untitled RR26*, 2009
mixed media on paper
60 × 36½ in. (152.4 × 91.4 cm)

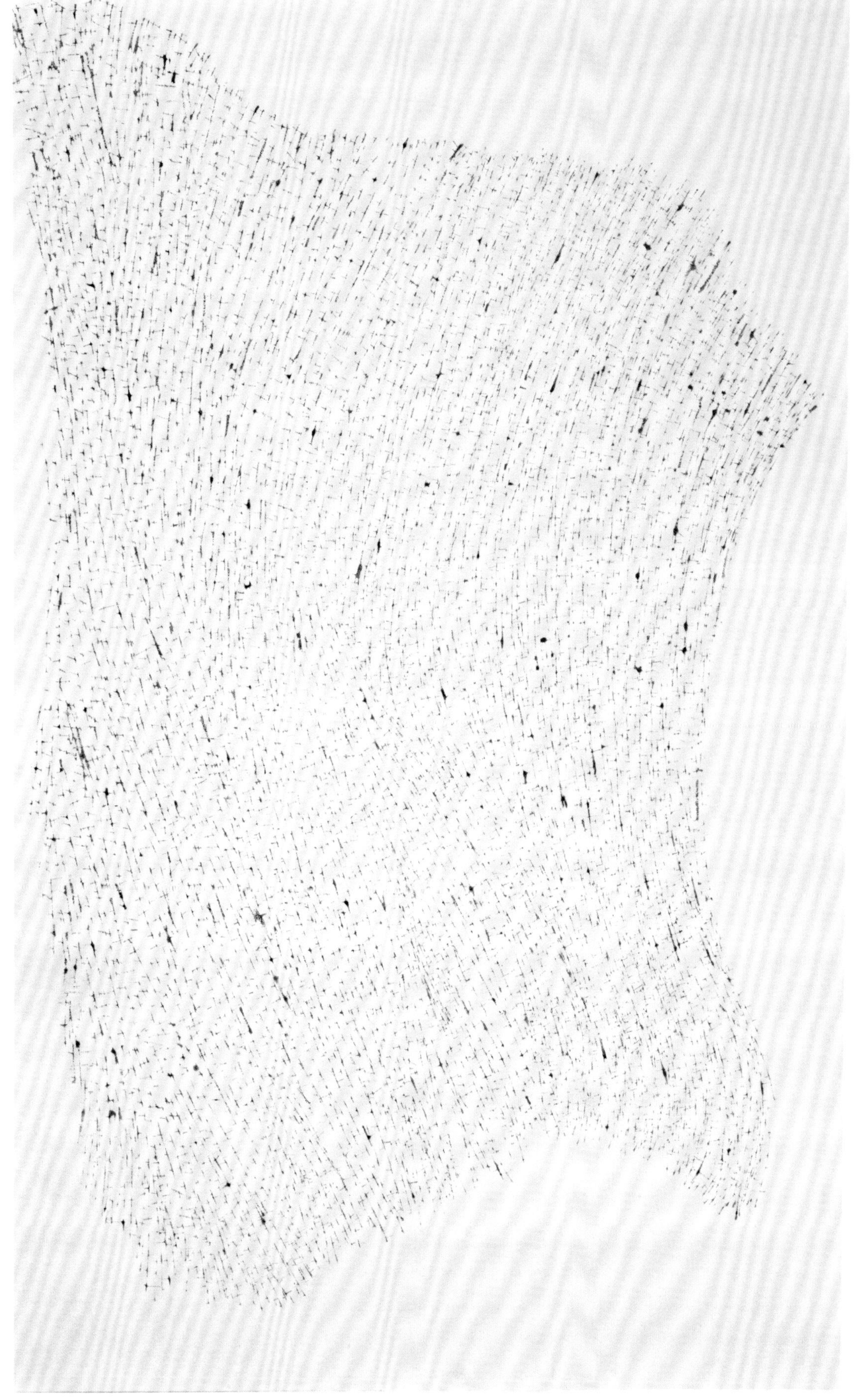

137 *Untitled 2009–32*, 2009
mixed media on paper
29⅛ × 17⅞ in. (74 × 45.5 cm)

138 *Untitled 2009–27*, 2009
mixed media on paper
29⅛ × 17⅞ in. (74 × 45.5 cm)
detail overleaf

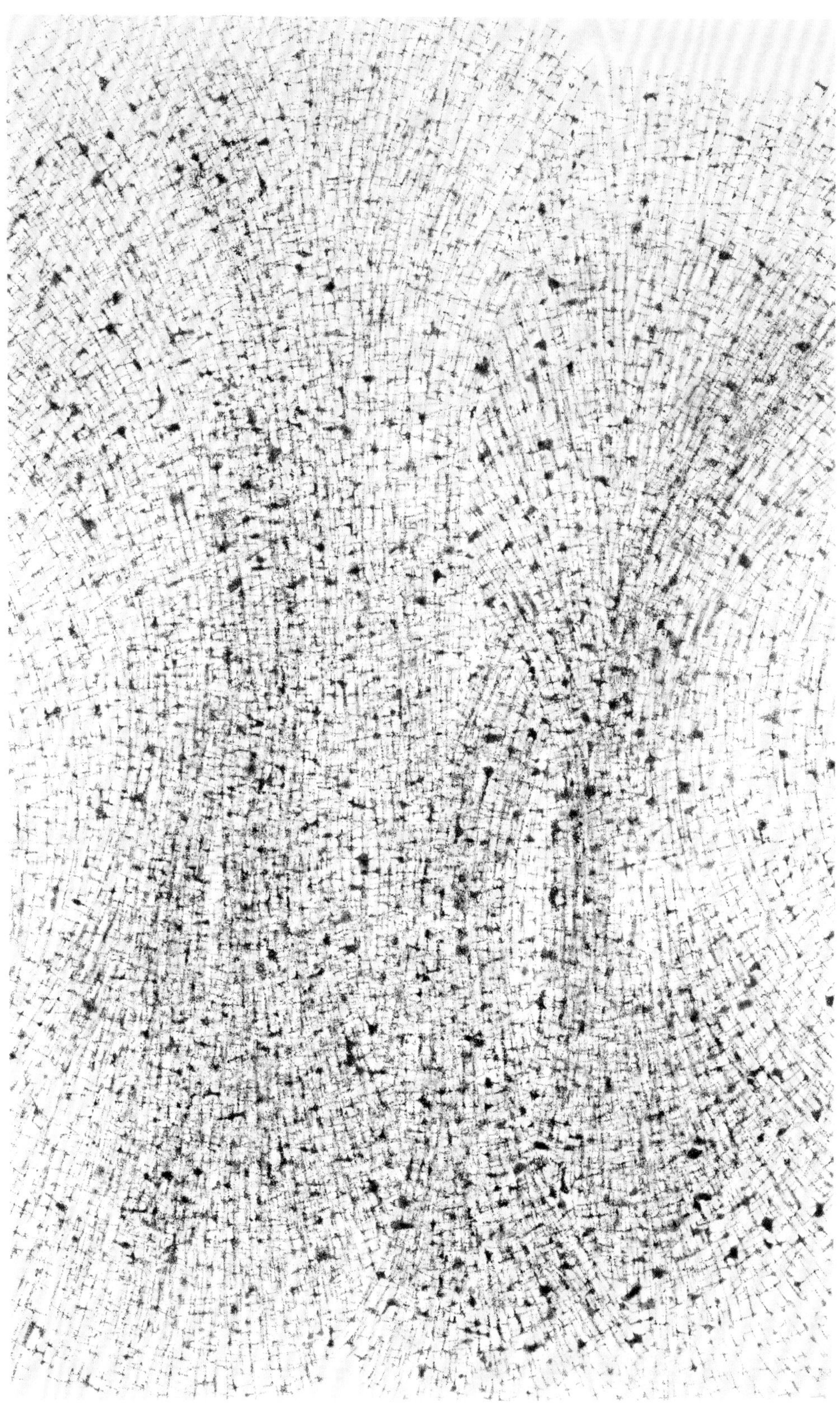

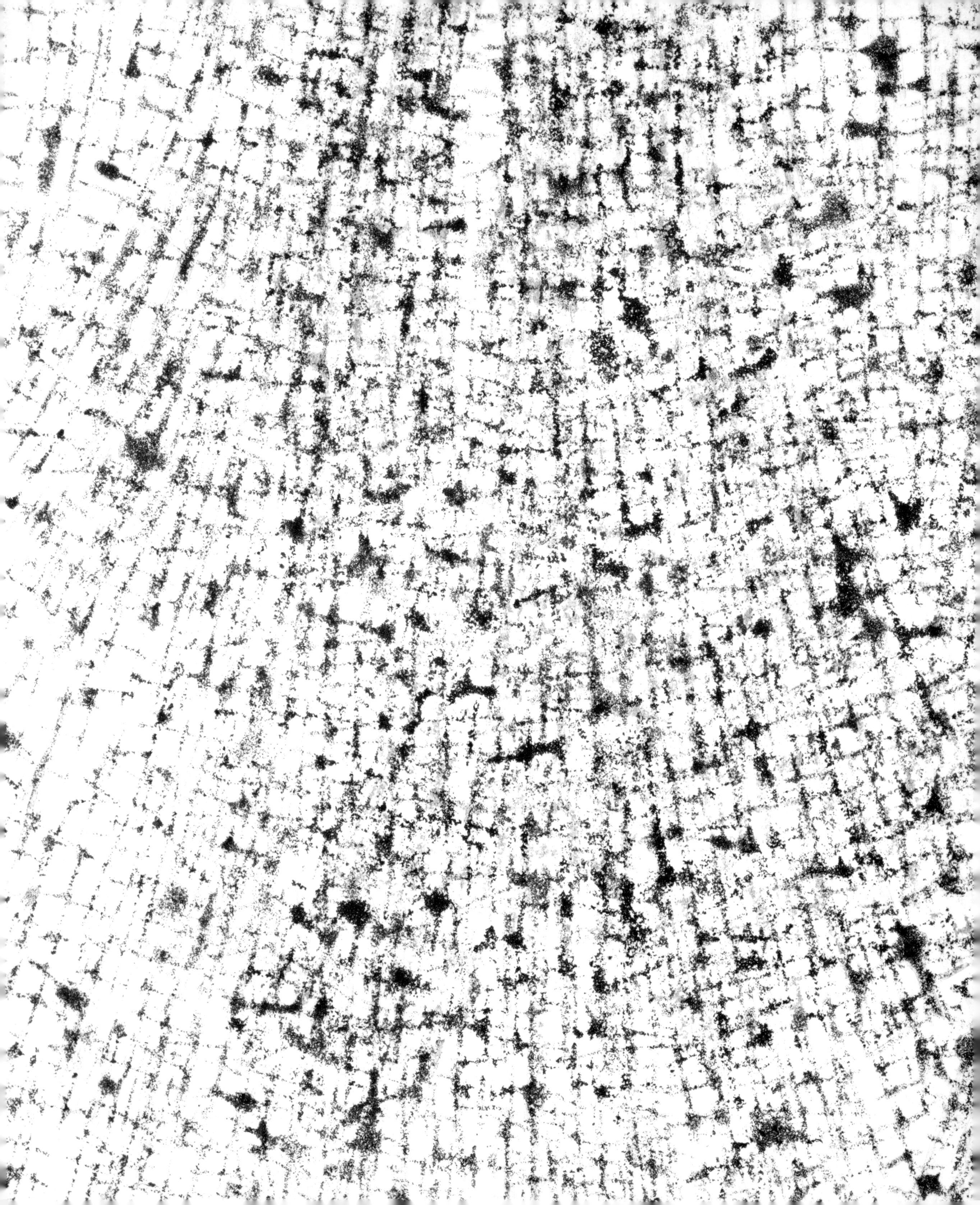

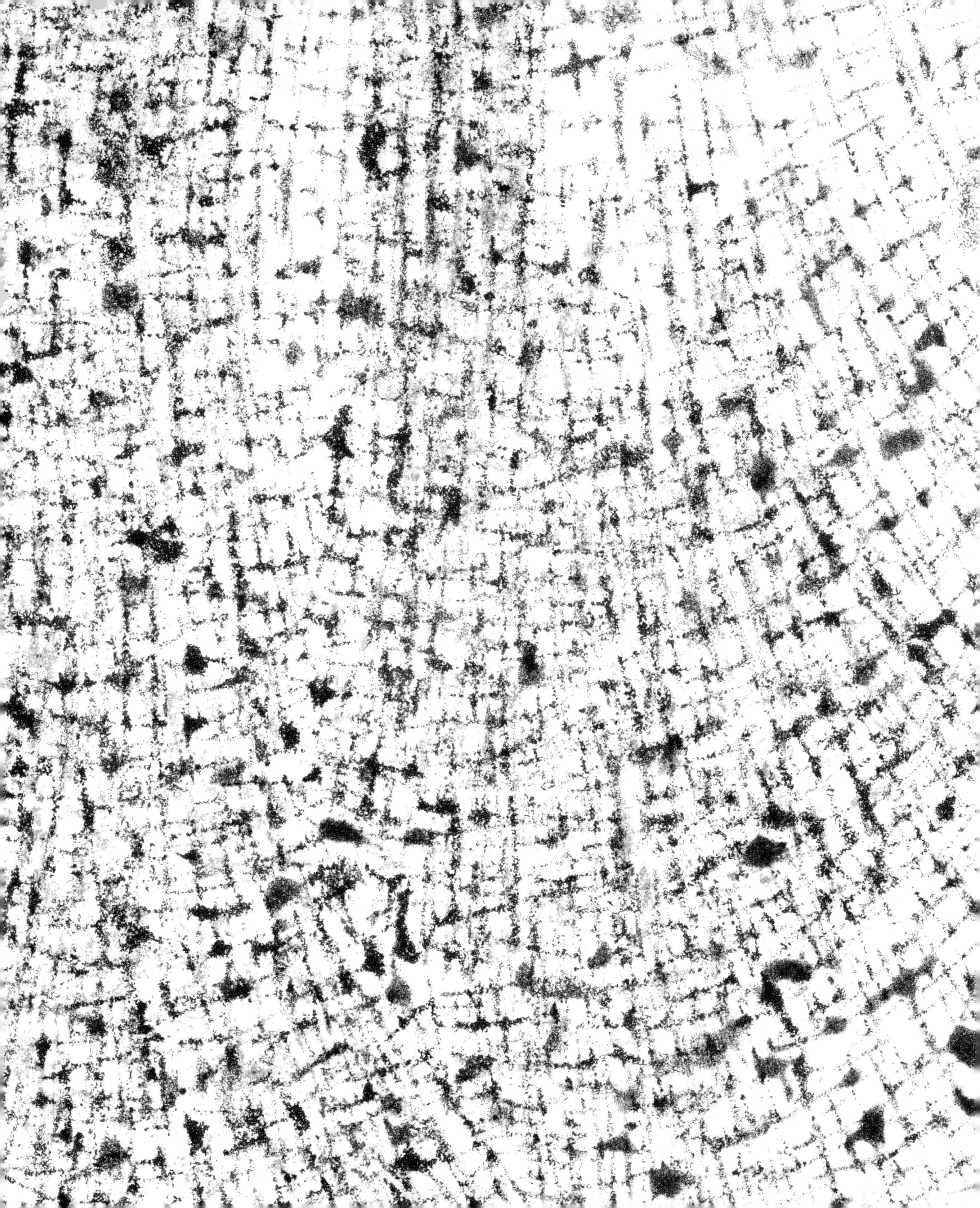

139 *Untitled RR1*, 2009
mixed media on linen
23⅝ × 17¾ in. (60 × 45 cm)

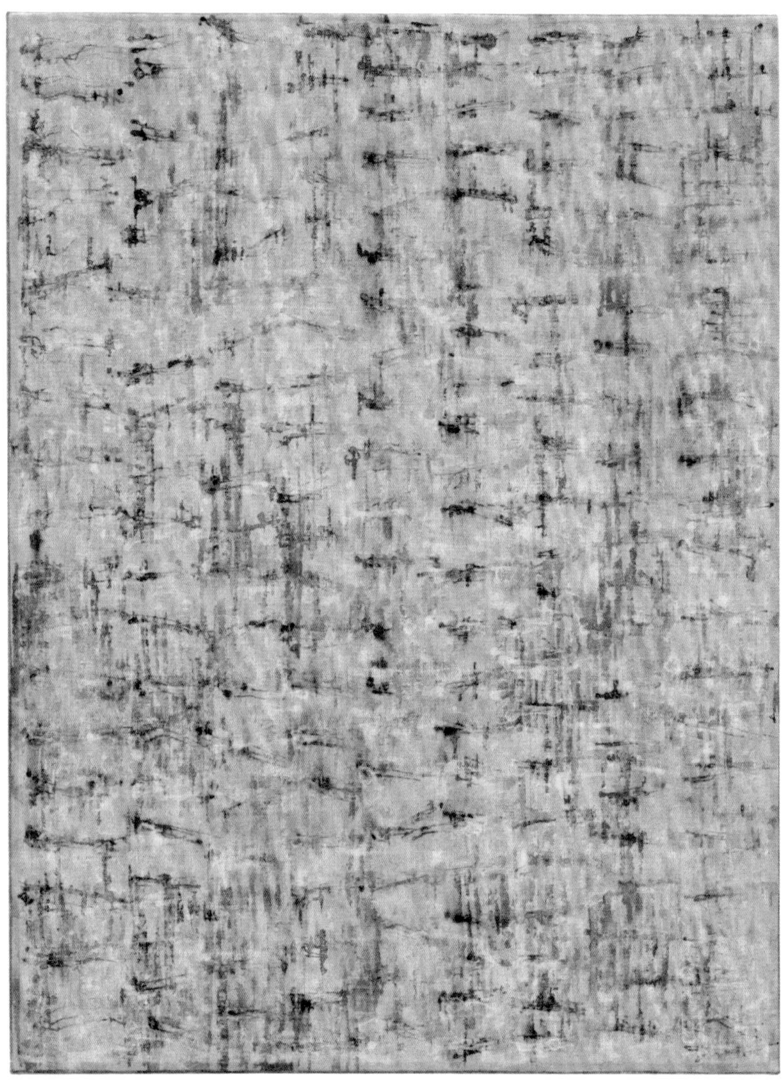

140 *Untitled AB2*, 2010
mixed media on linen
25½ × 19½ in. (64.8 × 49.5 cm)

141 *Lead 1*, 2009
mixed media on lead
12 × 10 in. (30.5 × 25.4 cm)

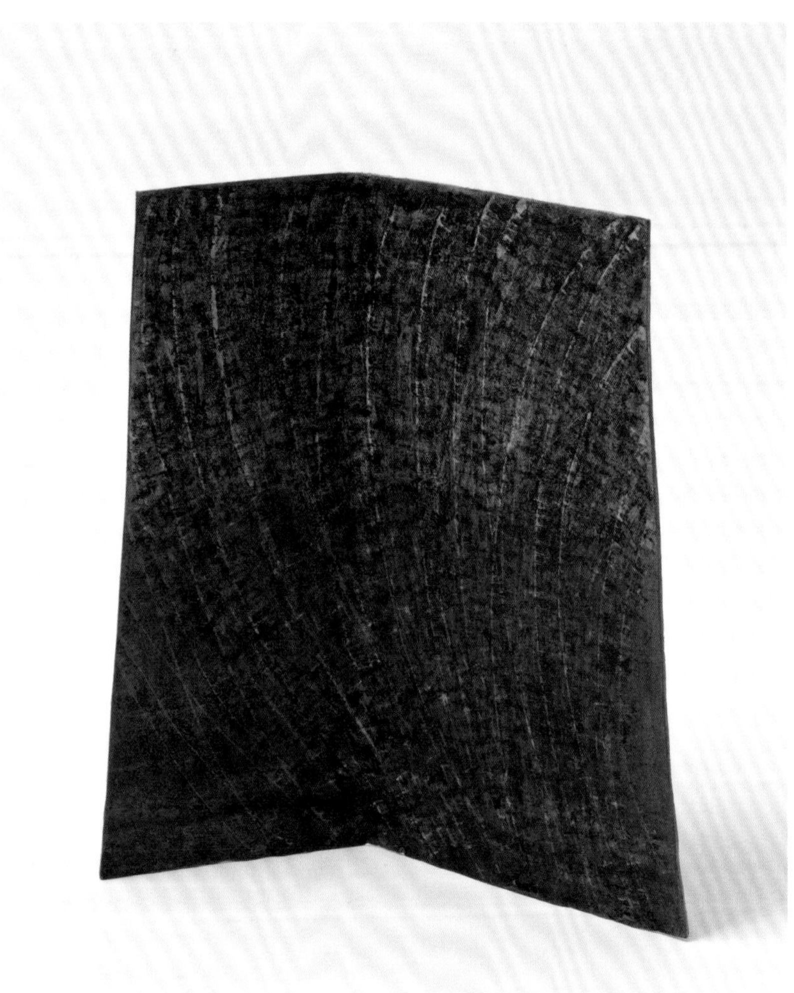

142 *Lead 3*, 2009
mixed media on lead
10 × 1½ in. (25.4 × 3.8 cm)

143 *Lead 6*, 2009
mixed media on lead
10 × 2½ in. (25.4 × 6.4 cm)

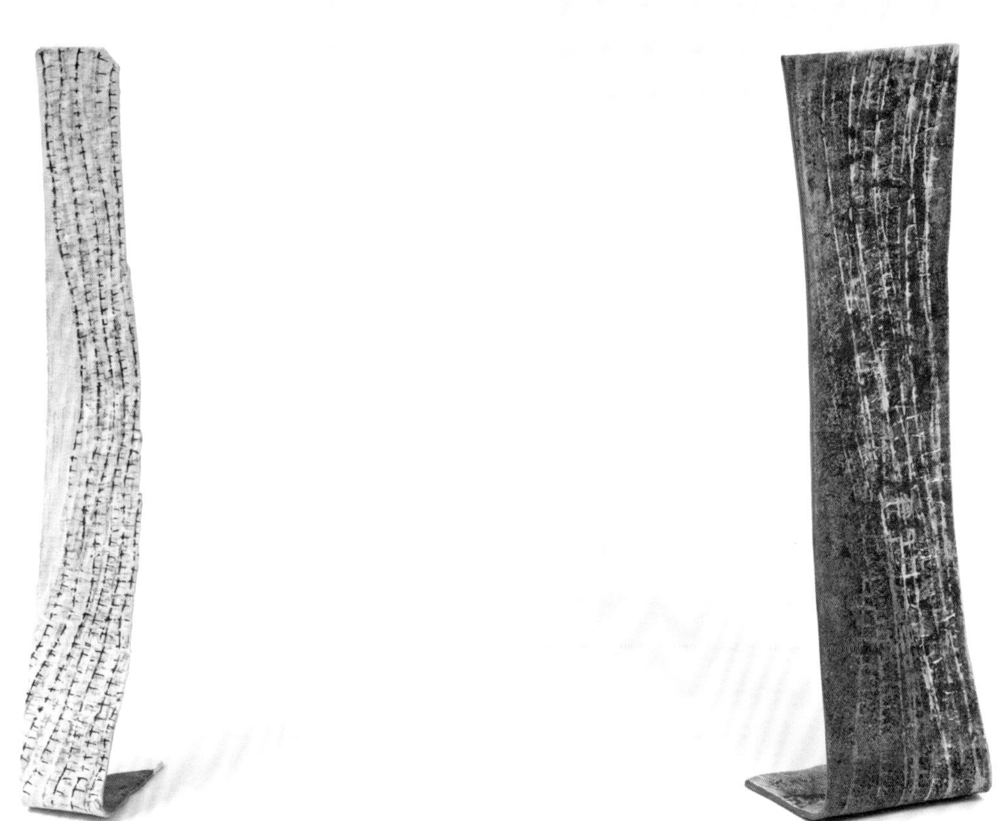

144 *Lead 9*, 2009
mixed media on lead
12¼ × 3½ in. (31.1 × 8.9 cm)

145 *Lead 12*, 2010
mixed media on lead
10 × 2½ in. (25.4 × 6.4 cm)

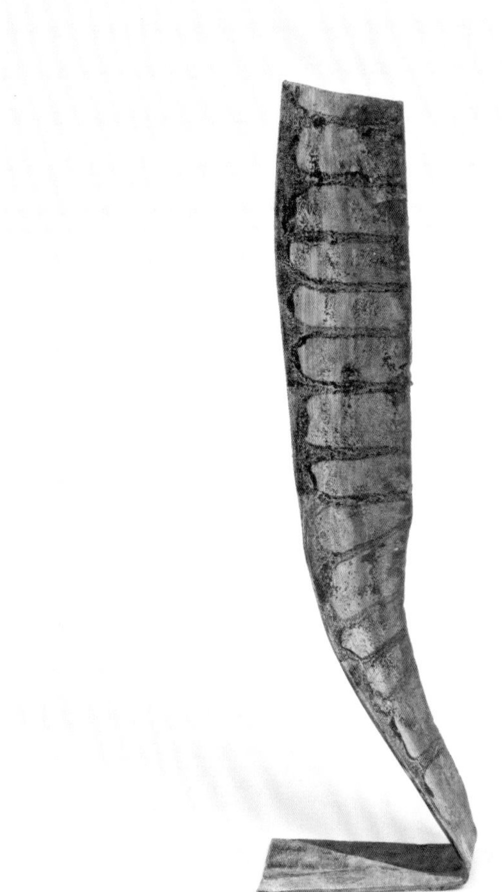

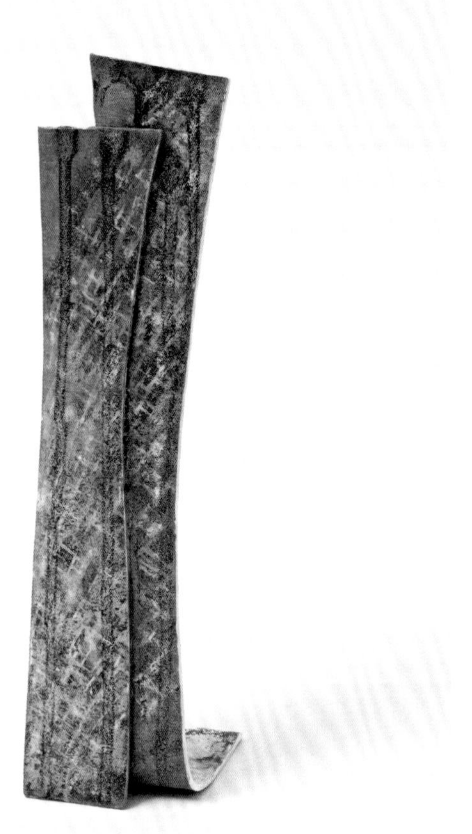

146 *Lead 13*, 2010
mixed media on lead
7 × 3½ in. (17.8 × 8.9 cm)

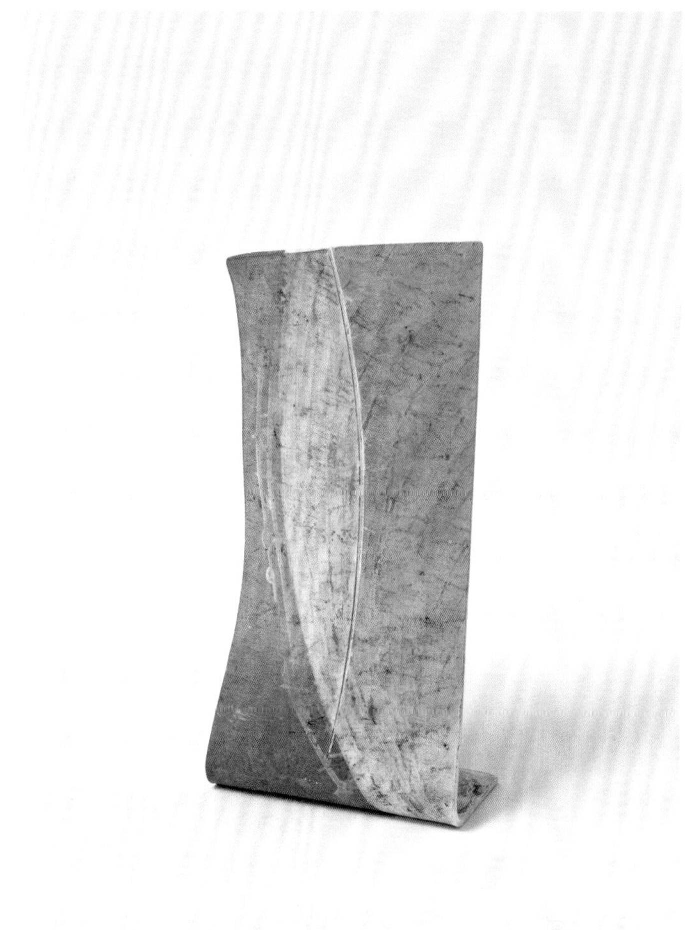

147 *Untitled RR31*, 2009
mixed media on linen
74¾ × 70⅞ in. (190 × 180 cm)

148 *Untitled RR21*, 2009
mixed media on linen
43¼ × 35⅜ in. (110 × 90 cm)

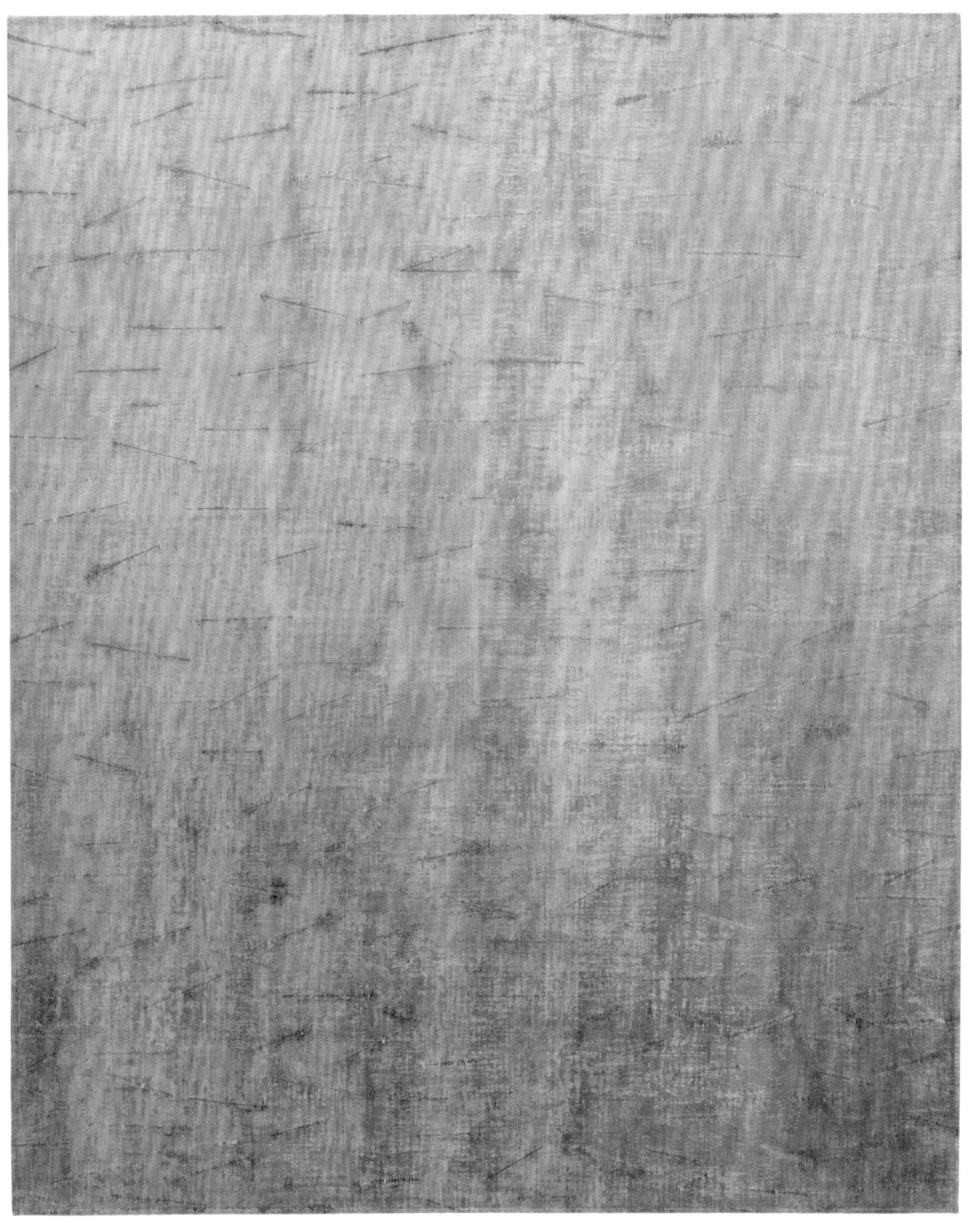

149 *Untitled RR35*, 2009
mixed media on paper
39⅜ × 59⅞ in. (100 × 152 cm)

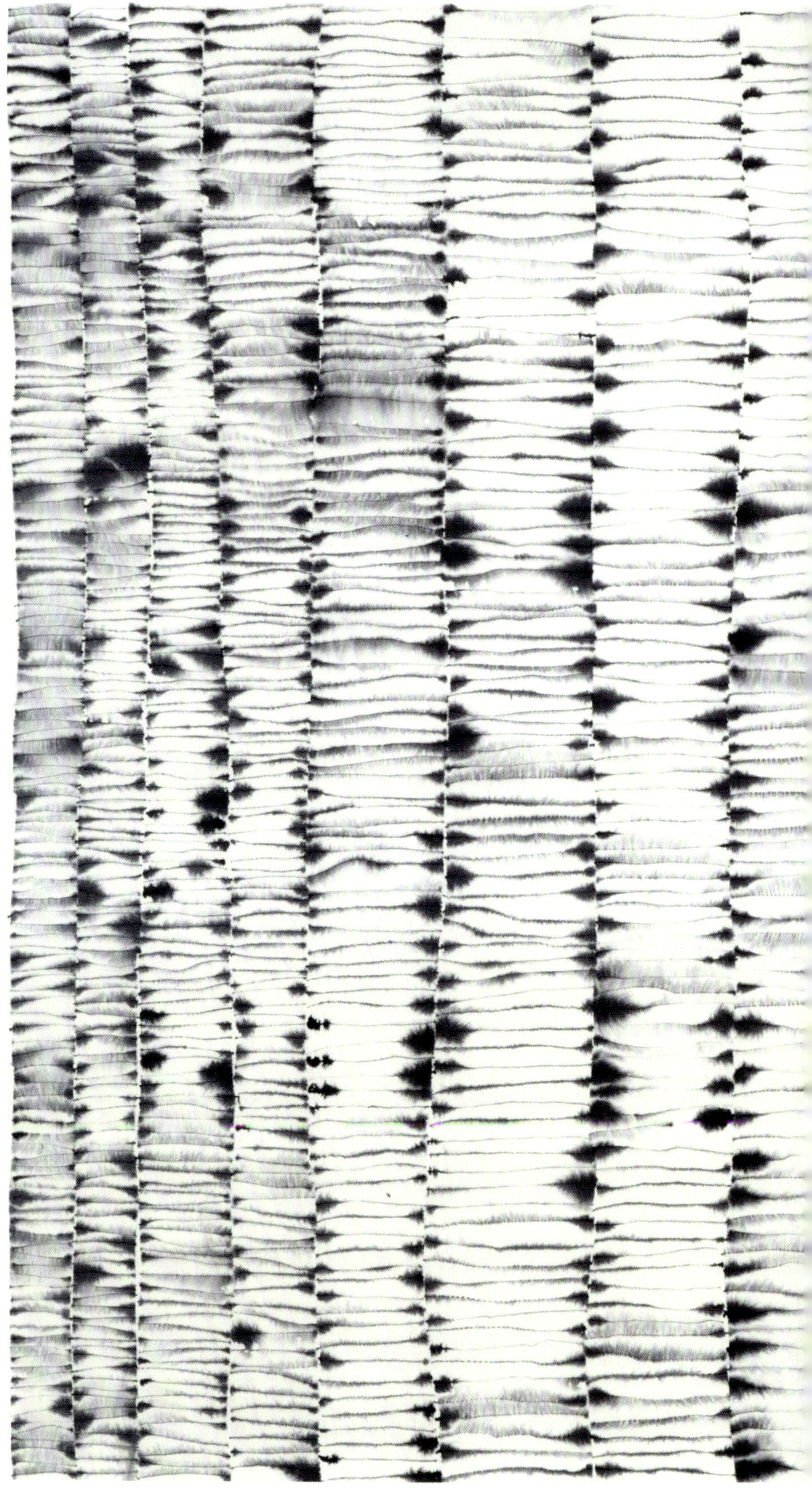

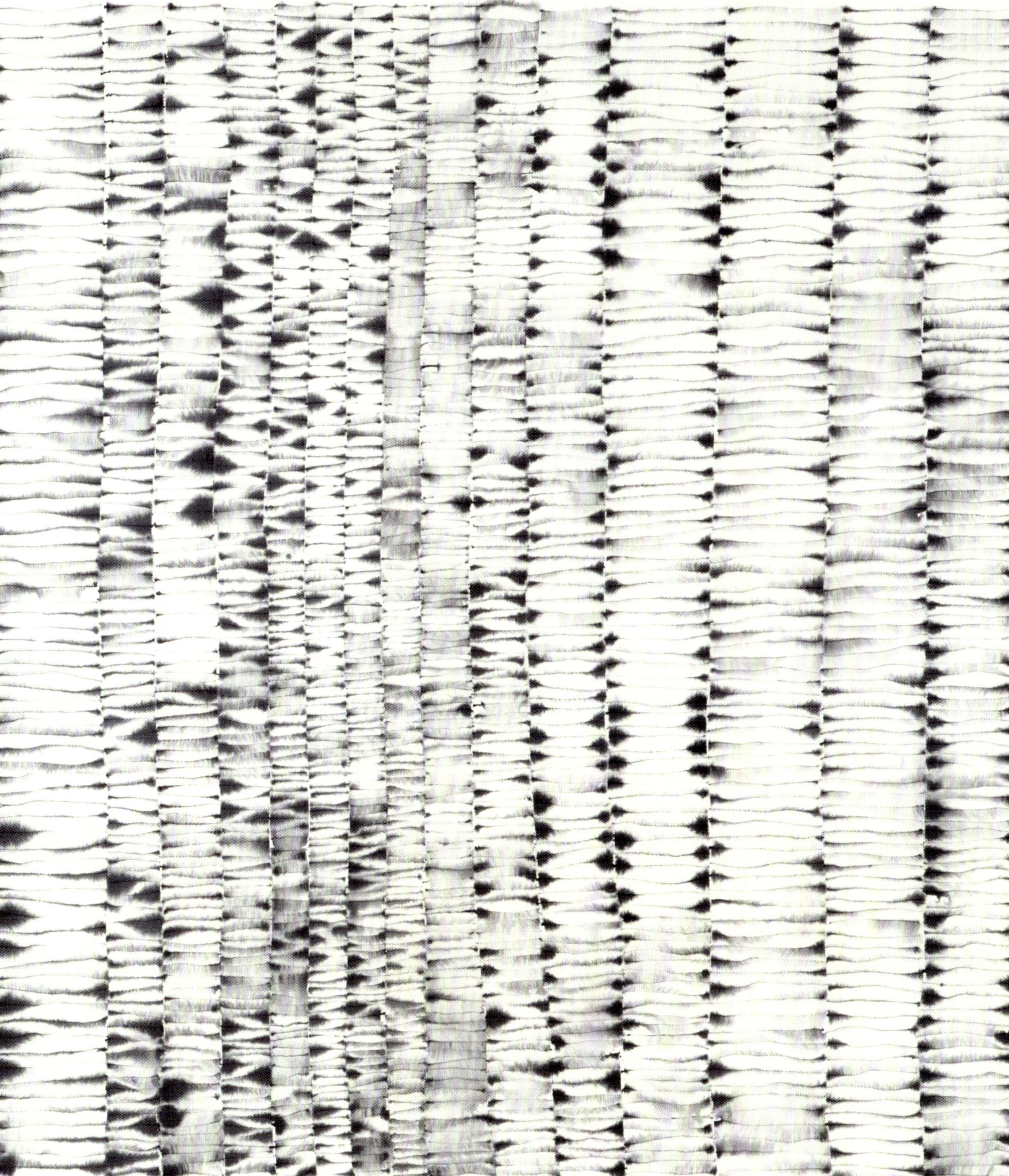

150 *Untitled AB19*, 2010
mixed media on linen
55 × 51 in. (140 × 130 cm)

151 *Untitled AB4*, 2010
mixed media on linen
30 × 27½ in. (76.2 × 69.9 cm)

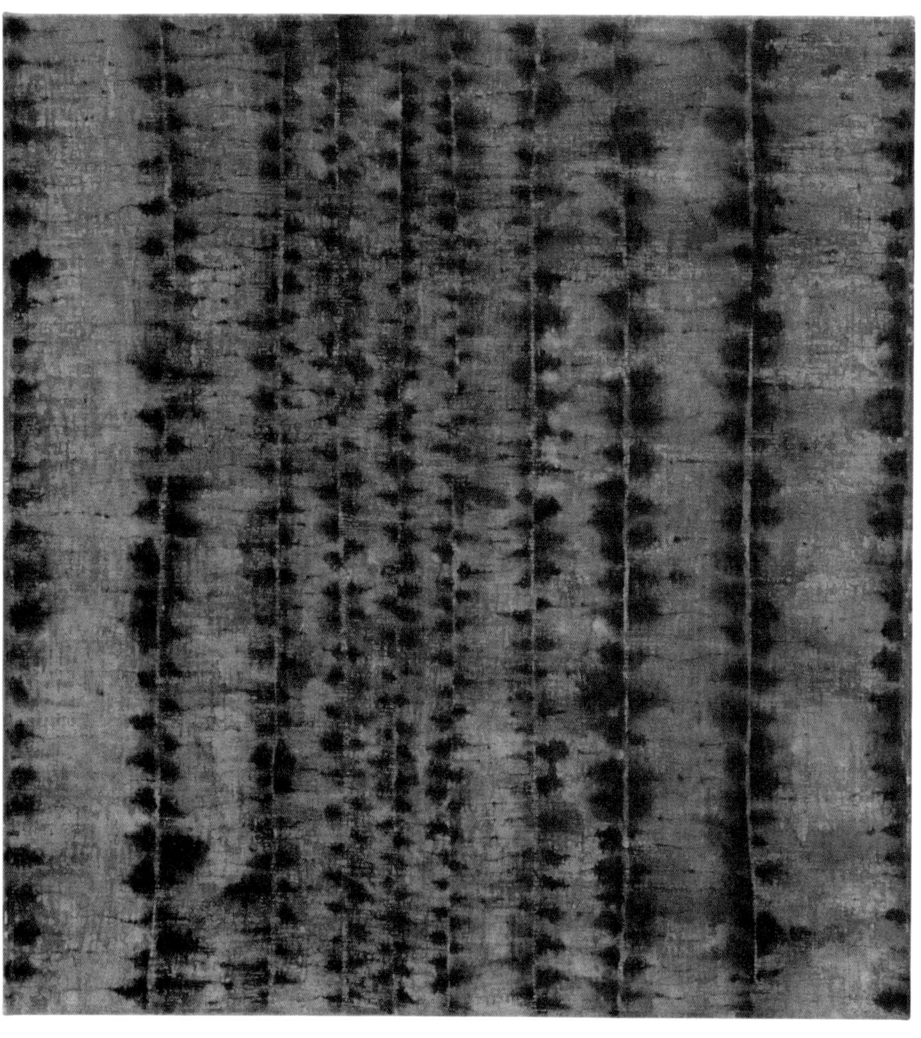

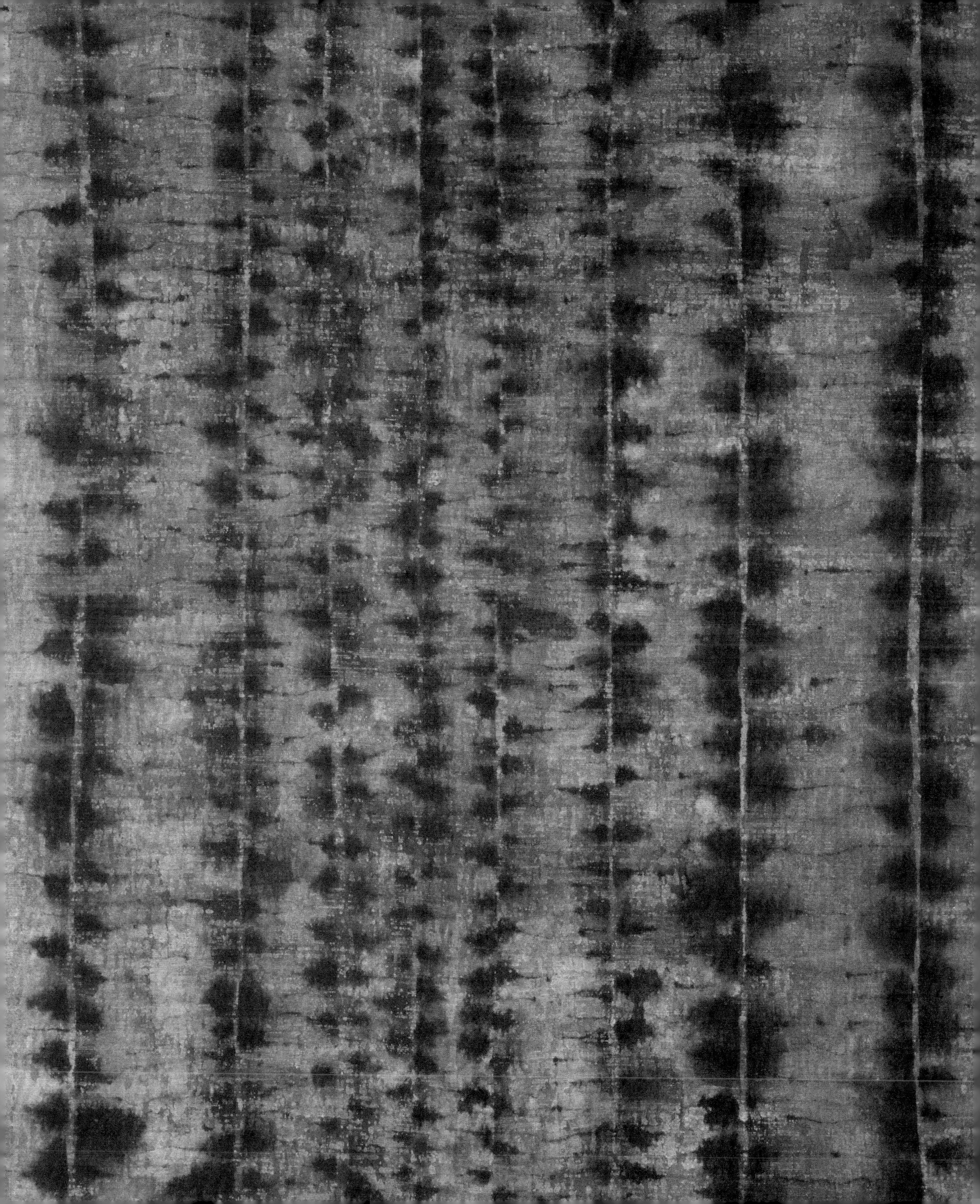

152 *Quadra Drawings 1–4*, 2010
watercolor on Somerset paper
11⅞ × 11⅞ in. (30 × 30 cm)

A suite of drawings translated into traditional Japanese woodblock printing (opposite) by Satō Woodblock Workshop, Kyoto

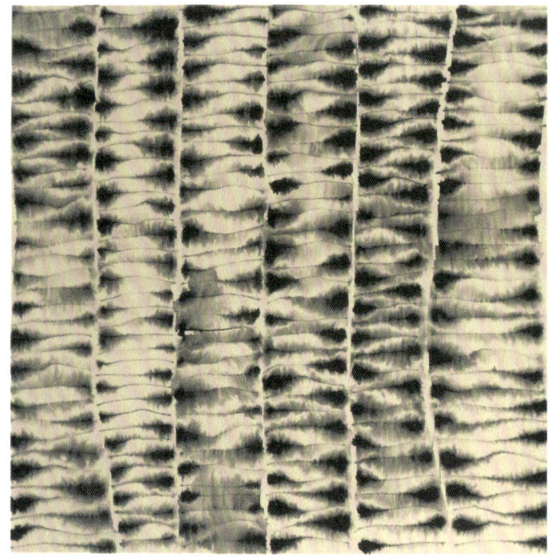

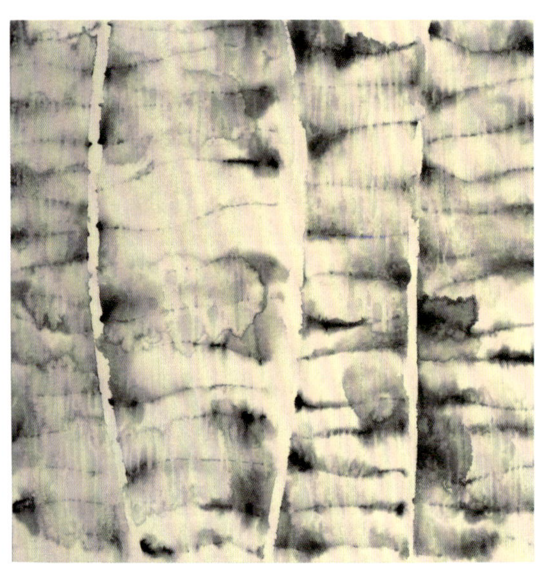

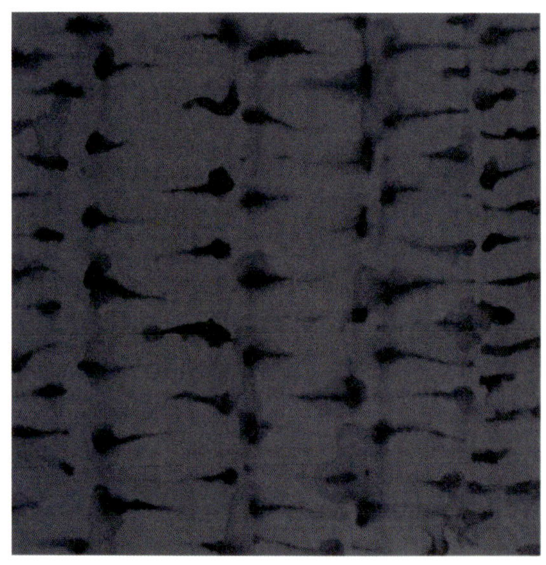

153 *Quadra 1–4, 2010*
woodblock on Torinoko paper
11⅞ × 11⅞ in. (30 × 30 cm)

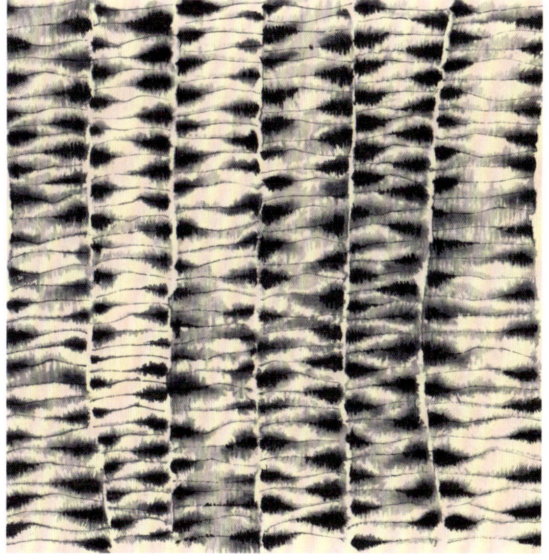

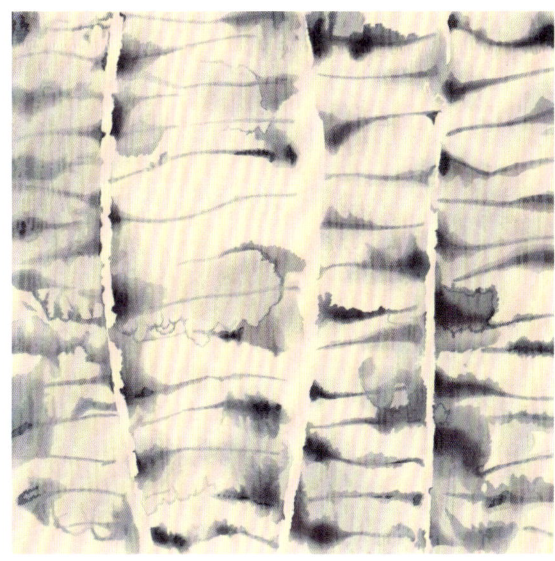

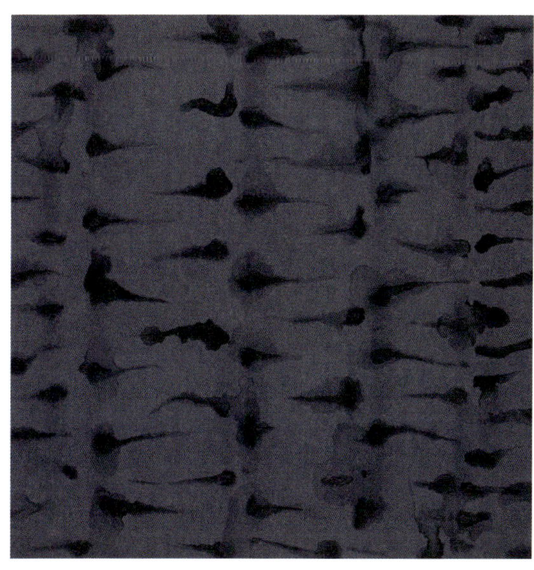

154 *into the light of things*, 2010
digital print on Japanese paper
edition of 4
2⅝ × 43¾ in. (6.7 × 111.1 cm)

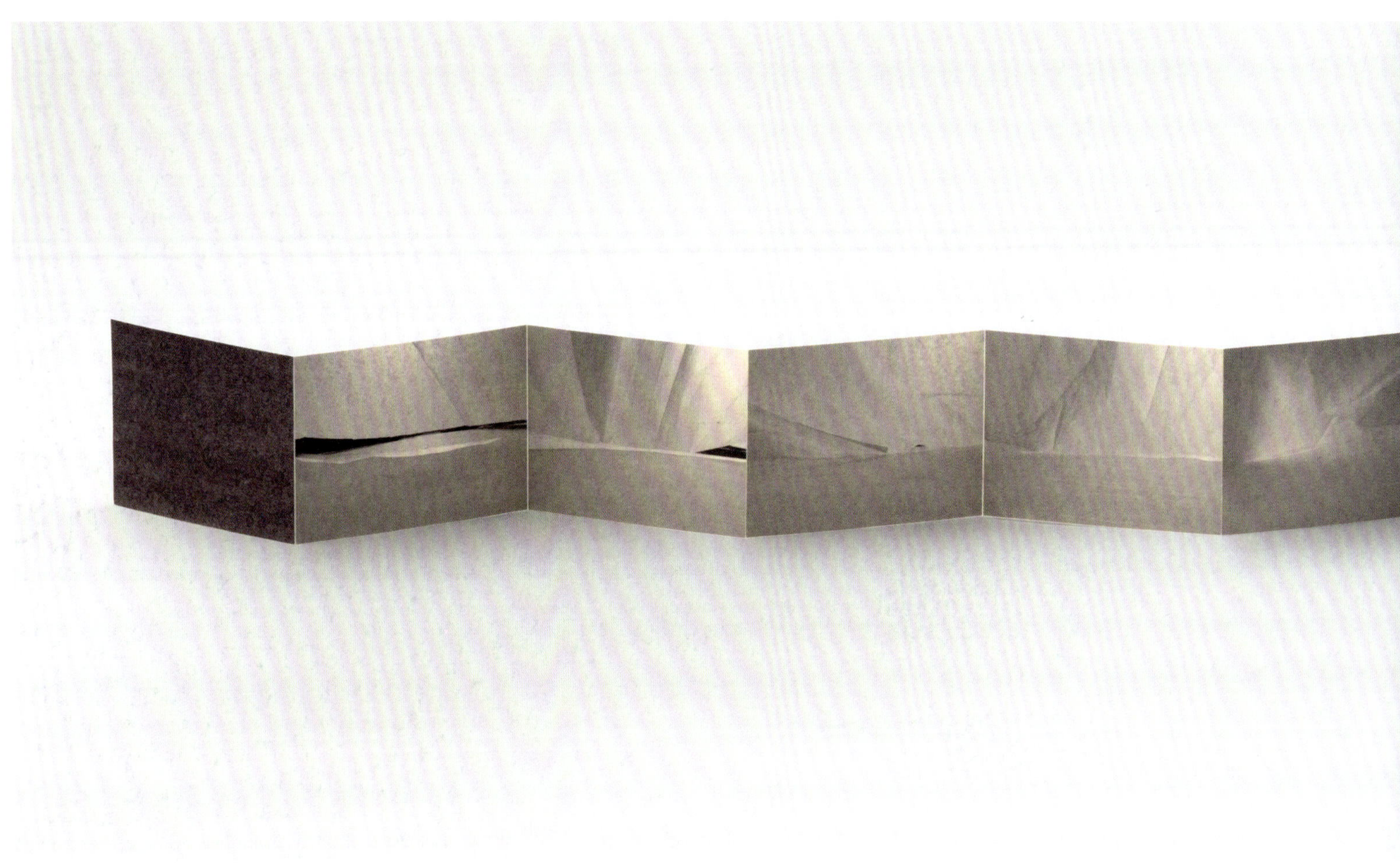

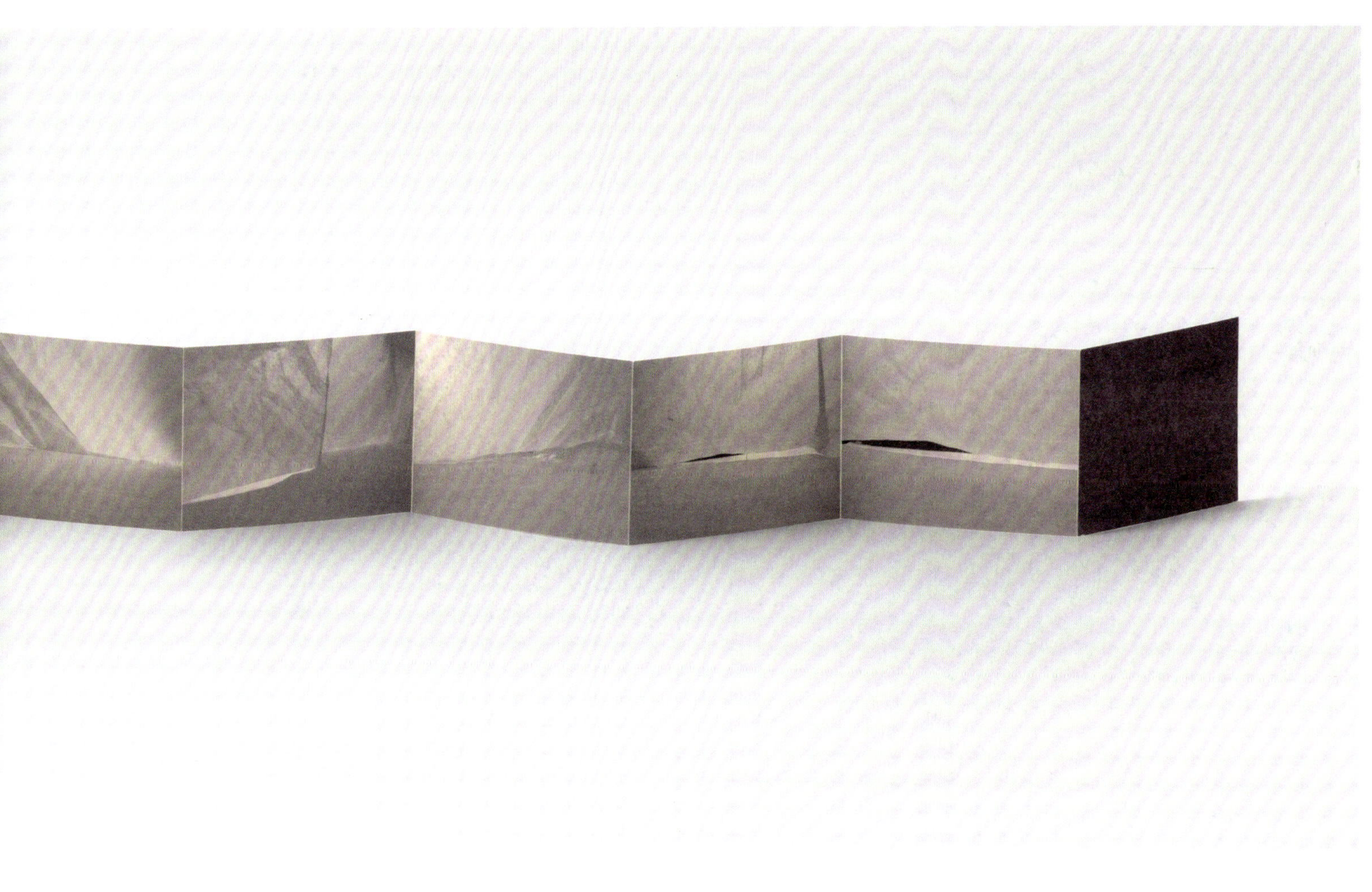

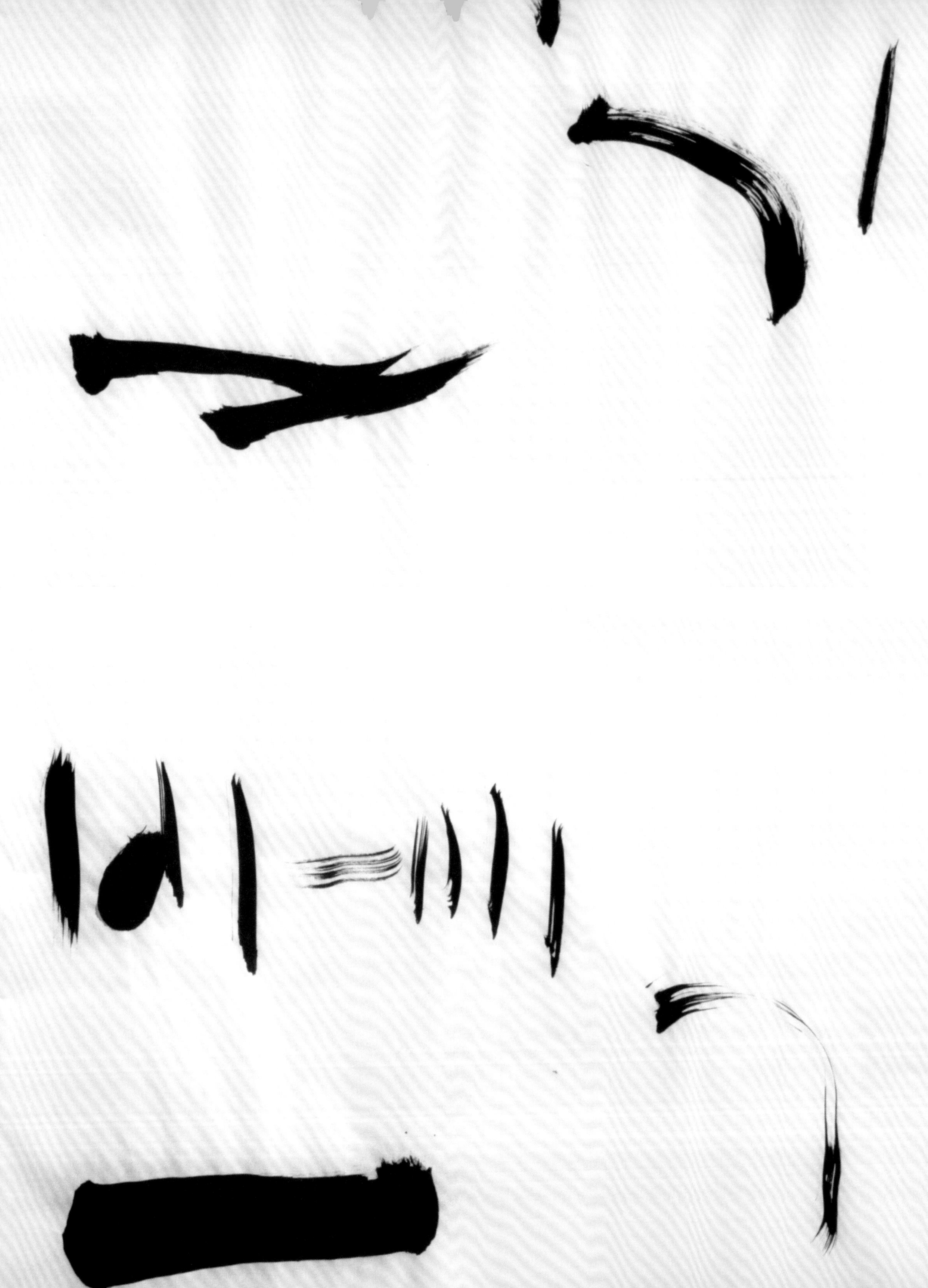

155 *Calligraphy of Light*
preliminary ink drawings, 2007
sumi ink on Japanese paper
each drawing: 9½ × 13¼ in.
(24.1 × 33.7 cm)

156 *Calligraphy of Light*
preliminary ink drawings, 2007
sumi ink on Japanese paper
each drawing: 9½ × 13¼ in.
(24.1 × 33.7 cm)

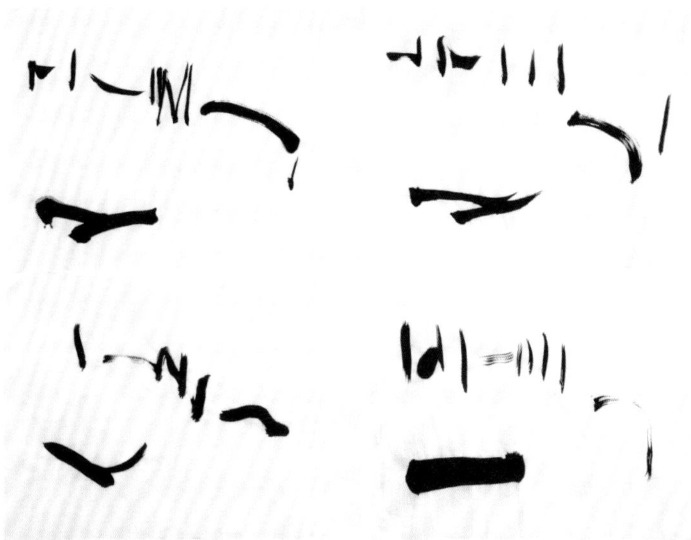

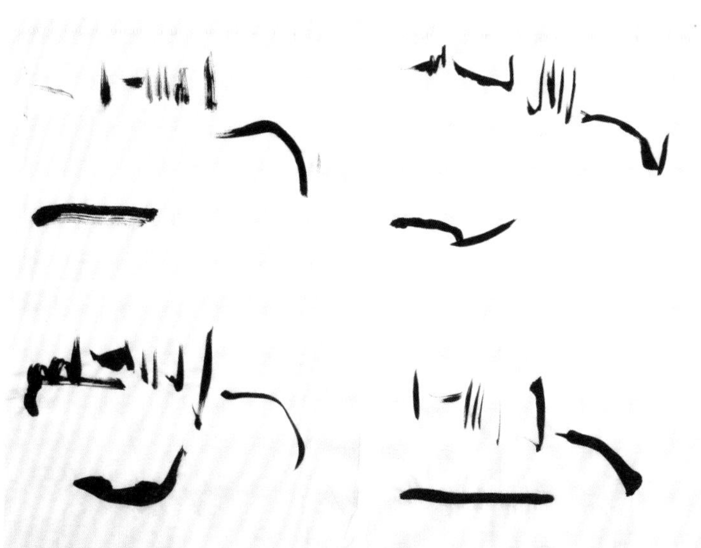

157 *Calligraphy of Light*
layout of scheme, 2007
collage on Japanese paper
2¾ × 37¾ in. (7 × 95.9 cm)

158 *Calligraphy of Light*
detailed layout of white/green
glass, 2007
gouache on paper
2 × 38½ in. (5 × 98 cm)

159 *Calligraphy of Light*
detailed layout of white/green
glass, 2007
gouache on paper
2¼ × 25¼ in. (5.7 × 64.2 cm)

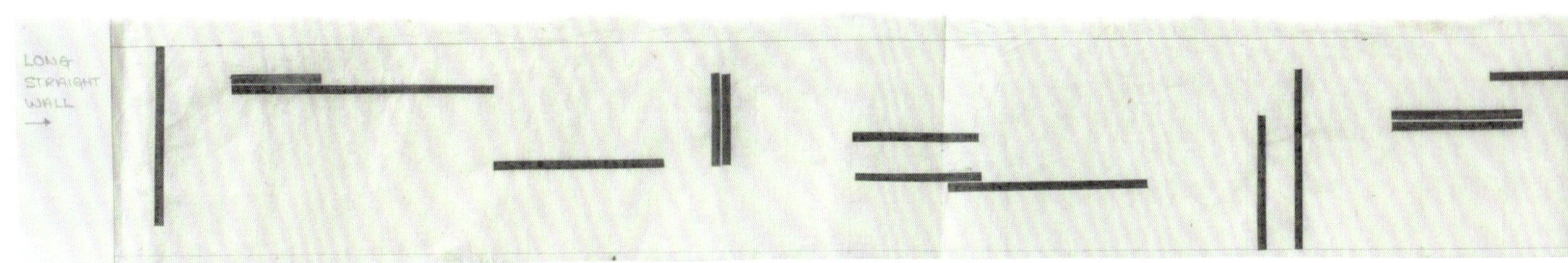

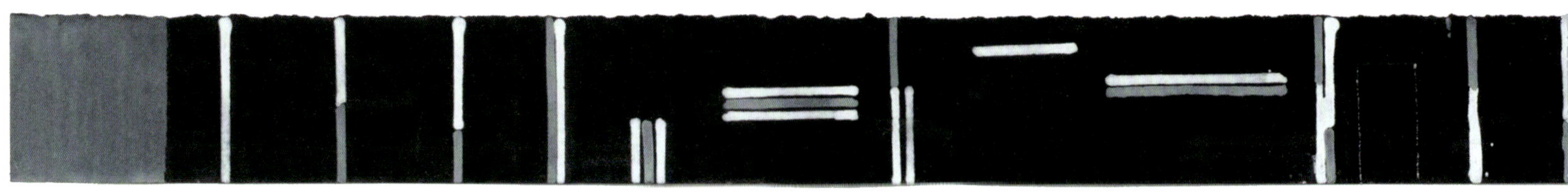

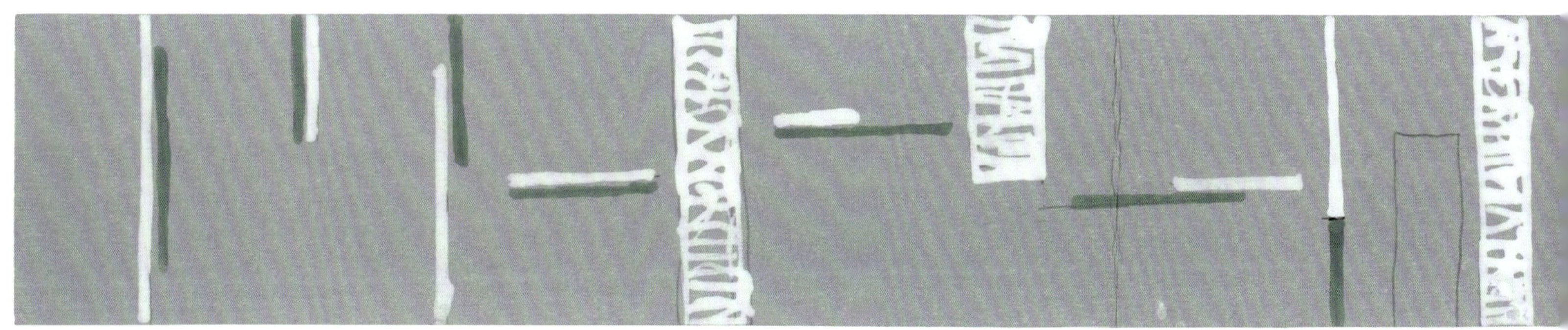

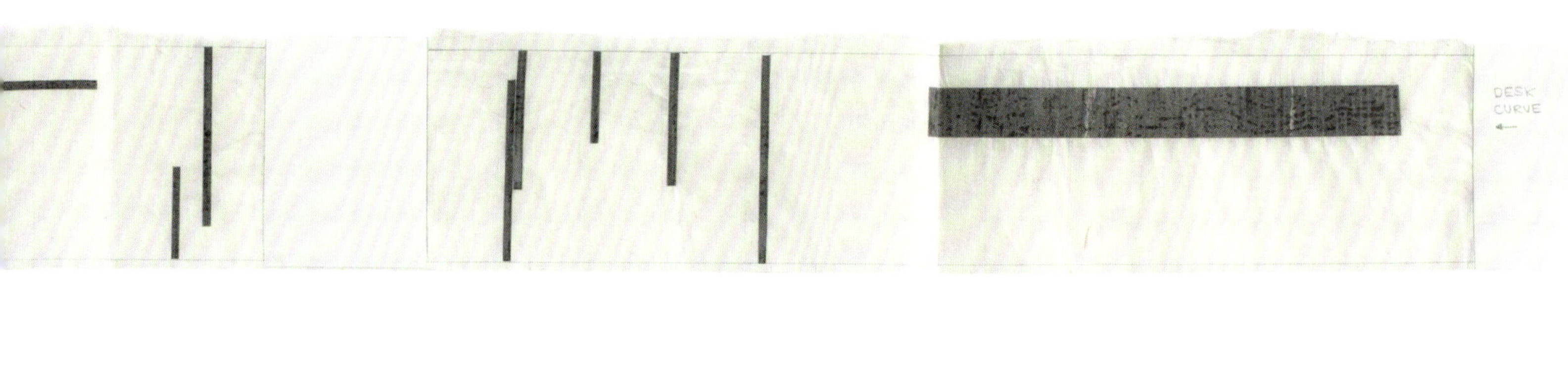
DESK
CURVE
←

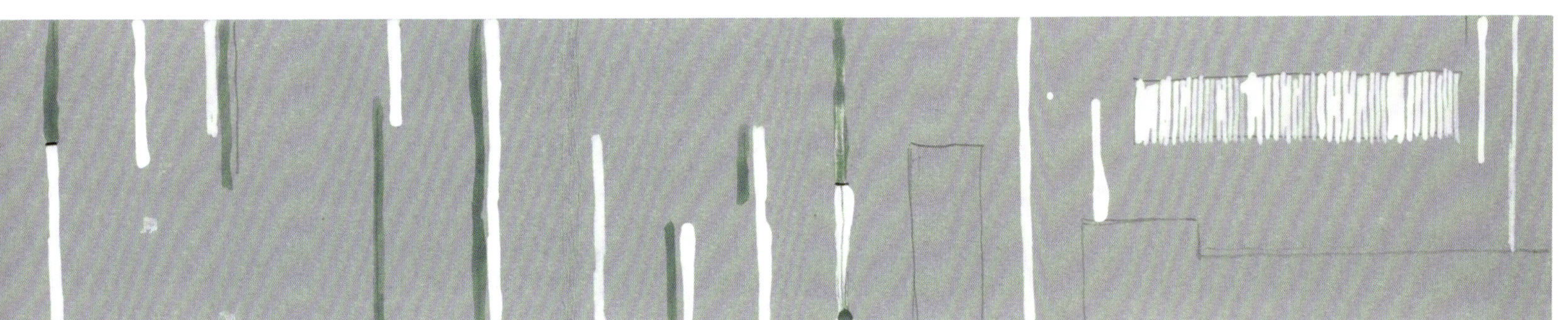

160 *Calligraphy of Light*
ink drawing of scheme, 2007
sumi ink on Japanese paper
9½ × 167 in. (24 × 425 cm)

161 *Calligraphy of Light*
final layout drawing, 2008
ink and gouache on paper
5⅛ × 82¾ in. (13 × 210.2 cm)

162 *Calligraphy of Light*, 2009
recycled glass and bamboo panel,
main entrance, St. George's
Hospital, London
(in collaboration with Gibberd,
architects)

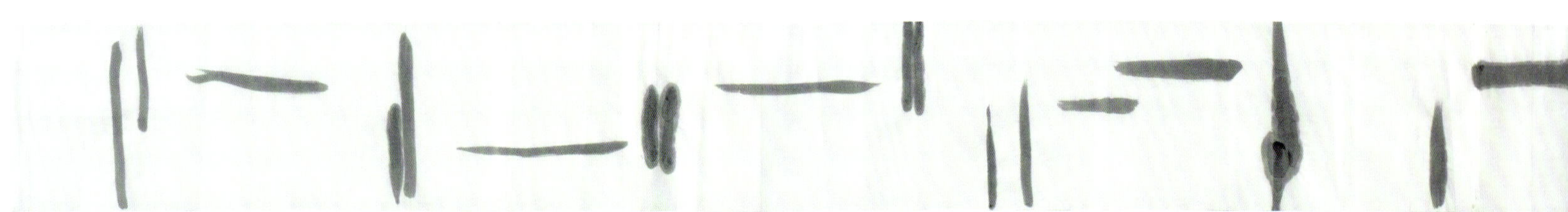

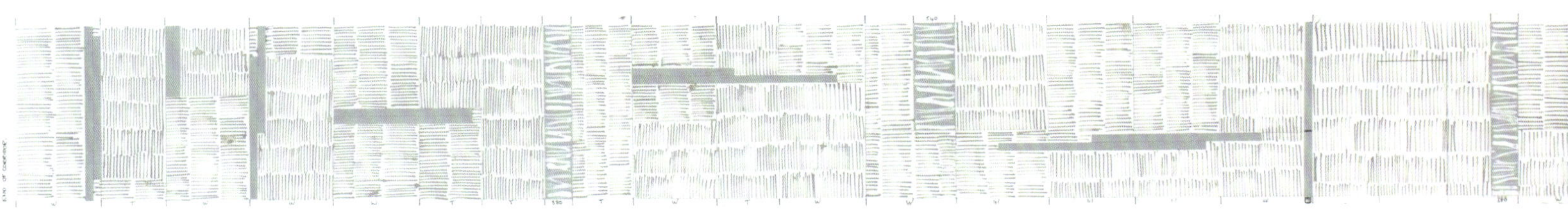

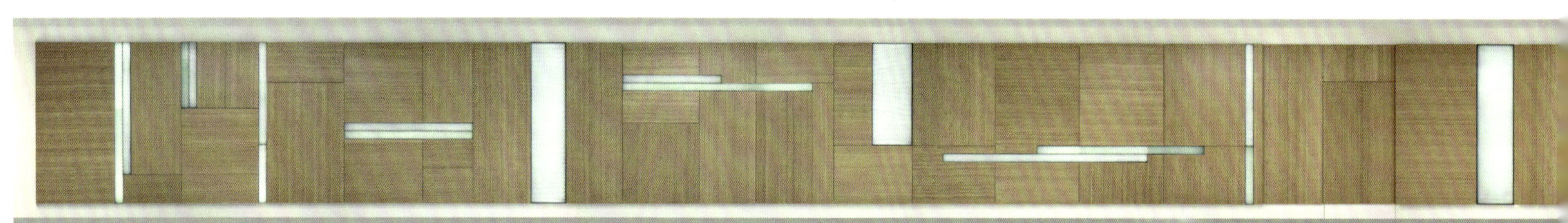

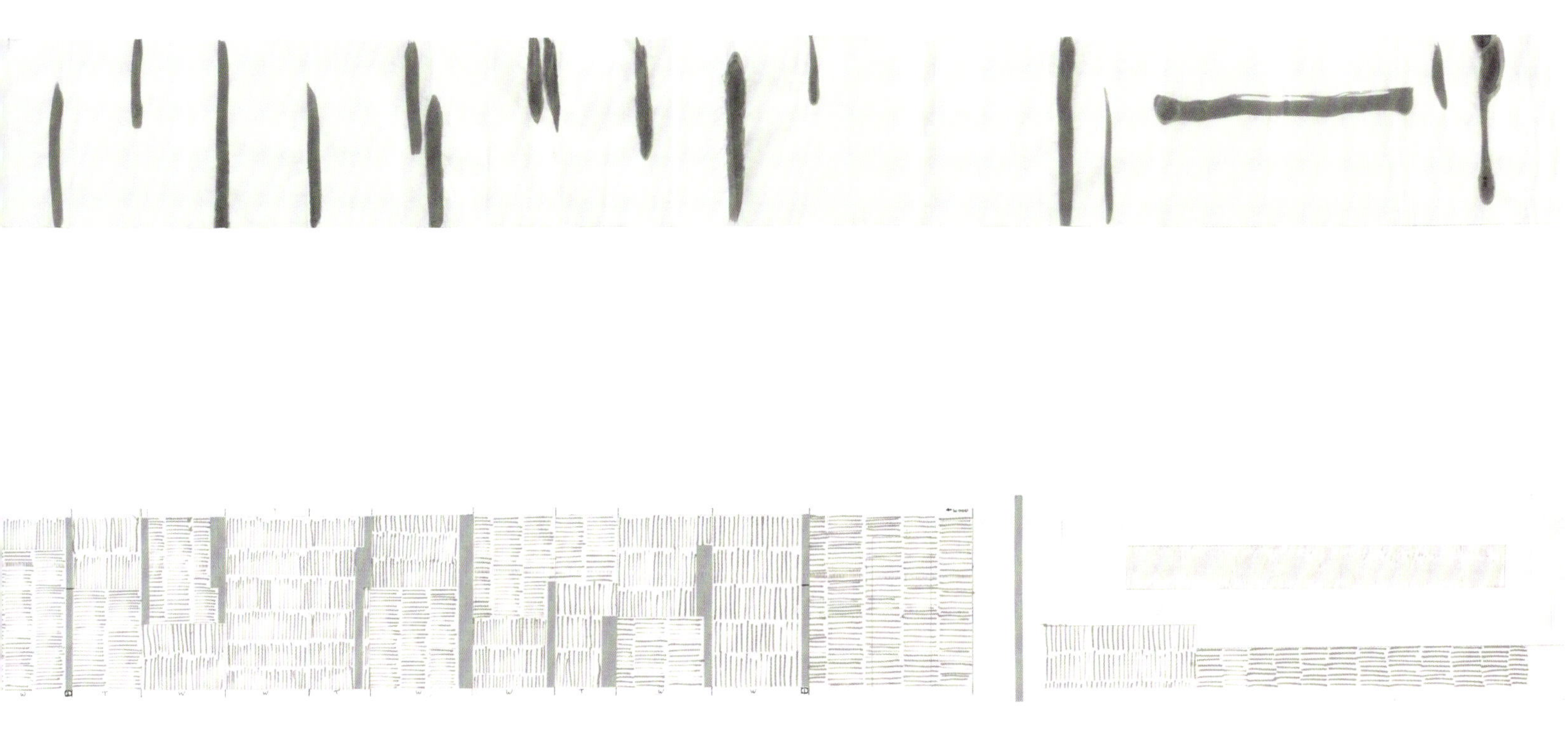

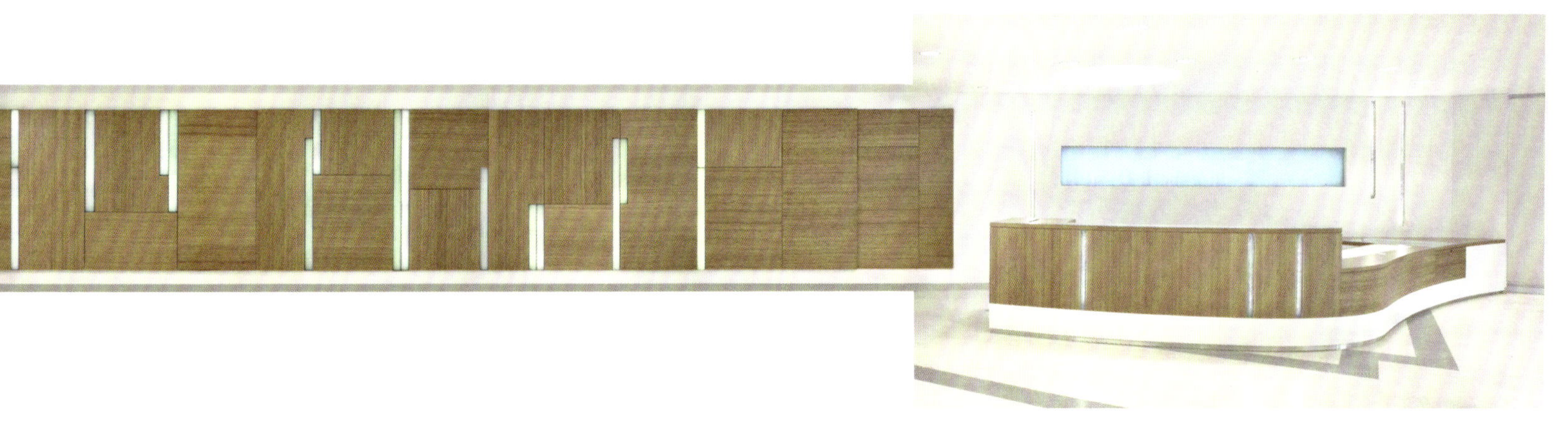

163 *Delta*
original design, 2008
woodblock on Japanese paper
2⅛ × 15 in. (5.4 × 38.1 cm)

164 *Delta*
color scheme, 2008
collage on Japanese paper
3⅛ × 10¼ in. (7.9 × 26 cm)

165 *Delta*, 2009
sandblasted glass with LED
backlighting, waiting area,
St. George's Hospital, London
(in collaboration with Gibberd,
architects)

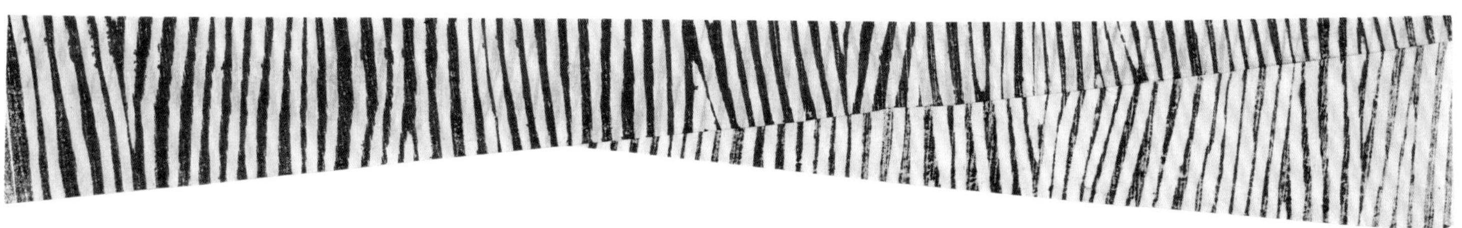

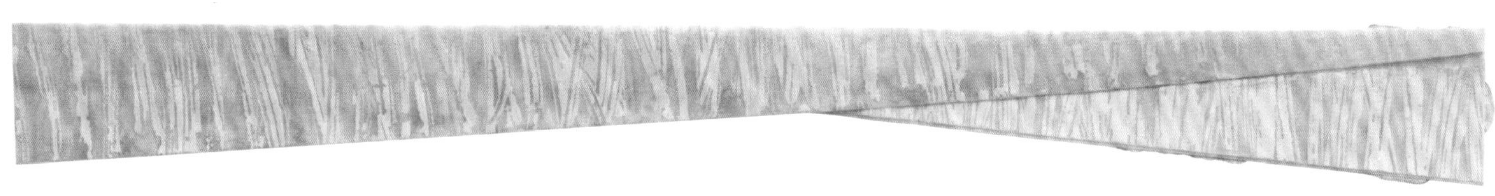

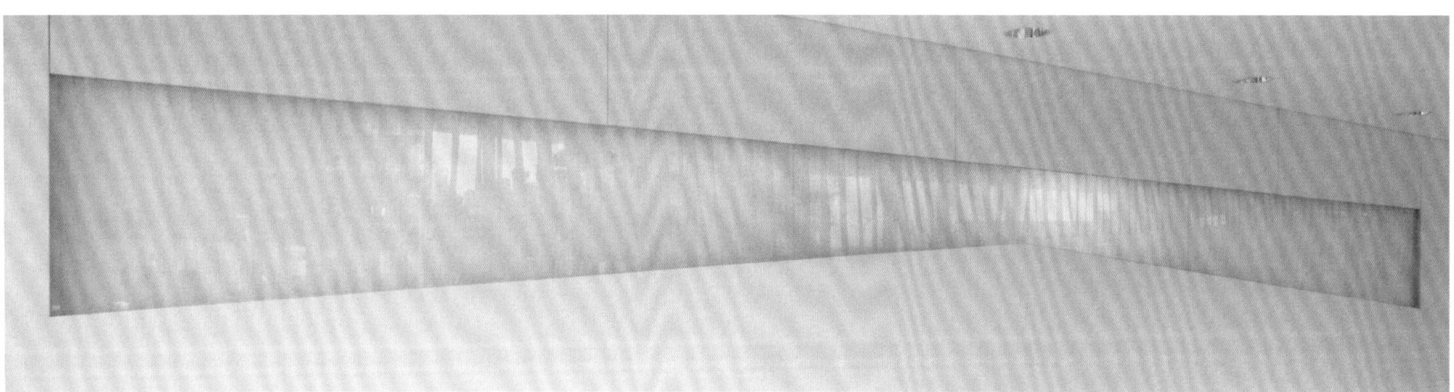

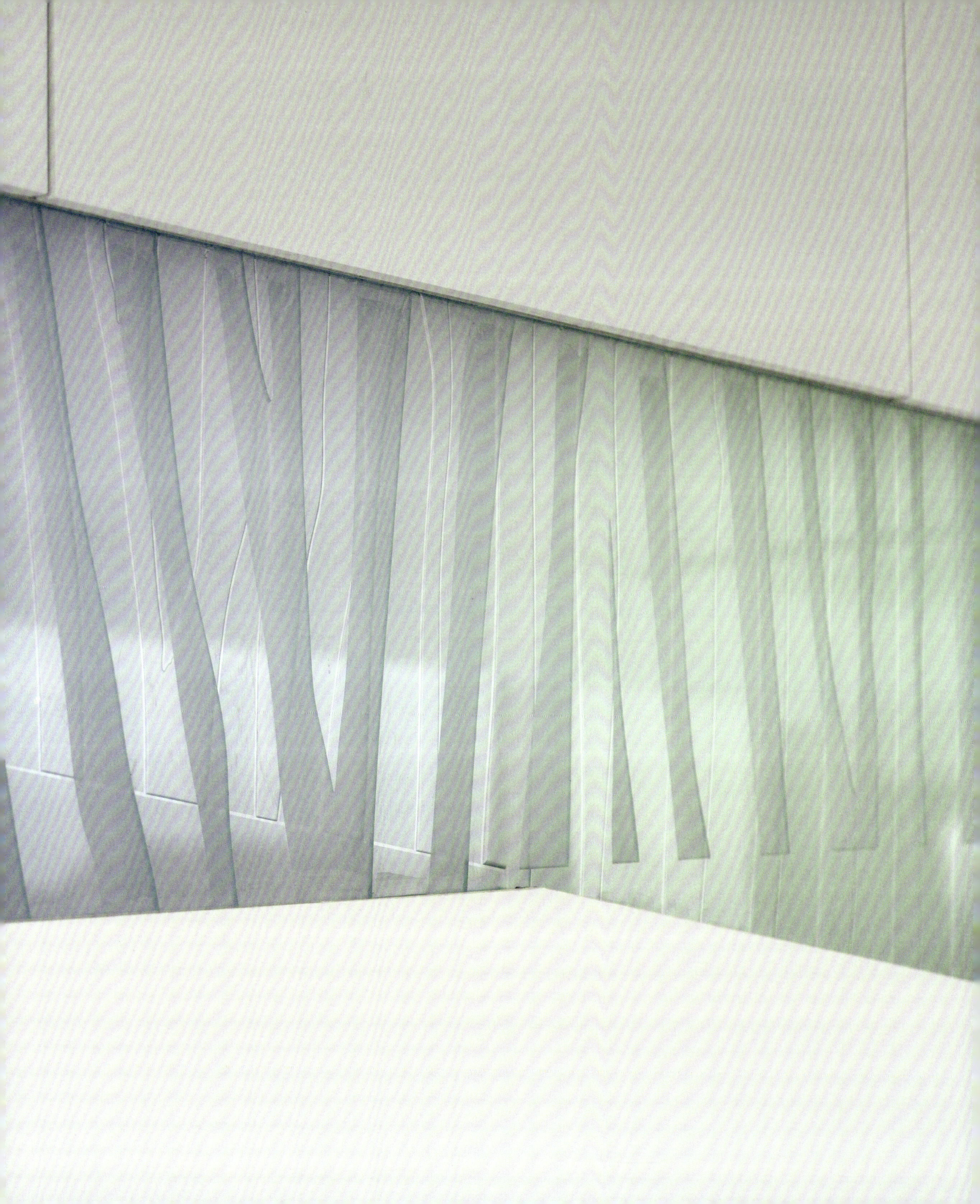

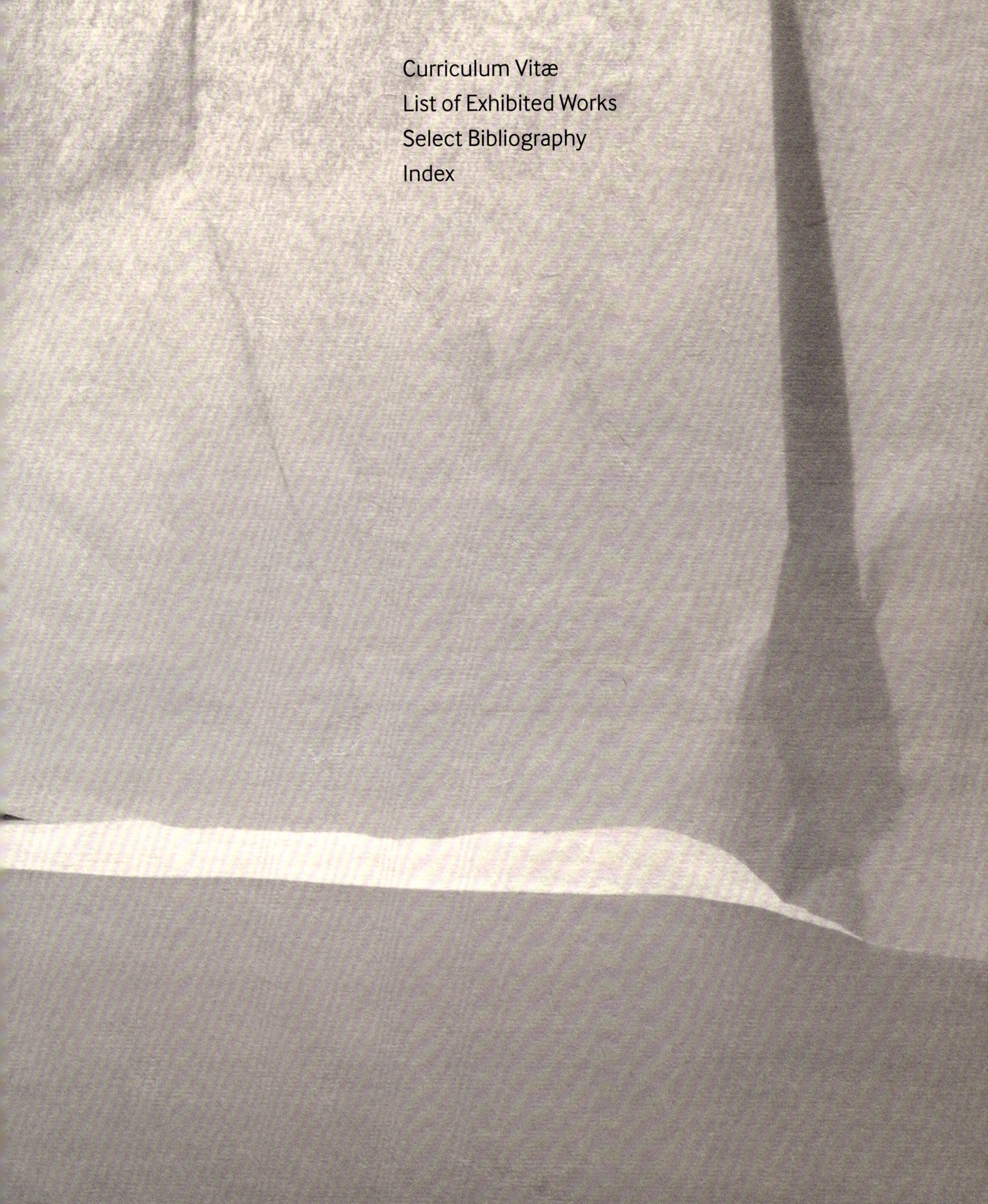

fig. 64: Rebecca Salter working in her London studio, 2009

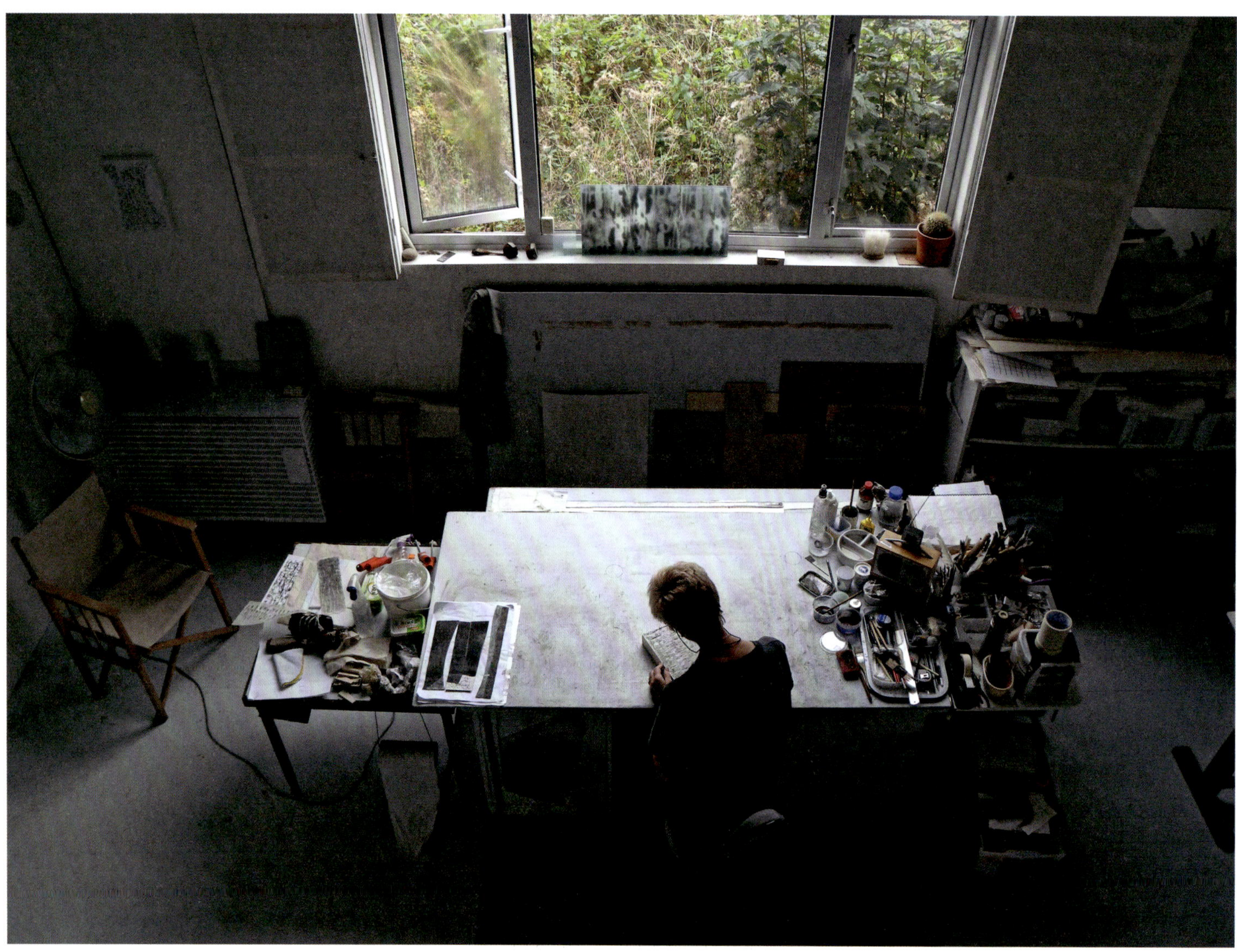

Curriculum Vitæ

Rebecca Salter

1977

Graduated from Bristol Polytechnic, Faculty of Art and Design

1979–81

Research student, Kyoto City University of Arts, Japan (Leverhulme Scholarship)

1981–85

Living and working in Japan

1985–

Living and working in London

1985

Greater London Arts Award

1989

Commission, EPR Architects, London

1993

Commission, United Airlines, London

1995

Pollock-Krasner Foundation award

1997

Cheltenham Open Drawing award (first prize)

2001

Published *Japanese Woodblock Printing*, A&C Black/Univ. of Hawai'i

2002–

Research Fellow, Camberwell College of Arts

2003

Artist in Residence, Josef and Anni Albers Foundation, Connecticut

Pollock-Krasner Foundation award

2006

Published *Japanese Popular Prints*, A&C Black/Univ. of Hawai'i

Commission, Vale Practice, London

2009

Architectural Commission, St. George's Hospital, London

2010

Commission, Guy's Hospital Haematology Unit, London

2011

Architectural Commission, Sackville Street, London

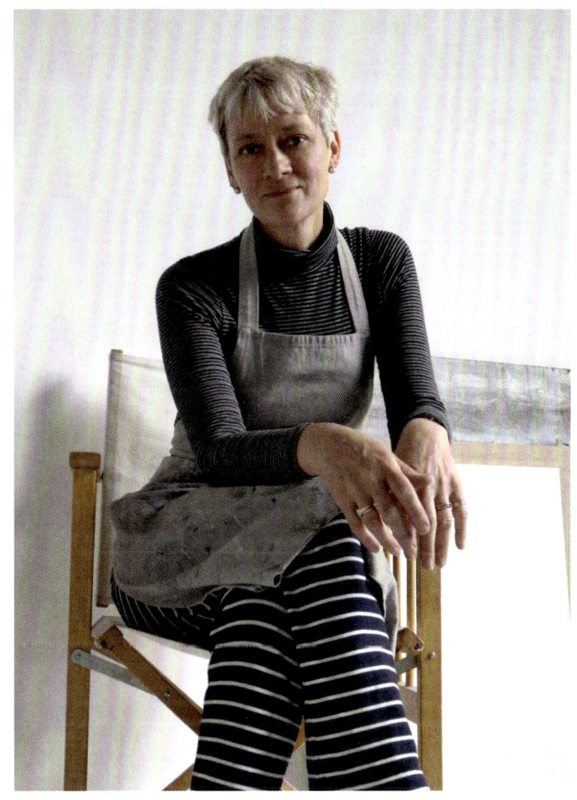

fig. 65: Rebecca Salter in her London studio, 2010

Solo exhibitions

2011

into the light of things, Yale Center for British Art, New Haven, Connecticut

2009

Pale Remembered, Beardsmore Gallery, London

2007

The Unquiet Gaze, Howard Scott Gallery, New York

2006

Bliss of Solitude, Beardsmore Gallery, London

2004

Line, Fosterart, London

Inward Eye, Howard Scott Gallery, New York

2002

Hirschl Contemporary Art, London

2001

Galerie Michael Sturm, Stuttgart

Galerie Pousse (Ichikawa), Tokyo

Gallery Sowaka, Kyoto

Howard Scott Gallery, New York

Russell-Cotes Art Gallery and Museum, Bournemouth

2000

Jill George Gallery, London

Wellcome Building, University of Cambridge

1999

Galerie Michael Sturm, Stuttgart (with Thomas Ruppel)

Feichtner & Mizrahi, Vienna

Gallery Sowaka, Kyoto

M-13/Howard Scott Gallery, New York

1998

Jill George Gallery, London

Galerie Pousse (Ichikawa), Tokyo

1997

M-13/Howard Scott Gallery, New York

1996

Jill George Gallery, London

1994

Jill George Gallery, London

1993

Ishiyacho Gallery, Kyoto

1992

Quay Arts Centre, Isle of Wight

Ichikawa Gallery, Tokyo

1991

Greene Gallery, Connecticut

1990

Ichikawa Gallery, Tokyo (with *UK '90* Festival)

1987

Curwen Gallery, London

1986

Miller/Brown Gallery, San Francisco

1985

CDR Fine Art, London

Art Forum, Singapore

World Print Council, San Francisco

1984

Ypsilon, Nishinomiya, Japan

Gallery Maronie, Kyoto

Gallery Te, Tokyo

Curwen Gallery, London

1983

Gallery Maronie, Kyoto

Gallery Suzuki, Kyoto (in conjunction with the International Paper Conference)

1982

Amano Gallery, Osaka

1981

Gallery Maronie, Kyoto

Craft International, Tokyo

Group exhibitions

2010

25 Years, Howard Scott Gallery, New York

2009

40 Artists–80 Drawings, The Drawing Gallery, London
Contemporary Prints, The Drawing Gallery, London
40 Artists–40 Drawings, Victoria and Albert Museum

2008

Water, Wood, Paper, Toronto
Drawing into Painting into Drawing, The Drawing Gallery, London
Chromophobia, OK Harris, New York

2007

Fils de…, Plouec de Trieux

2006

White II, Howard Scott Gallery, New York
Drawing, Galerie Michael Sturm, Stuttgart
Surface Tensions, Jagged Art, London
To the Edge, Beardsmore Gallery, London
16 Artists/32 Drawings, The Drawing Gallery, London
Drawing Inspiration, Abbot Hall Art Gallery, Cumbria
Individual Poetries, Howard Scott Gallery, New York
Drawing Breath, London; Bristol; Aberdeen; Sydney; Singapore

2005

Howard Scott Gallery, New York
Yale Center for British Art, New Haven
Larry Becker Contemporary Art, Philadelphia
40 Artists–40 Drawings, The Drawing Gallery, London

2004

Sublime Present, Musashino Art University, Tokyo; Nishio City, Aichi

2003

Untitled: work on paper, Hirschl Contemporary Art, London
20x5 Drawings, Eagle Gallery, London

2002

Working the Grid, Lafayette College, Easton, Pennsylvania
Works on Paper, Howard Scott Gallery, New York
Galerie Michael Sturm, Stuttgart

2000

Albright-Knox Museum, Buffalo, NY

1999

Preview, M-13/Howard Scott Gallery, New York

1998

5 artists, M-13/Howard Scott Gallery, New York
Abstraction, M-13/Howard Scott Gallery, New York

1997

12 British Artists, NORD/LB Forum, Hanover
Rubicon Gallery, Dublin
Cheltenham Open Drawing 1997

1996

White, M-13/Howard Scott Gallery, New York
Cheltenham Open Drawing 1996
Galerie Michael Sturm, Stuttgart
Monoprints, Gainsborough's House, Sudbury

1995

From Picasso to Woodrow, Tate Gallery, London
Cabinet Art, Jason & Rhodes Gallery, London
The Drawing Show III, Jill George Gallery, London
Turnpike Gallery, Leigh, Lancashire

1994

Recent Prints, Jill George Gallery, London

1993

Salama Caro Gallery, London

1991

Shimada Shigeru Gallery, Tokyo
Cleveland Drawing Biennale, Middlesborough

1990

Curwen Gallery, London
Runkel-Hue-Williams, London
A View of the New, Royal Over-Seas League (prizewinner)

1989

Critic's Space, Air Gallery, London
Drawing '89, Gallery Maronie, Kyoto/Wacoal Art Space, Tokyo
Curwen Gallery, London

fig. 66: *Light Lines*
artist's proof, 2009
woodblock on Japanese paper
7 × 9 in. (17.8 × 22.9 cm)

1988

Presentation '88, Curwen Gallery, London
Six Artists on Paper, Oxford Gallery, Oxford

1987

Eight by Eight, Curwen Gallery, London

1986

Norwegian International Print Biennale

1985

Curwen Gallery, London
Ljubljana International Print Biennale, Yugoslavia

1984

Curwen Gallery, London
Gallery Humanite, Nagoya
Norwegian International Print Biennale

1983

Ypsilon, Nishinomiya, Japan
International Impact Art Festival, Kyoto and Seoul
Ark, Kyoto
Miller/Brown Gallery, San Francisco
Woodworks, Plymouth Arts Centre
Frechen Kunstverein, Germany (prizewinner)
Cleveland Drawing Biennale, UK (prizewinner)
World Print Council IV, San Francisco (purchase prize)

1982

Rijeka Drawing Biennale, Yugoslavia
Ryu International Gallery, Tokyo
Gallery Maronie, Kyoto
Vallauris Ceramics Biennale, France (prizewinner)

1980

Gallery Soiree, Kyoto
Gallery Washio, Tokyo

Collections

UK

British Council
British Land
British Museum
Cambridge Institute for Medical Research
Gartmore
Government Art Collection
Leslie & Godwin Group
Linklaters
MSP Ltd
Pearl Assurance Ltd
Redcar and Cleveland Borough Council
Tate
Unilever
Victoria and Albert Museum
W. H. Smith

Germany

Kunstverein zu Frechen e.V.
Graphothek

USA

AT&T
California College of the Arts
The Heithoff Family Collection
J. P. Morgan
Lake Tower Collection, Chicago
Library of Congress
Portland Art Museum, Oregon
San Francisco Museum of Modern Art
World Print Council
Yale Center for British Art

List of Exhibited Works

Unless otherwise noted all works are in the collection of the artist. This checklist follows the order of the exhibition.

Untitled MM37
2008
mixed media on linen
63 × 66⅞ in. (160 × 170 cm)
Private collection
plate 134

Untitled MM2
2008
mixed media on paper
39¾ × 59⅞ in. (101 × 152 cm)
plate 131

Untitled RR2
2009
mixed media on paper
39¾ × 59⅞ in. (101 × 152 cm)
plate 135

Untitled RR35
2009
mixed media on paper
59⅞ × 39⅜ in. (152 × 100 cm)
plate 149

Untitled JJ11
2006
mixed media on paper
37⅜ × 57⅛ in. (95 × 145 cm)
plate 109

Quadra Drawings 1–4
2010
watercolor on Somerset paper
each 11⅞ × 11⅞ in. (30 × 30 cm)
plate 152

Satō Woodblock Workshop, Kyoto
Quadra 1–4
2010
woodblock on Torinoko paper
each 11⅞ × 11⅞ in. (30 × 30 cm)
plate 153

Satō Woodblock Workshop, Kyoto
Quadra 1–4
2010
original woodblocks
each 11⅞ × 11⅞ in. (30 × 30 cm)

Untitled RR31
2009
mixed media on linen
74¾ × 70⅞ in. (190 × 180 cm)
plate 147

Untitled AB19
2010
mixed media on linen
55 × 51 in. (140 × 130 cm)
plate 150

Untitled B161
1984
mixed media on Japanese paper
25¼ × 63 in. (64 × 160 cm)
plate 21

Calligraphy of Light
final layout drawing
2008
ink and gouache on paper
5⅛ × 82¾ in. (13 × 210.2 cm)
plate 161

Calligraphy of Light
detailed layout of white/green glass
2007
gouache on paper
2¼ × 25¼ in. (5.7 × 64.2 cm)
plate 159

Calligraphy of Light
preliminary ink drawings
2007
sumi ink on Japanese paper
each drawing: 9½ × 13¼ in.
(24.1 × 33.7 cm)
plate 155

Calligraphy of Light
preliminary ink drawings
2007
sumi ink on Japanese paper
each drawing: 9½ × 13¼ in.
(24.1 × 33.7 cm)
plate 156

Calligraphy of Light
layout of scheme
2007
collage on Japanese paper
2¾ × 37¾ in. (7 × 95.9 cm)
plate 157

Light Lines
artist's proof
2009
woodblock on Japanese paper
6¾ × 9 in. (17.1 × 22.9 cm)
fig. 66

Delta
original design
2008
woodblock on Japanese paper
2⅛ × 15 in. (5.4 × 38.1 cm)
plate 163

Delta
color scheme
2008
collage on Japanese paper
3⅛ × 10¼ in. (7.9 × 26 cm)
plate 164

Calligraphy of Light
2009
digital print on paper
plate 162

Delta
2009
digital print on paper
plate 165

Untitled A34
1982
mixed media on Japanese paper
25¼ × 37¾ in. (64 × 96 cm)
plate 3

Untitled A21
1982
mixed media on Japanese paper
25¼ × 37¾ in. (64 × 96 cm)
plate 4

Untitled A57
1982
mixed media on Japanese paper
25¼ × 37¾ in. (64 × 96 cm)
plate 5

Untitled A56
1982
mixed media on Japanese paper
25¼ × 37¾ in. (64 × 96 cm)
plate 6

Untitled B116
1982
mixed media on Japanese paper
16⅞ × 12⅝ in. (43 × 32 cm)
plate 1

Untitled B168
1982
mixed media on Japanese paper
16⅞ × 12⅝ in. (43 × 32 cm)
Klingensmith Collection, London
plate 2

Untitled B96
1983
mixed media on Japanese paper
39⅜ × 25¼ in. (100 × 64 cm)
plate 16

Untitled B117
1984
mixed media on Japanese paper
38⅛ × 25¼ in. (97 × 64 cm)
plate 17

Untitled B119
1984
mixed media on Japanese paper
25¼ × 37 in. (64 × 94 cm)
Promised gift, Nina Collection, Kyoto
plate 19

Untitled B120
1984
mixed media on Japanese paper
25¼ × 37 in. (64 × 94 cm)
Promised gift, Nina Collection, Kyoto
plate 20

Untitled B34
1984
mixed media on Japanese paper
33⅛ × 39⅜ in. (84 × 100 cm)
plate 18

Untitled B140
1983
mixed media on Japanese paper
36¼ × 58⅝ in. (92 × 149 cm)
plate 22

Untitled A39
1983
mixed media on Japanese paper
26⅜ × 38¼ in. (67 × 97 cm)
plate 13

Untitled A72
1983
mixed media on Japanese paper
25¼ × 37¾ in. (64 × 96 cm)
plate 12

Untitled A50
1982
mixed media on Japanese paper
25¼ × 37¾ in. (64 × 96 cm)
plate 14

Untitled A65
1983
mixed media on Japanese paper
25 × 38 in. (63.5 × 96.5 cm)
plate 15

Japan I
1982
woodblock on Japanese paper
7⅛ × 9 in. (18 × 23 cm)
plate 7

Japan II
1982
woodblock on Japanese paper
7⅛ × 9 in. (18 × 23 cm)
plate 8

Japan III
1982
woodblock on Japanese paper
7⅛ × 9 in. (18 × 23 cm)
plate 9

Japan IV
1982
woodblock on Japanese paper
7⅛ × 9 in. (18 × 23 cm)
plate 10

Japan V
1982
woodblock on Japanese paper
7⅛ × 9 in. (18 × 23 cm)
plate 11

Narumi Hiroyuki
Boku Sorekkiri (Just Me)
ぼく、それっきり
鳴海裕行
designed by Rebecca Salter
printed book, 1983
in plastic or paper slipcase
cover: 7¾ × 6 in. (19.8 × 15.2 cm)
plate 23

Elias Canetti
The Human Province
cover design incorporating a work
on paper by Rebecca Salter
printed book, 1986
cover: 7¾ × 5⅛ in. (19.7 × 12.9 cm)
plate 24

J. Marvin Spiegelman
Active Imagination
J.M. シュピーゲルマン
能動的想像法—内なる魂との対話
cover design by Nishioka Tsutomu
incorporating drawing by
Rebecca Salter
printed book, 1994
cover: 7⅝ × 5⅜ in. (19.4 × 13.7 cm)
plate 25

into the light of things
2010
digital print on Japanese paper
edition of four
2⅝ × 43¾ in. (6.7 × 111.1 cm)
plate 154

Untitled F48
1990
mixed media on paper
49¼ × 51⅛ in. (125 × 130 cm)
plate 26

Untitled G27
1991
mixed media on canvas
48 × 48 in. (121.9 × 121.9 cm)
plate 42

Untitled G14
1991
mixed media on canvas
48 × 48 in. (121.9 × 121.9 cm)
plate 39

Untitled F133
1990
mixed media on Japanese paper
32¼ × 35⅜ in. (82 × 90 cm)
plate 31

Untitled D90
1989
mixed media on Japanese paper
25¼ × 37⅜ in. (64 × 95 cm)
plate 33

Untitled G3
1990
mixed media on paper
11¾ × 11¾ in. (30 × 30 cm)
plate 28

Untitled F121
1990
mixed media on paper
14⅛ × 14⅛ in. (36 × 36 cm)
plate 29

Untitled D137
1989
mixed media on Japanese paper
63 × 28⅜ in. (160 × 72 cm)
plate 36

Untitled F68
1990
mixed media on paper
16 × 16 in. (40.6 × 40.6 cm)
plate 27

Untitled F112
1990
mixed media on canvas
30 × 30 in. (76.2 × 76.2 cm)
plate 38

Square 1
1991
mixed media on canvas
7⅛ × 7⅛ in. (18 × 18 cm)
plate 37

Square 2
1991
mixed media on canvas
7⅛ × 7⅛ in. (18 × 18 cm)
plate 37

Square 3
1991
mixed media on canvas
7⅛ × 7⅛ in. (18 × 18 cm)
plate 37

Square 4
1991
mixed media on canvas
7⅛ × 7⅛ in. (18 × 18 cm)
plate 37

Square 5
1991
mixed media on canvas
7⅛ × 7⅛ in. (18 × 18 cm)
plate 37

Square 6
1991
mixed media on canvas
7⅛ × 7⅛ in. (18 × 18 cm)
plate 37

Untitled D58
1988
mixed media on Japanese paper
18½ × 23¼ in. (47 × 59 cm)
plate 32

Untitled D86
1988
mixed media on Japanese paper
11⅜ × 52¾ in. (29 × 134 cm)
plate 35

Untitled D79
1988
mixed media on Japanese paper
18⅛ × 61⅜ in. (46 × 156 cm)
plate 34

Untitled J23
1995
mixed media on canvas
72 × 86 in. (182.9 × 218.4 cm)
plate 48

Untitled K20
1996
mixed media on canvas
29 × 39 in. (73.7 × 99.1 cm)
plate 55

Untitled G86
1992
mixed media on canvas
48 × 48 in. (121.9 × 121.9 cm)
Collection of Delia Smith
plate 40

Untitled H49
1993
mixed media on canvas
22 × 22 in. (55.9 × 55.9 cm)
plate 41

Untitled H32
1993
mixed media on canvas
60 × 60 in. (152.4 × 152.4 cm)
Private collection
plate 43

Untitled F40
1990
mixed media on paper
29⅛ × 29⅞ in. (74 × 76 cm)
plate 30

Untitled M43
1997
mixed media on canvas
32 × 36 in. (81.3 × 91.4 cm)
Lea Babcock Scherer and
Jeffrey A. Scherer
plate 67

London Sketchbook
1990
sumi ink on paper
12½ × 9½ in. (31.8 × 24.1 cm)
plate 44

Lake District Sketchbook
1990
pastel on paper
12½ × 9½ in. (31.8 × 24.1 cm)
Yale Center for British Art, Gift of
the artist in memory of her father,
Samuel Salter (1922–2005)
plate 45

Lake District Sketchbook
1994
watercolor on paper
12½ × 9½ in. (31.8 × 24.1 cm)
plates 49, 50

Lake District Sketchbook
2007
watercolor on paper
12½ × 9½ in. (31.8 × 24.1 cm)
plates 49, 50

Lake District Sketchbook
2008
watercolor on paper
12½ × 9½ in. (31.8 × 24.1 cm)
plates 49, 50

Lingmell/Kirkfell
2009
watercolor on paper
5 × 19¼ in. (12.7 × 48.9 cm)
plate 51

Lingmell/Kirkfell/Screes
2009
watercolor on paper
5½ × 43¾ in. (14 × 111.1 cm)
plate 52

Sana'a
1998
watercolor on paper
5¾ × 8¼ in. (14.6 × 21 cm)
plate 70

Sana'a
1998
watercolor on paper
5½ × 6⅛ in. (14 × 15.6 cm)
plate 71

Shibam
1998
watercolor on paper
5¼ × 7 in. (13.3 × 17.8 cm)
plate 72

Shibam
1998
watercolor on paper
5¼ × 7 in. (13.3 × 17.8 cm)
plate 73

Tools used by Rebecca Salter: scrapers, reed pen, syringe, spatulas, plastic brush, blade, brush, line scraper, mud stick, feather, wooden clothes-pin, wooden chopsticks, dip pen, comb

Graphite shavings in plastic box

Pigments scraped from paintings

Untitled K37
1996
mixed media on canvas
72 × 96 in. (182.9 × 243.8 cm)
plate 59

Untitled M20
1997
mixed media on canvas
48 × 54 in. (121.9 × 137.2 cm)
plate 61

Untitled J1
1994
mixed media on canvas
72 × 108 in. (182.9 × 274.3 cm)
plate 47

Arc I
1995
woodblock on Japanese paper
22½ × 22½ in. (57 × 57 cm)
plate 56

Arc II
1995
woodblock on Japanese paper
22½ × 22½ in. (57 × 57 cm)
plate 57

Untitled K78
1996
mixed media on paper
44⅛ × 60⅝ in. (112 × 154 cm)
plate 60

Untitled J72
1994
mixed media on canvas
72 × 73 in. (182.8 × 185.4 cm)
Diane Davies
plate 58

Untitled J47
1995
mixed media on paper
29½ × 28¾ in. (75 × 73 cm)
plate 53

Untitled J48
1995
mixed media on paper
29⅞ × 44⅛ in. (76 × 112 cm)
plate 54

Untitled M72
1997
mixed media on canvas
68 × 72 in. (172.7 × 182.9 cm)
Private collection
plate 68

Untitled M134
1998
mixed media on canvas
66 × 66 in. (167.6 × 167.6 cm)
Plate 69

Untitled H43
1993
mixed media on paper
48⅞ × 46½ in. (124 × 118 cm)
plate 46

Untitled AA29
2000
mixed media on canvas
22½ × 19 in. (57.2 × 48.3 cm)
Gill and Steve Evans
plate 76

Untitled R76
1999
mixed media on canvas
21 × 18 in. (53.3 × 45.7 cm)
Private collection
plate 75

Untitled R72
1999
mixed media on canvas
72 × 61 in. (182.9 × 154.9 cm)
plate 74

Untitled BB32
2001
mixed media on canvas
28 × 7 in. (71.1 × 17.8 cm)
Private collection, Cambridge, UK
plate 79

Untitled BB45
2001
mixed media on canvas
28 × 7 in. (71.1 × 17.8 cm)
Private collection, Cambridge, UK
plate 80

Untitled AA3
2000
mixed media on canvas
96 × 34 in. (243.8 × 86.4 cm)
plate 77

Untitled AA52
2000
mixed media on canvas
96 × 34 in. (243.8 × 86.4 cm)
plate 78

Untitled BB47
2001
mixed media on canvas
50 × 72 in. (127 × 182.9 cm)
Collection of Holly Lennihan and Robert Cox, Washington, DC
plate 81

Untitled M108
1998
mixed media on canvas
32 × 34 in. (81.3 × 86.4 cm)
plate 66

Untitled CC43
2002
mixed media on canvas
14 × 56 in. (35.6 × 142.2 cm)
plate 87

Untitled CC40
2002
mixed media on paper
20⅞ × 31⅛ in. (53 × 79 cm)
plate 86

Untitled CC45
2002
mixed media on canvas
36 × 35½ in. (91.4 × 90.2 cm)
plate 88

Satō Woodblock Workshop, Kyoto, after drawings by Rebecca Salter
Sky I–IV
2008
woodblock on Japanese paper
each 11 × 16⅞ in. (28 × 43 cm)
commissioned by
ANA InterContinental Tokyo
plates 127–130

Eminence I
2005
woodblock on tissue on Japanese paper
10⅝ × 15 in. (27 × 38 cm)
plate 105

Eminence II
2005
woodblock on tissue on Japanese paper
10⅝ × 15 in. (27 × 38 cm)
plate 106

Eminence III
2005
woodblock on tissue on Japanese paper
10⅝ × 15 in. (27 × 38 cm)
plate 107

Eminence IV
2005
woodblock on tissue on Japanese paper
10⅝ × 15 in. (27 × 38 cm)
plate 108

March Monoprint I
1998
woodblock on Japanese paper
15 × 16⅛ in. (38 × 41 cm)
plate 62

March Monoprint III
1998
woodblock on Japanese paper
15 × 16⅛ in. (38 × 41 cm)
plate 63

March Monoprint VII
1998
woodblock on Japanese paper
15 × 16⅛ in. (38 × 41 cm)
plate 64

March Monoprint IX
1998
woodblock on Japanese paper
15 × 16⅛ in. (38 × 41 cm)
plate 65

Untitled BB60
2001
mixed media on canvas
10 × 9 in. (25.4 × 22.9 cm)
Private collection
plate 82

Untitled JJ2
2006
mixed media on canvas
10 × 9 in. (25.4 × 22.9 cm)
plate 85

Untitled DD14
2003
mixed media on canvas
10 × 9 in. (25 × 23 cm)
plate 84

Untitled DD25
2003
mixed media on canvas
10 × 9 in. (25.4 × 22.9 cm)
plate 83

Untitled EE12
2004
mixed media on canvas
36 × 36 in. (91.4 × 91.4 cm)
plate 100

Untitled EE13
2004
mixed media on canvas
36 × 36 in. (91.4 × 91.4 cm)
plate 101

Untitled DD21
2003
mixed media on canvas
24 × 20 in. (61 × 50.8 cm)
plate 91

Untitled DD8
2003
mixed media on canvas
24 × 22 in. (61 × 55.9 cm)
plate 90

Untitled 2006-27
2006
mixed media on paper
10⅝ × 12⅝ in. (27 × 32 cm)
Andrew Lambirth Collection
plate 110

Untitled 2006-35
2006
mixed media on paper
11¼ × 11¾ in. (28.5 × 30 cm)
plate 113

Untitled JJ46
2006
mixed media on paper
11 × 11¾ in. (28 × 30 cm)
plate 114

Untitled JJ47
2006
mixed media on paper
11 × 11¾ in. (28 × 30 cm)
plate 111

Untitled 2007-18
2007
mixed media on paper
14⅛ × 18⅞ in. (36 × 48 cm)
plate 121

Untitled 2007-19
2007
mixed media on paper
14⅛ × 18⅞ in. (36 × 48 cm)
plate 122

Untitled KK40
2007
mixed media on paper
14⅛ × 18⅞ in. (36 × 48 cm)
plate 125

Untitled KK33
2007
mixed media on paper
14⅛ × 18⅞ in. (36 × 48 cm)
plate 124

Untitled 2006-28
2006
mixed media on paper
14¼ × 15 in. (36 × 38 cm)
plate 112

Untitled 2006-29
2006
mixed media on paper
14¼ × 15 in. (36 × 38 cm)
plate 115

Untitled JJ21
2006
mixed media on paper
14¼ × 15 in. (36 × 38 cm)
plate 116

Bethany Squares
2003
mixed media on paper mounted on aluminum
each square: 9 × 9 in. (23 × 23 cm)
Yale Center for British Art, Friends of British Art Fund
plate 99

Bethany 6
2003
mixed media on paper
19¼ × 25¼ in. (49 × 64 cm)
plate 94

Bethany 11
2003
mixed media on paper
19¼ × 25¼ in. (49 × 64 cm)
plate 97

Bethany 21
2003
mixed media on paper
19¼ × 25¼ in. (49 × 64 cm)
plate 98

Bethany 10
2003
mixed media on paper
19¼ × 25¼ in. (49 × 64 cm)
plate 96

Bethany 8
2003
mixed media on paper
19¼ × 25¼ in. (49 × 64 cm)
plate 95

Bethany 3
2003
mixed media on paper
19¼ × 25¼ in. (49 × 64 cm)
plate 93

Untitled DD1
2003
mixed media on paper
39 × 59⅞ in. (99 × 152 cm)
plate 89

Untitled DD20
2003
mixed media on canvas
43 × 67 in. (109.2 × 170.2 cm)
Collection of Vicky and Terry Klein
plate 92

Untitled HH4
2005
mixed media on linen
30 × 30 in. (76.2 × 76.2 cm)
Private collection
plate 102

Untitled KK14
2007
mixed media on linen
21⅝ × 23⅝ in. (55 × 60 cm)
Private collection, New York
plate 120

Untitled KK5
2007
mixed media on linen
23⅝ × 23⅝ in. (60 × 60 cm)
Howard Foote
plate 118

Untitled HH37
2005
mixed media on linen
10 × 9 in. (25.4 × 22.9 cm)
Bruce and Aphrodite Garrison
plate 119

Untitled KK12
2007
mixed media on linen
10 × 10 in. (25.4 × 25.4 cm)
Collection of Richard and
Janet Caldwell
plate 117

Untitled HH30
2005
mixed media on linen
72 × 96 in. (182.9 × 243.8 cm)
plate 104

Untitled HH5
2005
mixed media on linen
38 × 44 in. (96.5 × 111.8 cm)
Private collection
plate 103

Untitled KK37
2007
mixed media on linen
33½ × 47¼ in. (85 × 120 cm)
plate 126

Untitled AB2
2010
mixed media on linen
25½ × 19½ in. (64.8 × 49.5 cm)
plate 140

Untitled MM8
2008
mixed media on linen
51⅛ × 55⅛ in. (130 × 140 cm)
plate 132

Untitled RR1
2009
mixed media on linen
23⅝ × 17¾ in. (60 × 45 cm)
plate 139

Untitled KK25
2007
mixed media on linen
47¼ × 66⅞ in. (120 × 170 cm)
Collection of Fanchon and
Howard Hallam
plate 123

Untitled MM33
2008
mixed media on linen
41⅜ × 43¼ in. (105 × 110 cm)
Private collection, London
plate 133

Untitled AB4
2010
mixed media on linen
30 × 27½ in. (76.2 × 69.9 cm)
plate 151

Untitled 2009-27
2009
mixed media on paper
29⅛ × 17⅞ in. (74 × 45.5 cm)
plate 138

Untitled 2009-32
2009
mixed media on paper
29⅛ × 17⅞ in. (74 × 45.5 cm)
plate 137

Lead 1
2009
mixed media on lead
12 × 10 in. (30.5 × 25.4 cm)
plate 141

Lead 3
2009
mixed media on lead
10 × 1½ in. (25.4 × 3.8 cm)
plate 142

Lead 6
2009
mixed media on lead
10 × 2½ in. (25.4 × 6.4 cm)
plate 143

Lead 9
2009
mixed media on lead
12¼ × 3½ in. (31.1 × 8.9 cm)
plate 144

Lead 12
2010
mixed media on lead
10 × 2½ in. (25.4 × 6.4 cm)
plate 145

Lead 13
2009
mixed media on lead
7 × 3½ in. (17.8 × 8.9 cm)
plate 146

Works by Rebecca Salter included in *Rebecca Salter and Japan*, a companion exhibition at the Yale University Art Gallery (February 3 – May 1, 2011):

Untitled RR26
2009
mixed media on paper
60 × 36½ in. (152.4 × 91.4 cm)
plate 136

Untitled RR21
2009
mixed media on linen
43¼ × 35⅜ in. (110 × 90 cm)
plate 148

Select Bibliography

Beckett, Sister Wendy. *Sister Wendy Beckett on Art and the Sacred.* London: Rider, 1992: 126.

Klonk, Charlotte. "Towards the Problem of an Aesthetic of Nature in Contemporary Art: Nicky Hirst and Rebecca Salter." *Scroope – Cambridge Architecture Journal 3* (June 1991): 11–14.

Lambirth, Andrew. *Rebecca Salter: Paintings and Works on Paper.* London: Jill George Gallery, 1996 (unpaginated).

Lambirth, Andrew. "Illusory Images of Universal Order." *Rebecca Salter: Recent Paintings and Works on Paper.* New York: M-13, Howard Scott Gallery, 1999 (unpaginated).

Lynton, Norbert. *Rebecca Salter: Paintings, Drawings, Prints.* London: Jill George Gallery, 1998 (unpaginated).

Mozynska, Anna. *Rebecca Salter: Paintings and Works on Paper.* London: Jill George Gallery, 2000 (unpaginated).

Mozynska, Anna. "Rebecca Salter: the Bliss of Solitude." *Rebecca Salter: Bliss of Solitude.* London: Beardsmore Gallery, 2006 (unpaginated).

Mullins, Charlotte. "Painting Shadows." *Rebecca Salter.* London: Hirschl Contemporary Art, 2002 (unpaginated).

Petherbridge, Deanna. *Drawing Inspiration: Contemporary British Drawing.* Kendal, UK: Lakeland Arts Trust and Abbot Hall Art Gallery, 2006.

Robinson, Fiona. "Rebecca Salter." *Axis Journal 7* (November 2007 – February 2008). www.axisweb.org/dlFULL.aspx?ESSAYID=96

Scott, Howard and John Goodrich. *Working the Grid: Diana Cooper, Francisco Castro Lenero, Lance Letscher, Rebecca Salter, Boris Viskin, Stephan Westfall.* Easton, PA: Lafayette College, 2002.

Smith, Alison, ed. *Watercolour.* London: Tate Publishing, 2011.

Weber, Nicholas Fox. "Inward Eye." New York: Howard Scott Gallery, 2004 (unpaginated).

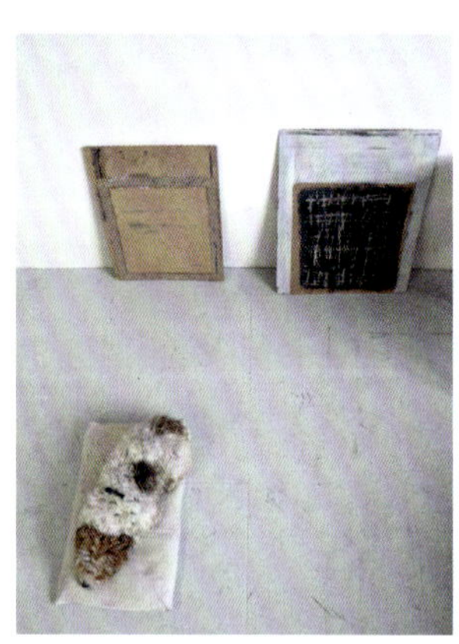

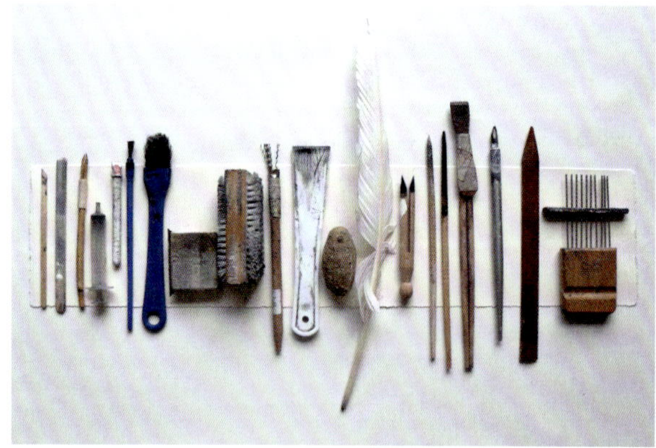

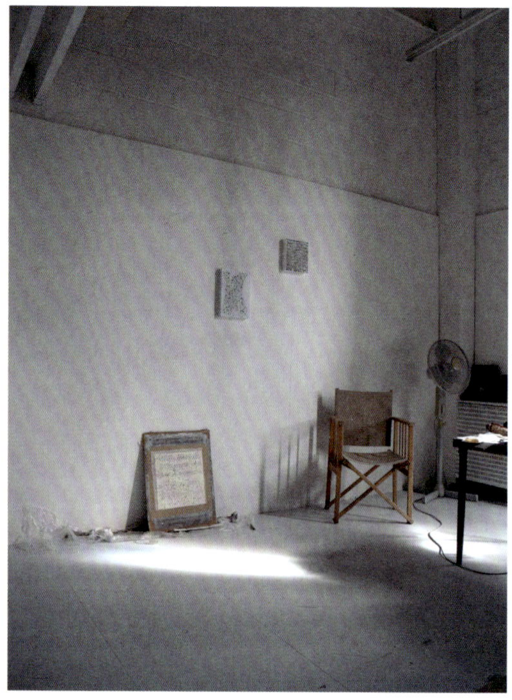

figs. 67–69: Rebecca Salter's studio and tools, London, 2009–10

Index

Note
Page numbers in italics refer to illustrations.
For a full list of works by Rebecca Salter included in the exhibition, see the List of Exhibited Works. All works listed are reproduced on pp. 84–257.
St. is indexed as though spelled in full as Saint.
RS = Rebecca Salter.